Essential Chemistry
for Cambridge IGCSE®

2nd Edition

Roger Norris
Series Editor: Lawrie Ryan

Oxford excellence for Cambridge IGCSE®

OXFORD

OXFORD
UNIVERSITY PRESS

Great Clarendon Street, Oxford, OX2 6DP, United Kingdom

Oxford University Press is a department of the University of Oxford. It furthers the University's objective of excellence in research, scholarship, and education by publishing worldwide. Oxford is a registered trade mark of Oxford University Press in the UK and in certain other countries

© Oxford University Press 2015

The moral rights of the authors have been asserted

First published in 2015

All rights reserved. No part of this publication may be reproduced, stored in a retrieval system, or transmitted, in any form or by any means, without the prior permission in writing of Oxford University Press, or as expressly permitted by law, by licence or under terms agreed with the appropriate reprographics rights organization. Enquiries concerning reproduction outside the scope of the above should be sent to the Rights Department, Oxford University Press, at the address above.

You must not circulate this work in any other form and you must impose this same condition on any acquirer

British Library Cataloguing in Publication Data
Data available

978-0-19-839923-0

10 9 8 7 6

Paper used in the production of this book is a natural, recyclable product made from wood grown in sustainable forests. The manufacturing process conforms to the environmental regulations of the country of origin.

Typeset by GreenGate Publishing Services, Tonbridge, Kent

Illustrations include artwork drawn by GreenGate Publishing Services

Printed in China by Golden Cup

Acknowledgements

The publishers would like to thank the following for permissions to use their photographs:

Cover image: © Stan Fellerman/Corbis; p2: © Jorge Salcedo, 2007; p9: © sciencephotos / Alamy; p10: © sciencephotos / Alamy; p13: Andrew Lambert; p14: © Itani / Alamy; p16: Martyn Chillmaid; p18: © Roger Hutchings / Alamy; p25: CATHERINE POUEDRAS/SCIENCE PHOTO LIBRARY; p26: © ITAR-TASS Photo Agency / Alamy; p28: CHARLES D. WINTERS/ SCIENCE PHOTO LIBRARY; p30: Martyn Chillmaid; p34: CREDIT: CHARLES D. WINTERS/SCIENCE PHOTO LIBRARY; p38: © Patrick Gijsbers / iStock; p41: Martyn Chillmaid; p42: © RIA Novosti / Alamy; p44: STEVE ALLEN/SCIENCE PHOTO LIBRARY; p48: SHEILA TERRY/SCIENCE PHOTO LIBRARY; p48: SCIENCE PHOTO LIBRARY; p54: Martyn Chillmaid; p59: STEPHEN AUSMUS/US DEPARTMENT OF AGRICULTURE/ SCIENCE PHOTO LIBRARY; p59: © Fearnstock / Alamy; p60: Getty Images; p62: DigiStu/Istock; p65: © nano / iStock; p67: © kirza / iStock; p68: kaband / Shutterstock; p70: Martyn Chillmaid; p74: © PITCHAL FREDERIC/CORBIS SYGMA; p77: Martyn Chillmaid; p79: SHEILA TERRY/SCIENCE PHOTO LIBRARY; p80: CREDIT: SIMON FRASER/NORTHUMBRIA CIRCUITS/SCIENCE PHOTO LIBRARY; p82: © Hugh Sitton/ zefa/Corbis; p84: © Howard Davies/CORBIS; p87: Ehrman Photographic / Shutterstock; p92: CREDIT: ALEXANDER LOWRY/SCIENCE PHOTO LIBRARY; p90: CAN BALCIOGLU / Shutterstock; p94: Angelo Giampiccolo / Shutterstock; p96: Martyn Chillmaid; p98: ENVIRONMENTAL IMAGES/ UIG/SCIENCE PHOTO LIBRARY; p102: © John Fryer / Alamy; p104: © Peter Johnson/CORBIS; p107: CREDIT: ASTRID & HANNS-FRIEDER MICHLER/SCIENCE PHOTO LIBRARY; p108: SUE PRIDEAUX/SCIENCE PHOTO LIBRARY; p111: © bjones27 / iStock; p112: © Nigel Cattlin / Alamy; p116: MARTYN F. CHILLMAID/SCIENCE PHOTO LIBRARY; p118: Jane September / Shutterstock; p120: NASA/SCIENCE PHOTO LIBRARY; p122: Ysbrand Cosijn; p127: ANDREW LAMBERT PHOTOGRAPHY/ SCIENCE PHOTO LIBRARY; p128: DmitriMaruta / Shutterstock; p130: © guy harrop / Alamy; p133: Martyn Chillmaid; p134: © Wayne HUTCHINSON / Alamy; p139: MARTYN F. CHILLMAID/SCIENCE PHOTO LIBRARY; p143: Kekyalyaynen / Shutterstock; p144: Martyn Chillmaid; p146: ANDREW LAMBERT PHOTOGRAPHY/SCIENCE PHOTO LIBRARY; p148: ANDREW LAMBERT PHOTOGRAPHY/SCIENCE PHOTO LIBRARY; p154: ANDREW LAMBERT PHOTOGRAPHY/SCIENCE PHOTO LIBRARY; p154: Philip Evans/Visuals Unlimited/ Corbis; p156: By Ian Miles-Flashpoint Pictures / Alamy; p156: © Lester V. Bergman/CORBIS; p156: CREDIT: ANDREW LAMBERT PHOTOGRAPHY/SCIENCE PHOTO LIBRARY; p159: Vietnam Art / Shutterstock; p160: tratong / Shutterstock; p165: pixinoo / Shutterstock; p167: Martyn Chillmaid; p167: Martyn Chillmaid; p167: Martyn Chillmaid; p169: PETICOLAS/ MEGNA/FUNDAMENTAL PHOTOS/SCIENCE PHOTO LIBRARY; p171: kaspargallery / iStock; p172: LARRY W. SMITH / Corbis; p177: © All Canada Photos / Alamy; p181: © brozova / iStock; p183: © mozcann / iStock; p186: © Boaz Rottem / Alamy; p188: GEORGETTE DOUWMA/SCIENCE PHOTO LIBRARY; p190: © NHPA/Photoshot; p192: imantsu / Shutterstock; p194: eClick / Shutterstock; p194: L.F / Shutterstock; p196: 1000 Words / Shutterstock; p199: Getty Images/National Geographic; p199: Mike Volk; p203: © Nigel Cattlin / Alamy; p204: TonyV3112 / Shutterstock; p206: CREDIT: ANDREW LAMBERT PHOTOGRAPHY/SCIENCE PHOTO LIBRARY; p208: CREDIT: MARTIN BOND/SCIENCE PHOTO LIBRARY; p211: © nagelestock.com / Alamy; p214: Martyn Chillmaid; p216: CHAIWATPHOTOS / Shutterstock; p218: © claudiodivizia / iStock; p220: Gerard Koudenburg / Shutterstock; p224: Martyn Chillmaid; p227: KAJ R. SVENSSON/SCIENCE PHOTO LIBRARY; p228: Martyn Chillmaid; p230: bioraven / Shutterstock; p233: © christophe_cerisier / iStock; p236: Martyn Chillmaid; p238: Jeffrey Coolidge / Corbis; p240: ©2007 Rex Features; p245: gnomeandi / Shutterstock; p247: szefei / Shutterstock; p248: satit_srihin/ Shutterstock; CHARLES D. WINTERS/SCIENCE PHOTO LIBRARY.

Artwork by GreenGate Publishing Services and OUP.

Although we have made every effort to trace and contact all copyright holders before publication this has not been possible in all cases. If notified, the publisher will rectify any errors or omissions at the earliest opportunity.

All questions, example answers, marks awarded and comments that appear in this book were written by the author. In examination, the way marks are awarded to answers like these may be different. Cambridge International Examinations bears no responsibility for the example answers which are contained in this publication.

® IGCSE is the registered trademark of Cambridge International Examinations

Seperating techniqus (handwritten annotation)

Contents

Introduction ... 1

Unit 1 Particles and purification ... 2
1.1 Solids, liquids and gases ... 2
1.2 Changing states ... 4
1.3 The kinetic particle model ... 6
1.4 Diffusion ... 8
1.5 Apparatus for measuring ... 10
1.6 Paper chromatography ... 12
1.7 Is that chemical pure? ... 14
1.8 Methods of purification ... 16
1.9 More about purification ... 18
Summary and Practice questions ... 20

Unit 2 Atoms, elements and compounds ... 22
2.1 Inside the atom ... 22
2.2 Isotopes ... 24
2.3 Electronic structure and the Periodic Table ... 26
2.4 Elements, compounds and mixtures ... 28
2.5 Metals and non-metals ... 30
Summary and Practice questions ... 32

Unit 3 Structure and bonding ... 34
3.1 Ionic bonding ... 34
3.2 Covalent bonding (1): simple molecules ... 36
3.3 Covalent bonding (2): more complex molecules ... 38
3.4 Ionic or covalent? ... 40
3.5 Giant covalent structures ... 42
3.6 Metallic bonding ... 44
Summary and Practice questions ... 46

Unit 4 Formulae and equations ... 48
4.1 Chemical formulae ... 48
4.2 Working out the formula ... 50
4.3 Chemical equations ... 52
4.4 More about equations ... 54
Summary and Practice questions ... 56

Unit 5 Chemical calculations ... 58
5.1 Reacting masses ... 58
5.2 Chemical calculations ... 60
5.3 How much product? ... 62
5.4 Percentages and volumes ... 64
5.5 Yield and purity ... 66
5.6 More chemical calculations ... 68
5.7 Titrations ... 70
Summary and Practice questions ... 72

Unit 6 Electricity and chemistry ... 74
6.1 Electrolysis ... 74
6.2 More about electrolysis ... 76
6.3 Explaining electrolysis ... 78
6.4 Purifying copper ... 80
6.5 Electroplating ... 82
6.6 Extracting aluminium ... 84
6.7 Conductors and insulators ... 86
Summary and Practice questions ... 88

Unit 7 Chemical changes ... 90
7.1 Physical and chemical changes ... 90
7.2 Energy transfer in chemical reactions ... 92
7.3 Fuels and energy production ... 94
7.4 Energy from electrochemical cells ... 96
7.5 Fuel cells ... 98
Summary and Practice questions ... 100

Unit 8 Rate of reaction ... 102
8.1 Investigating rate of reaction ... 102
8.2 Interpreting data ... 104
8.3 Surfaces and reaction rate ... 106
8.4 Concentration and rate of reaction ... 108
8.5 Temperature and rate of reaction ... 110
8.6 Light-sensitive reactions ... 112
Summary and Practice questions ... 114

Unit 9 Chemical reactions ... 116
9.1 Reversible reactions ... 116
9.2 Shifting the equilibrium ... 118
9.3 Redox reactions ... 120
9.4 More about redox reactions ... 122
Summary and Practice questions ... 124

Unit 10 Acids and bases ... 126
10.1 How acidic? ... 126
10.2 Properties of acids ... 128
10.3 Bases ... 130
10.4 More about acids and bases ... 132
10.5 Oxides ... 134
Summary and Practice questions ... 136

Unit 11 Making and identifying salts ... 138
11.1 Making salts (1) ... 138
11.2 Making salts (2): Titration method ... 140
11.3 Making salts (3): Precipitation ... 142
11.4 What's that gas? ... 144
11.5 Testing for cations ... 146
11.6 Testing for anions ... 148
Summary and Practice questions ... 150

Contents

Unit 12	**The Periodic Table**	**152**
12.1	The Periodic Table	152
12.2	The Group I metals	154
12.3	The Group VII elements	156
12.4	The noble gases and more	158
12.5	Transition elements	160
	Summary and Practice questions	162
Unit 13	**Metals and reactivity**	**164**
13.1	Alloys	164
13.2	The metal reactivity series	166
13.3	More about metal reactivity	168
13.4	From metal oxides to metals	170
13.5	Thermal decomposition	172
	Summary and Practice questions	174
Unit 14	**Metal extraction**	**176**
14.1	Metals from their ores	176
14.2	Extracting iron	178
14.3	Iron into steel	180
14.4	Uses of metals	182
	Summary and Practice questions	184
Unit 15	**Air and water**	**186**
15.1	Water	186
15.2	Air	188
15.3	Air pollution	190
15.4	The nitrogen oxide problem	192
15.5	Global warming	194
15.6	The carbon cycle	196
15.7	Preventing rust	198
	Summary and Practice questions	200
Unit 16	**The chemical industry**	**202**
16.1	Fertilisers	202
16.2	Making ammonia	204
16.3	Sulfur and sulfuric acid	206
16.4	Manufacturing sulfuric acid	208
16.5	The limestone industry	210
	Summary and Practice questions	212
Unit 17	**Organic chemistry and petrochemicals**	**214**
17.1	Organic chemistry	214
17.2	Hydrocarbons	216
17.3	Fuels	218
17.4	Petroleum	220
	Summary and Practice questions	222
Unit 18	**The variety of organic chemicals**	**224**
18.1	Alkanes	224
18.2	Cracking alkanes	226
18.3	Alkenes	228
18.4	Alcohols	230
18.5	Carboxylic acids	232
	Summary and Practice questions	234
Unit 19	**Polymers**	**236**
19.1	What are polymers?	236
19.2	More about polymer structure	238
19.3	Polyamides and polyesters	240
	Summary and Practice questions	242
Unit 20	**Biological molecules**	**244**
20.1	Natural macromolecules	244
20.2	Mainly carbohydrates	246
20.3	Fermentation	248
	Summary and Practice questions	250

Alternative to Practical section	252
C1 Using and organising techniques, apparatus and material	252
C2 Observing, measuring and recording	254
C3 Handling experimental observations and data	256
C4 Planning investigations	258
Revision checklist	260
Glossary	271
Index	277

Support website:
www.oxfordsecondary.com/9780198399230

Introduction

This book is designed specifically for Cambridge IGCSE® Chemistry 0620. Experienced teachers have been involved in all aspects of the book, including detailed planning to ensure that the content gives the best match possible to the syllabus.

Using this book will ensure that you are well prepared for studies beyond the IGCSE level in pure sciences, in applied sciences or in science-dependent vocational courses. The features of the book outlined below are designed to make learning as interesting and effective as possible:

LEARNING OUTCOMES
- These are at the start of each spread and will tell you what you should be able to do at the end of the spread.
- Some outcomes will be needed only if you are taking a supplement paper and these are clearly labelled, as is any content in the spread that goes beyond the syllabus.

DID YOU KNOW?
These are not needed in the examination but are found throughout the book to stimulate your interest in chemistry.

PRACTICAL
These show the opportunities for practical work. The results are included to help you if you do not actually tackle the experiment or are studying at home.

Extra resources, including answers:
www.oxfordsecondary.com/9780198399230

STUDY TIP...
Experienced teachers give you suggestions on how to avoid common errors or give useful advice on how to tackle questions.

KEY POINTS
These summarise the most important things to learn from the spread.

SUMMARY QUESTIONS
These questions are at the end of each spread and allow you to test your understanding of the work covered in the spread.

At the end of each unit there is a double page of examination-style questions written by the author.

At the end of the book you will also find:

'Alternative to Practical' section – this provides guidance if you are doing this examination paper instead of coursework or the practical examination.

A useful Revision Checklist to help you prepare for the examination in four written papers.

Assessment structure

Paper 1: Multiple Choice (Core)

Paper 2: Multiple Choice (Supplement)

Paper 3: Theory (Core)

Paper 4: Theory (Supplement)

Paper 5: Practical Test

Paper 6: Alternative to Practical

1 Particles and purification

1.1 Solids, liquids and gases

LEARNING OUTCOMES

- State the distinguishing properties of solids, liquids and gases
- Describe the structures of solids, liquids and gases in terms of particle separation, arrangement and motion
- Describe the pressure and temperature of a gas in terms of the motion of its particles

The three states of matter

Substances can be solids, liquids or gases. These are the three **states** of matter. Most substances can exist in all three states. For example, water can exist as ice, liquid water or steam.

All matter is made up of **particles**. Three types of particles make up most matter – atoms, molecules and ions.

An **atom** is the smallest particle that cannot be broken down by chemical means.

A **molecule** is a particle of two or more atoms joined together.

An **ion** is an atom or group of atoms that carries an electrical charge.

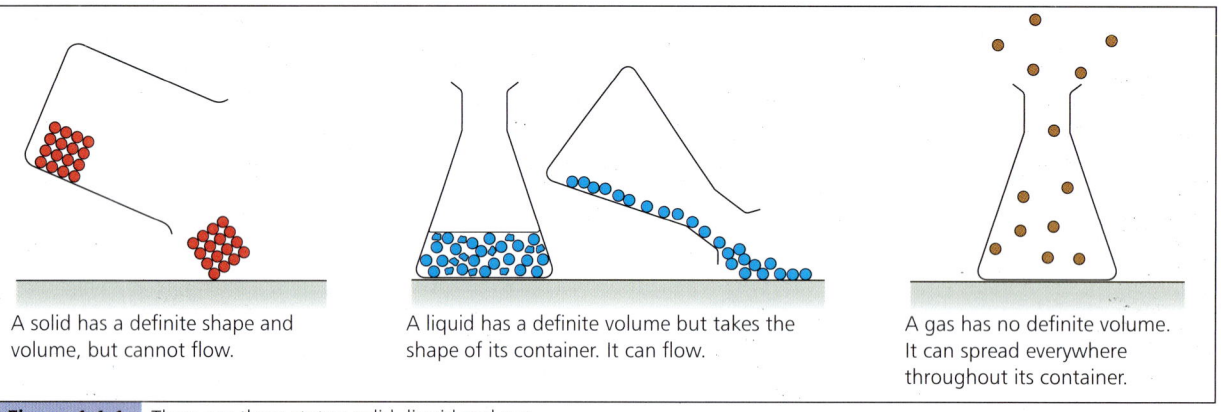

A solid has a definite shape and volume, but cannot flow.

A liquid has a definite volume but takes the shape of its container. It can flow.

A gas has no definite volume. It can spread everywhere throughout its container.

Figure 1.1.1 There are three states: solid, liquid and gas

We can explain the properties of solids, liquids and gases by looking at their particles' arrangement, movement and proximity – how close they are together.

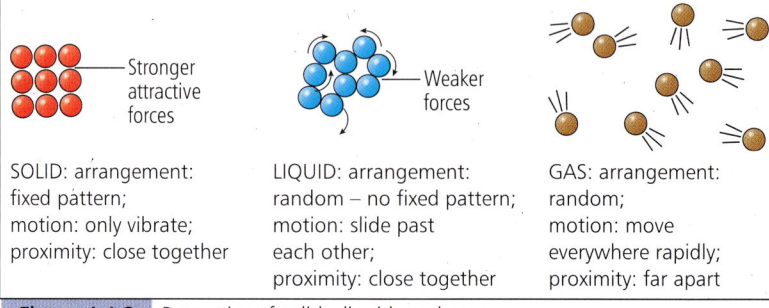

Steam is water in the gaseous state

DID YOU KNOW?

The heat from your hand is enough to change the metal gallium from a solid to a liquid.

SOLID: arrangement: fixed pattern; motion: only vibrate; proximity: close together — Stronger attractive forces

LIQUID: arrangement: random – no fixed pattern; motion: slide past each other; proximity: close together — Weaker forces

GAS: arrangement: random; motion: move everywhere rapidly; proximity: far apart

Figure 1.1.2 Properties of solids, liquids and gases

Compressing gases

We can picture a gas as a collection of randomly moving particles (molecules or atoms) which are continuously colliding with each other and with the walls of their container. The gas particles exert a force on the walls of their container, causing pressure. When we decrease the volume of a gas, the molecules get closer together and hit the wall of the container more often, so the pressure increases (see Figure 1.1.3). The higher the pressure, the closer the particles are to each other.

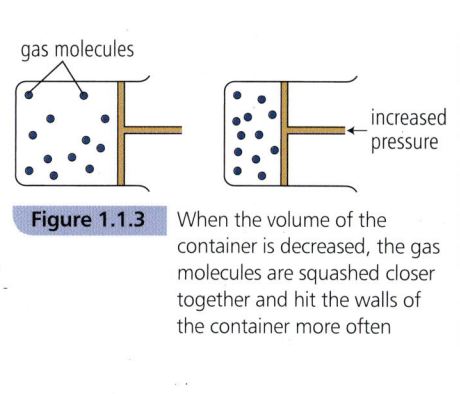

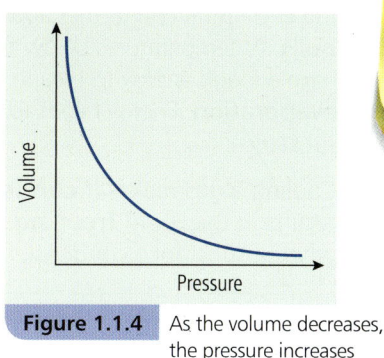

Figure 1.1.3 When the volume of the container is decreased, the gas molecules are squashed closer together and hit the walls of the container more often

Figure 1.1.4 As the volume decreases, the pressure increases

> **STUDY TIP**
>
> Note that increasing temperature increases the average speed of the gas particles, but increasing the pressure at constant temperature has no effect on the speed of the gas particles.

Heating gases

A closed container has a fixed volume. If we heat a gas in a closed container, as the temperature increases, the gas particles move faster and hit the walls of the container with increased force. We say that the molecules have greater kinetic energy (energy associated with movement) at a higher temperature. The higher the temperature, the higher the pressure will be in the container. If the volume of the gas is not fixed, for example in a gas syringe, the volume of gas increases as the temperature increases. This is because the higher the temperature, the faster the gas molecules move – they have more kinetic energy. The greater force exerted by the gas molecules on the syringe plunger pushes the plunger out until the pressure is balanced by the pressure of the atmosphere.

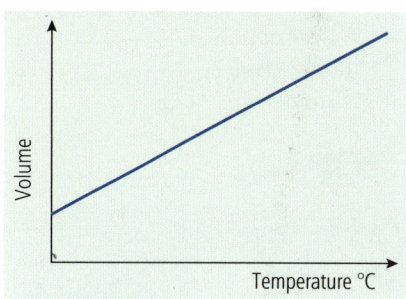

Figure 1.1.5 As pressure increases at constant temperature, the volume of gas decreases

SUMMARY QUESTIONS

1. Which of these phrases refers to
   a gases
   b liquids
   c solids?
   - The particles are close together.
   - The particles are randomly arranged.
   - They have the strongest attractive forces between the particles.

2. Explain why it is dangerous to heat a closed glass jar full of gas.

3. How does the motion and distance between the gas particles change in a closed container when **a** the temperature increases and **b** the pressure increases?

KEY POINTS

1. Solids, liquids and gases can be distinguished by their shapes and how easily they flow or spread out.

2. The properties of solids, liquids and gases can be explained in terms of the closeness, arrangement and motion of their particles.

3. The closer the particles in a gas, the higher the pressure.

4. Increasing the temperature of a gas increases the speed at which the particles move.

1.2 Changing states

LEARNING OUTCOMES

- Describe the changes of state in terms of melting, boiling, evaporation, freezing, condensation and sublimation
- Explain changes of state in terms of the kinetic theory

STUDY TIP

Remember that most of the particles in liquids are touching one another. It is a common error to think that they are well separated.

Changes of state

When we heat a solid, its particles gain energy and vibrate more vigorously. The **forces of attraction** between the particles are weakened and the solid **melts**. Heating to a higher temperature weakens these forces of attraction further and the liquid turns into a gas – the liquid **boils**. At temperatures below the boiling point of a liquid, some particles have enough energy to escape and form a vapour. This process is called **evaporation**. Energy needs to be supplied to melt, boil or evaporate a substance.

Cooling a gas makes it **condense** into a liquid. Further cooling results in the liquid **freezing**. Energy is given out to the surroundings when a substance condenses or freezes.

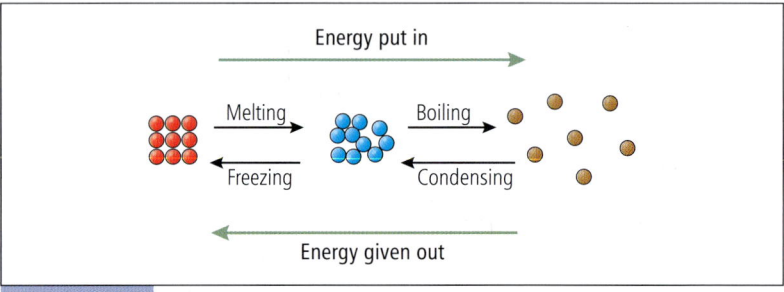

Figure 1.2.1 The changes of state. Energy must be put in to melt and boil a substance. Energy is given out on condensing and freezing.

A few substances change directly from a solid to a gas when heated. No liquid state is seen. On cooling, the gas turns back to the solid state. This change of state is known as sublimation.

$$\text{solid} \xrightleftharpoons[\text{sublimation}]{\text{sublimation}} \text{gas}$$

Explaining changes of state

We can explain changes of state by using the idea that particles are constantly in motion. This is called the **kinetic particle theory**. We can explain a heating curve using ideas about the energy and motion of the particles.

- A–B: Increasing heat energy increases the vibrations of the particles in the solid. So the temperature of the solid increases.
- B–C: The forces of attraction between the particles are weakened enough so that the particles begin to slide over each other. The temperature is constant because the energy supplied is going to overcome the forces between the particles instead of raising the temperature. The substance melts.

Figure 1.2.2 A substance is heated at a constant rate. The flat parts of the curve show where the substance is melting (B-C) and boiling (D-E).

- C–D: Increasing the energy increases the motion of the particles in the liquid. So the temperature of the liquid increases.
- D–E: The forces of attraction between the particles are weakened enough so that the particles move well away from each other. The temperature is constant because the energy supplied is going in to overcome the forces between the particles instead of raising the temperature. The substance boils.
- E–F: Increasing the energy increases the speed of the gas particles. So the temperature increases. The gas particles are far away from each other.

PRACTICAL

A heating curve for salicylic acid

1. A solid called salicylic acid is heated at a constant rate. We can record its temperature at intervals. A graph of temperature against time shows that the temperature does not increase steadily all the way.
2. We can also plot a heating curve directly using a temperature probe attached to a data logger and computer.

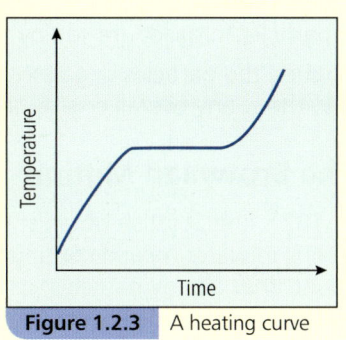

Figure 1.2.3 A heating curve

In the heating curve in Figure 1.2.3, the horizontal part of the graph is where the solid is changing to liquid. This is its melting point. There is no temperature rise here. So the energy supplied is not raising the temperature. The energy is being absorbed to overcome the attractive forces holding the particles of solid in position. A similar thing happens at the boiling point.

KEY POINTS

1. The terms melting, boiling, condensing and freezing are used for specific changes of state.
2. Sublimation is the direct change of state from solid to gas, or gas to solid without a liquid being formed.
3. In melting, boiling and evaporation energy is absorbed (put in).
4. In condensing and freezing, energy is released.
5. When a change of state occurs, there is a change in the type of motion of the particles which can be explained using the kinetic theory.

SUMMARY QUESTIONS

1. Give the names of these changes of state:
   a liquid to gas
   b solid to liquid
   c gas to liquid

2. Copy and complete using the words below:

   **absorbed   energy
   flat   forces   melting**

   When we heat a solid, _____ is absorbed and raises the temperature of the solid. At the _____ point, the energy is _____ to overcome the attractive _____ between the particles rather than raising the temperature. That is why there is a _____ part to the heating curve.

3. What does the term *sublimation* mean?

4. Explain, using the kinetic particle theory, what happens to the motion and closeness of particles when:
   a bromine vapour cools and forms liquid bromine
   b liquid sulfur cools and forms solid sulfur

5. Sketch a graph to show how temperature changes when a gas is cooled slowly over time to form a liquid. Indicate the boiling point of the liquid on your graph.

1.3 The kinetic particle model

LEARNING OUTCOMES

- Show an understanding of Brownian motion as evidence for the kinetic particle model of matter
- Describe diffusion
- Describe and explain Brownian Motion in terms of random molecular bombardment
- Describe the evidence for Brownian Motion

The kinetic particle model

In liquids and gases, the particles are constantly moving and changing directions as they hit other particles. We say that they move in a **random** way. The idea that particles are constantly in motion is called the kinetic particle theory.

In Topic 1.2 you used the kinetic particle theory to explain the properties of solids, liquids and gases. The kinetic particle theory states that:

- particles in gases and liquids move randomly
- particles in gases do not attract each other
- particles in gases are so tiny that their volumes can be ignored
- when the particles in gases collide they bounce off each other without any overall energy change.

The Brownian Motion

In 1827, the Scottish botanist Robert Brown, noticed that pollen grains in water moved continually in a zig-zag way. Later, it was found that all tiny particles suspended in a liquid or gas show the same irregular movement. The type of movement is called **Brownian Motion** and it gives us evidence for the kinetic particle theory.

Brownian Motion is caused by the random, irregular bombardment (hitting) of visible particles, by even smaller particles of the liquid or gases they are suspended in, which cannot be seen. These smaller particles can be molecules, atoms or ions. When the collisions on one side of the particle are stronger, or greater in number, than those on the other sides, the particle moves slightly. Because the bombardment is random and irregular, the particle also moves irregularly. Brownian Motion gives us evidence for the kinetic particle model.

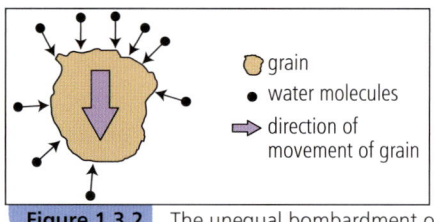

Figure 1.3.2 The unequal bombardment of water molecules on a tiny grain causes it to move in the direction shown

Other evidence for Brownian Motion includes:

- The random motion of tiny particles of graphite or toothpaste suspended in water.
- The random motion of dust or smoke particles in the air.

Larger, heavier particles do not show Brownian Motion. This is because the difference in force of the collisions of molecules on each side of the heavier particles is not large enough to move them.

DID YOU KNOW?

You usually need a microscope to be able to see tiny grains moving in an irregular manner. An exception to this is the movement of dust particles seen in bright sunlight when the air is still. The dust particles seem to dance about in a random way.

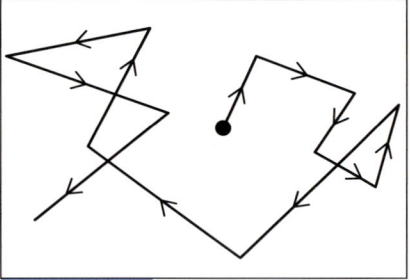

Figure 1.3.1 Brownian Motion is the irregular, random motion of tiny particles suspended in a liquid or gas

Introducing diffusion

When we are cooking, the smell can spread throughout the house. If we put a drop of ink into some water and leave it for a day, the colour of the ink will spread throughout the water. Why is this?

The random movement of different particles so they get mixed up is called **diffusion**. Diffusion results in the particles spreading throughout the space available. The overall direction of the movement is from where the particles are more concentrated to where the particles are less concentrated. However, the particles are moving randomly, so some are moving from less concentrated to more concentrated regions as well!

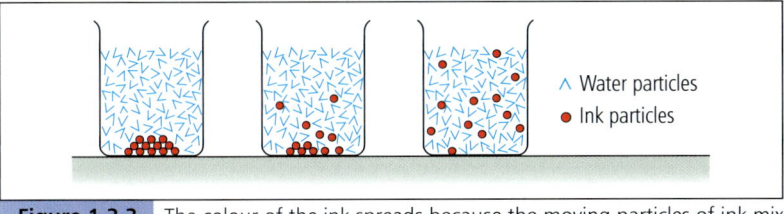

Figure 1.3.3 The colour of the ink spreads because the moving particles of ink mix with the moving water particles

Generally, diffusion occurs only in liquids and gases because the particles are able to move. Diffusion in gases is faster than in liquids. This is because in gases the particles move rapidly but in liquids they move slowly. Diffusion does not occur in solids because the particles are packed tightly together and, although they vibrate, they cannot move around.

Speed of diffusion and molecular mass

The speed at which a gas diffuses depends on how heavy its molecules are. We compare the mass of molecules with each other by using their relative molecular masses (see Topic 5.1). The greater the relative molecular mass, the heavier the molecule.

Molecules that have a lower mass move faster than those with a higher mass. If the light and heavy molecules have the same amount of energy when they collide, the lighter ones will bounce off the heavier ones quicker. So lighter molecules diffuse faster than heavier molecules.

KEY POINTS

1. The kinetic particle model states that the particles in liquids and gases are in constant motion and that when the particles in gases collide, they bouince off each other.
2. Brownian Motion gives us evidence for the kinetic particle model of matter.
3. Diffusion is the random movement of molecules, so that they get mixed up.
4. Brownian Motion can be explained in terms of random molecular bombardment of large visible particles by molecules, atoms or ions.
5. Evidence for Brownian Motion comes from the observations of the motion of tiny visible particles suspended in liquids or gases.

DID YOU KNOW?

The durian fruit found in Malaysia and Indonesia is so smelly that the fruit is not allowed inside most hotels!

SUMMARY QUESTIONS

1. Copy and complete using the words below.

   **different   diffusion
   gases   particles   random**

   The kinetic particle theory states that the _____ in liquids and _____ are in constant _____ motion. When freely moving particles collide, they bounce off each other in _____ directions. If the particles are different they get _____ up. This process is called _____.

2. a Describe Brownian Motion. Include a diagram in you answer.
   b Copy and complete this sentence:

   The movement of particles in Brownian Motion is evidence for the _____ particle model of matter.

3. Tiny pieces of clay suspended in water show a random irregular movement. Explain why by referring to the kinetic particle theory.

1.4 Diffusion

LEARNING OUTCOMES

- Describe and explain diffusion
- Describe and explain how molecular mass affects the rate of diffusion

Explaining diffusion

We have seen in Topic 1.3 that Brownian Motion provides evidence for the kinetic particle theory. Diffusion also provides evidence for the kinetic particle theory. Diffusion occurs in gases because the molecules in gases are constantly moving, colliding with each other and changing directions. This results in the gases spreading out. If we put a gas jar containing (colourless) oxygen above a gas jar of bromine vapour (brown) the molecules of the bromine vapour and oxygen gradually mix because the particles are moving and colliding randomly. The collisions occurring are:

- oxygen molecules with oxygen molecules
- oxygen molecules with bromine molecules
- bromine molecules with bromine molecules.

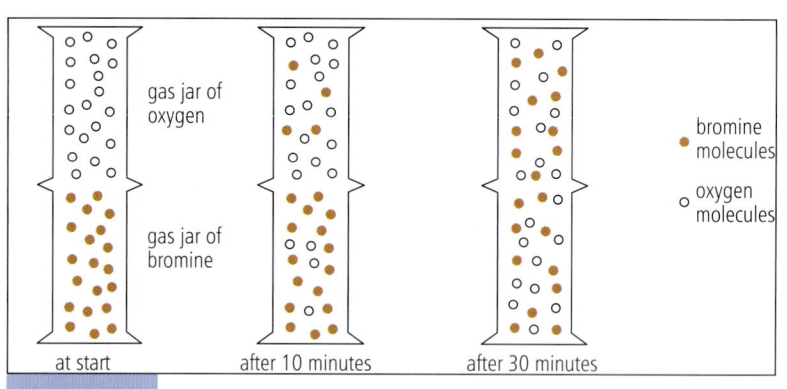

Figure 1.4.1 The molecules in the gas jars collide and move randomly. This leads to the gases mixing by diffusion.

DID YOU KNOW?

Diffusion can also occur in solids but it is extremely slow at room temperature. Atoms of gold and lead will diffuse into each other if the surfaces of the two metals are in close contact. If left for years in close contact, the lead and gold appear to be completely 'stuck together'. Some of the carbon atoms in iron alloys (see Topics 13.1 and 14.4) can diffuse into another metal when the two metals are placed in contact. An important application of diffusion in solids involves adding 'doping' metal atoms to a silicon microchip.

Diffusion can also occur in liquids which are miscible (are able to mix together) or when solids dissolve in liquids. When sugar is added to water, it first dissolves. The dissolved sugar then diffuses through the water because of the random collisions of the molecules. Eventually, the sugar molecules are spread throughout the water. The rate of diffusion is slower in liquids than in gases because there are weak attractive forces between the liquid particles, but the kinetic particle theory assumes that there are no forces of attraction between gas particles. So, particles in a liquid move less rapidly than the particles in a gas.

The speed at which a gas diffuses depends on the mass of its molecules. In Figure 1.4.2 the white ring is nearer the hydrogen chloride end of the tube. This shows that hydrogen chloride has heavier molecules than ammonia. We compare the masses of molecules by using their relative molecular masses (see Topic 5.1). The higher the relative mass, the heavier are the molecules. So, hydrogen chloride is heavier than ammonia.

> At the same temperature, molecules that have a lower mass move, on average, faster than those with a higher mass. If the light and heavy molecules have the same amount of energy when they collide, the lighter ones will bounce off the heavier ones quicker. So, lighter molecules diffuse faster than heavier molecules.

DEMONSTRATION

Comparing speeds of diffusion

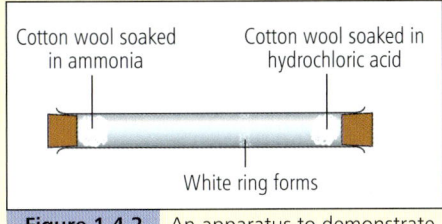

Figure 1.4.2 An apparatus to demonstrate diffusion of gases

A long glass tube is set up as shown.

Concentrated hydrochloric acid gives off fumes of a colourless gas called hydrogen chloride. Concentrated ammonia solution gives off colourless ammonia gas. These gases diffuse along the tube.

After a few minutes a white ring is seen nearer one end of the tube. The molecules of ammonia and hydrogen chloride have diffused. When the molecules of ammonia and hydrogen chloride collide with each other they react and form a white solid, ammonium chloride. The white ring is nearer the hydrogen chloride end of the tube. This shows that hydrogen chloride is a heavier molecule than ammonia.

The bromine gradually diffuses throughout the space available

SUMMARY QUESTIONS

1. Explain the following using the kinetic particle theory:
   a. A layer of water is placed carefully over a layer of red dye solution so that they do not immediately mix. After two days, the red colour has spread throughout the water.
   b. A bottle of perfume is opened at the front of the classroom. After a little while, you can smell the perfume at the back of the classroom.
2. Explain why diffusion in gases is faster than diffusion in liquids.
3. The relative molecular masses of four gases are: carbon dioxide 44; methane 16; nitrogen 28; oxygen 32. Put these gases in order of their rate of diffusion, with the fastest first.

KEY POINTS

1. Diffusion is the random movement of particles. This random movement leads to mixing up of the particles of different gases.
2. Diffusion in two miscible liquids is slower than diffusion in gases.
3. The rate of diffusion depends on the relative molecular mass. Substances with a higher relative molecular mass diffuse more slowly than those with a lower relative molecular mass.

1.5 Apparatus for measuring

LEARNING OUTCOMES

- Name appropriate apparatus for the measurement of time, temperature, mass and volume

DID YOU KNOW?

The Indus Valley civilisation in what is now Pakistan made very accurate measurements of mass, length and time 4600 years ago. Their smallest unit of length was equivalent to 1.704 metres.

Using a volumetric (graduated) pipette

Mass, time and temperature

In chemistry we often have to take accurate measurements of mass, time and temperature.

The standard unit for mass is the kilogram (abbreviation kg). Chemists, however, often find it more convenient to work in grams (abbreviation g). We usually use a top pan balance accurate to two decimal places to weigh out chemicals in the laboratory. Since many balances nowadays have a digital readout, measuring mass is fairly straightforward as long as you remember to set the balance to zero with the weighing boat on it.

Sometimes we need to know how quickly a reaction proceeds. In order to do this, we use a stop-clock. Electronic stop-clocks often read seconds to two decimal places. We can achieve this precision, however, only if we connect the stop-clock to another electronic measuring system, for example a data logger and computer.

Temperature is measured in degrees Celsius, written °C. The thermometer you use in the laboratory can sometimes be read to an accuracy of one-tenth of a degree. Very often we do not need this accuracy and a reading to the nearest degree is acceptable. When you take a thermometer reading, remember **not** to take it out of the liquid when you read it. If you do this it will cool down and so give an incorrect reading.

Measuring volumes

Volumes of liquids

In chemistry, we generally measure volumes in centimetres cubed (cm^3) or decimetres cubed (dm^3).

Since $1\,dm^3 = 1000\,cm^3$ we can easily change cm^3 into dm^3 by dividing by 1000.

Four pieces of apparatus are often used in chemistry for measuring volumes of liquids. These are shown below.

Measuring cylinder — Volumetric (graduated) pipette — Burette — Volumetric flask

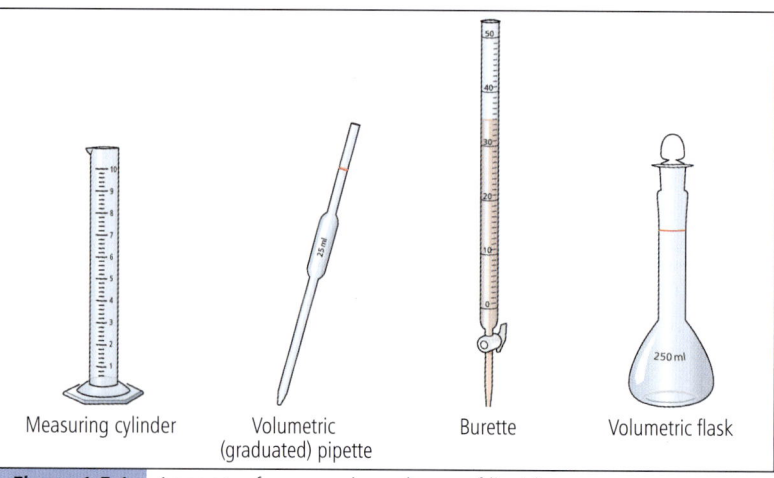

Figure 1.5.1 Apparatus for measuring volumes of liquid

There are various sizes of **measuring cylinders** and **volumetric flasks** ranging from 5 cm^3 to 1 dm^3. **Volumetric pipettes** are made in only a few sizes, the most common being 10 cm^3 and 25 cm^3. A **burette** is used to accurately deliver up to 50 cm^3 of liquid.

When we measure out volumes of liquids, we need to think about the accuracy required. A burette or a volumetric pipette is much more accurate than a measuring cylinder. The scale divisions on a burette can be read to the nearest 0.1 cm^3 but a 100 cm^3 measuring cylinder might have scale divisions only every 2 cm^3.

A volumetric pipette is used to measure out a single fixed volume of liquid very accurately. A burette is used if you want to measure out volumes more accurately than by using a measuring cylinder. It is also used in titrations (see Topic 11.2) when you are not sure of the exact volume of the solution you will be adding.

Measuring cylinders are most useful for making up fairly large volumes of solution or in cases where accuracy is not as important. If we want to make up a solution of a solid dissolved in a liquid accurately we use a volumetric flask. We add the weighed solid to the volumetric flask, then add the liquid until it reaches the line scratched on the glass.

Volumes of gases

Gas volumes can be measured using either a **gas syringe** or an upturned measuring cylinder. When a measuring cylinder is used in this way it is completely filled with water and then turned upside down in a bowl of water. The gas pushes the water down – we say it displaces the water. A burette can also be used in this way for greater accuracy.

Accuracy

Accurate measurements are very close to the true value. You are more likely to get accurate results for your experiments if you:

- repeat your measurements in the same way each time
- use apparatus with small scale divisions
- use the apparatus carefully.

> **STUDY TIP**
>
> When measuring out volumes, think about the accuracy required. A burette or volumetric pipette is far more accurate than a measuring cylinder.

Figure 1.5.2 When reading the level of a liquid accurately, your eye should be in line with the bottom of the meniscus

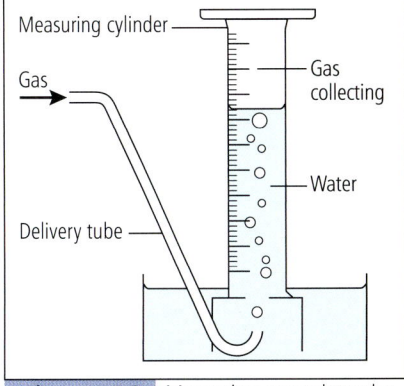

Figure 1.5.3 Measuring gas volume by displacement of water

SUMMARY QUESTIONS

1 What piece of apparatus would you use to:
   a measure mass
   b measure out 500 cm^3 of water
   c measure out 5 cm^3 of a liquid very accurately?

2 State the name of the units and their abbreviations for:
   a mass
   b volume
   c temperature.

3 State three ways you can improve the accuracy of results for an experiment.

KEY POINTS

1 In chemistry, mass is measured in grams, temperature in °C and volume in cm^3 or dm^3.

2 The apparatus you select for an experiment depends on the accuracy required in your experiment.

3 Volumes of gases can be measured using a gas syringe or by displacement of water.

1.6 Paper chromatography

LEARNING OUTCOMES

- Demonstrate knowledge and understanding of paper chromatography
- Interpret simple chromatograms

Supplement
- Interpret simple chromatograms using R_f values
- Outline how chromatography techniques can be applied to colourless substances using locating agents

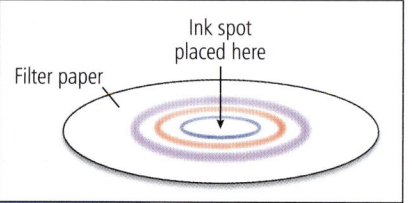

Figure 1.6.1 This blue ink contains three different dyes which are separated on the filter paper

DID YOU KNOW?

The Russian botanist Mikhail Tsvet in 1901 was the first to separate colours from plant leaves. His name was well matched – in the Russian language *Tsvet* means colour.

STUDY TIP

When drawing chromatography apparatus, you must draw the base line or origin line on the chromatogram so that it is above the starting level of the solvent.

Simple chromatography

We make use of many coloured chemicals in our lives. Ink and food colourings are just two examples. These are often mixtures of several different dyes. We can demonstrate this by placing a drop of ink on a piece of filter paper. If you then add a few drops of water to the ink, the colour spreads out from the ink drop and separates into several different colours.

This method of separating pigments (coloured substances) using filter paper is called paper **chromatography**. The colours separate if:

- the pigments have different solubilities in the **solvent,** and/or
- the pigments have different degrees of attraction for the filter paper.

These two factors determine how fast the pigments move across the filter paper. If the mixture of pigments is not soluble in water, other solvents such as ethanol or propanone can be used. The type of solvent also affects how far the pigment moves across the paper.

We can get more information about the substances present in a mixture of dyes by using a special chromatography apparatus.

PRACTICAL

Making a chromatogram

1. First draw a pencil base line across a piece of chromatography paper.
2. Then place a spot of the concentrated dye mixture, M, on the base line using a very fine pipette.
3. Then put some spots of some pure dyes that you think the mixture might contain, A, B and C, on the line as well. The chromatography paper is put in a jar with the solvent. Make sure that the solvent level is below the level of the spots otherwise the dye will wash off into the solvent. As the solvent moves up the paper, the dyes in the mixture separate from each other.

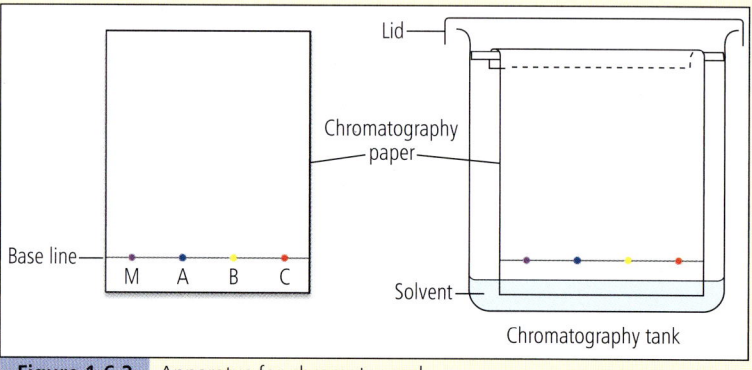

Figure 1.6.2 Apparatus for chromatography

How can we interpret the chromatogram shown here?

The mixture, M, has separated into three dyes. Two of the pure dyes have risen to the same height as two of the dyes in the mixture. So the mixture M contains dyes A and C but not dye B.

From this we can see that chromatography has two uses:
- identifying the substances in a mixture
- separating out substances in a mixture and thereby purifying them.

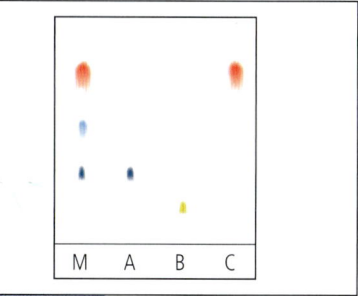

Figure 1.6.3 M contains dyes A and C

More about chromatography

Chromatography can also be used to identify colourless substances. We do this by carrying out chromatography in the same way but this time we mark a line near the top of the paper to show where the solvent has reached – this is the solvent front. The chromatography paper is then dried and sprayed with a chemical called a **locating agent**. The locating agent reacts with the chemicals in the colourless spot and a coloured compound is formed. The colour is usually developed by warming the paper in an oven. Different locating agents are used for different types of compounds. For example, amino acids form coloured spots on the paper when sprayed with ninhydrin. We can identify the substances on the chromatogram by comparing how far the spots have moved from the base line compared with the solvent front. This is called the R_f **value**.

$$R_f \text{ value} = \frac{\text{distance from base line to the centre of the spot}}{\text{distance of solvent front from the base line}}$$

For example, here:

R_f of A $= \frac{4}{6} = 0.67$

R_f of B $= \frac{1.5}{6} = 0.25$

When R_f values have been calculated they can be compared with tables of known R_f values and the compounds present identified. R_f values vary with the solvent used.

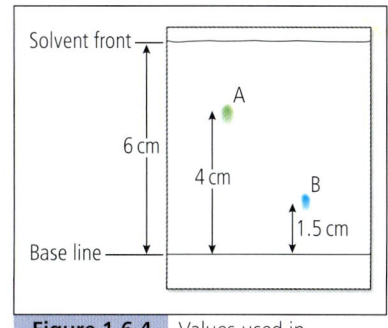

Figure 1.6.4 Values used in calculating R_f

A chromatogram of different coloured dyes

SUMMARY QUESTIONS

1. Copy and complete using the words below:

   **filter   mixture
   solubility   solvent**

   Chromatography can be used to separate a _____ of dyes. The _____ of the dyes in the _____ determines how far they travel up the _____ paper.

2. When setting up a chromatogram, suggest why the base line where the dyes are placed is not drawn with a pen.

3. The R_f values of four amino acids are shown below:

   alanine 0.38; lysine 0.14; serine 0.27; valine 0.60

   Put these amino acids in order of how far they would move up the chromatography paper. Put the one that would move farthest first.

KEY POINTS

1. Chromatography is a method for separating and purifying coloured compounds using filter paper and a solvent.
2. Chromatography can be used to identify compounds.
3. Locating agents are used to make colourless compounds visible on a chromatogram.
4. The compounds on a chromatogram can be identified using their R_f values.

1.7　Is that chemical pure?

LEARNING OUTCOMES

- Identify substances and assess their purity from melting point and boiling point information
- Understand the importance of purity of substances used in everyday life

Figure 1.7.1 Pure and impure water

Pure water contains only water molecules

Tap water is not pure; it contains minerals and chlorine, as well as water

It is important that each substance present in baby food is pure

DID YOU KNOW?

Helium has the lowest melting point of all the elements. It solidifies under pressure only at a temperature slightly lower than −272 °C.

Why is purity important?

In everyday life we use the idea of purity in a very inexact way. We might read 'pure orange juice' on the side of a drink bottle. When chemists talk about **purity** they mean that there is only one substance present. In orange juice there are hundreds of different compounds present so it can never be pure! It is very difficult to make substances absolutely pure. Even distilled water is not pure. It contains tiny amounts of material from the distillation apparatus and dissolved gases from the air.

It does not always matter if a substance is impure. But sometimes purity is important. Many chemical companies make medical drugs and food additives. Because we take these into our bodies, they must be very pure. If a substance has a small amount of unwanted substance mixed with it, the unwanted substance is called an **impurity**. Impurities in food additives or medical drugs may have harmful effects on health. The medical drugs are often mixed with other substances to form a tablet but these substances must also be very pure.

The computer industry needs very pure silicon to make computer chips. If there are tiny particles of unwanted substances in the silicon, the chips will not work properly.

How do we know if a substance is pure?

In the last topic you saw that chromatography can be used to separate coloured substances. This method is also valuable in checking for purity. If only a single spot is seen on the chromatogram, the substance is likely to be pure. If several spots are seen, the substance is impure. We cannot use chromatography to test every substance for purity but we can use another method – we can measure the melting and boiling points of a substance.

Most pure substances have distinct melting and boiling points. The **melting point** is the temperature at which a solid changes to a liquid. The **boiling point** is the temperature at which a substance changes from a liquid to a gas. Melting points and boiling points can be used to identify a substance. We can use tables of melting and boiling points to identify a pure substance.

Pure substances have a sharp melting point and boiling point. For example, the boiling point of water is exactly 100 °C.

If a substance is impure, the impurities will have an effect on the melting and boiling points.

The presence of an impurity in a substance has several effects on the melting and boiling points:

- The melting and boiling points are not sharp – the substance melts and boils over a narrow range of temperatures. The more impurities, the wider the range of temperatures over which melting or boiling occurs.

- The boiling point is increased by impurities. For example, adding salt to water can increase the boiling point of 'water' to 102 °C. The greater the amount of salt present, the greater the increase in the boiling point.
- The melting point is decreased by impurities. Adding salt to water decreases its melting point. The more salt added, the greater the decrease in melting point. This has its uses: salt can be put on the roads in icy weather so that ice is less likely to form.

How do we purify mixtures?

In most chemical reactions several products are made. These products often need to be separated from unused reactants or other impurities. We can use several simple methods to purify a mixture depending on the state of the substance we want to obtain. None of these results in an absolutely pure substance but they are suitable for general school laboratory use.

We can separate an undissolved solid from a liquid or a solution by several methods. These include filtration, decanting and centrifugation. Once separated, the solid is washed to remove any solution that might be between the solid particles.

Evaporation of solvent can be used to separate a dissolved solid from a liquid. The solution is heated in an evaporating basin. The solvent boils off leaving the solid behind. Leaving the solvent to evaporate at room temperature has the same effect. This will not produce a pure solid, however, if there is more than one dissolved solid in the original solution.

Fractional distillation is used to separate a mixture of liquids that have different boiling points.

You will learn more about these methods in the next two topics.

> **STUDY TIP**
>
> Remember that pure substances have definite sharp melting points and boiling points. Impure substances melt and boil over a range of temperatures.

Pure water boils at 100 °C exactly

Impure water boils above 100 °C depending on the concentration of the dissolved salt

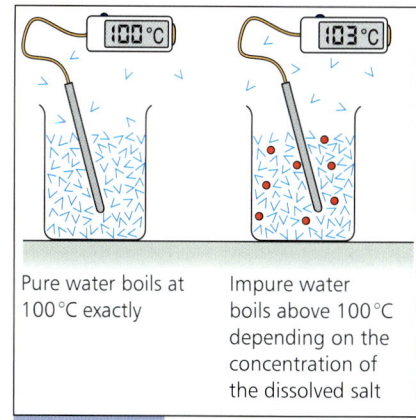

Figure 1.7.2 Boiling pure and impure water

SUMMARY QUESTIONS

1. Copy and complete using the words below:

   lower   pure   sharp   solution   wider

   Pure water has a _sharp_ melting point. A _solution_ of salt in water melts over a _wide_ range of temperature and has a _lower_ melting point than _pure_ water.

2. Give two examples where purity is important in everyday life.

3. Which of the following are pure and which are impure?
   a tap water — I
   b a crystal of sulfur — P
   c a mint-flavoured sweet — I
   d seawater — I

KEY POINTS

1. Medical drugs and food additives must be pure to avoid harming people.
2. Melting points and boiling points can be used to identify pure substances.
3. A pure substance melts and boils at definite temperatures. An impure substance melts and boils over a range of temperatures.

1.8 Methods of purification

LEARNING OUTCOMES

- Describe and explain methods of purification by filtration and crystallisation
- Describe and explain methods of purification by the use of a suitable solvent

Separating a solid from a solution

Filtration

A **solution** contains a solid dissolved in a solvent. The dissolved solid is called the **solute**. The liquid that dissolves the solid is the **solvent**.

An undissolved solid can be separated from a solution or liquid by passing it through a piece of filter paper in a filter funnel. This is called **filtration**. The solution which passes through the filter paper is called the **filtrate**. The solid that stays on the filter paper is called the **residue**. The solid should be washed with distilled water to remove any solution between the solid particles. The solid is then dried in an oven.

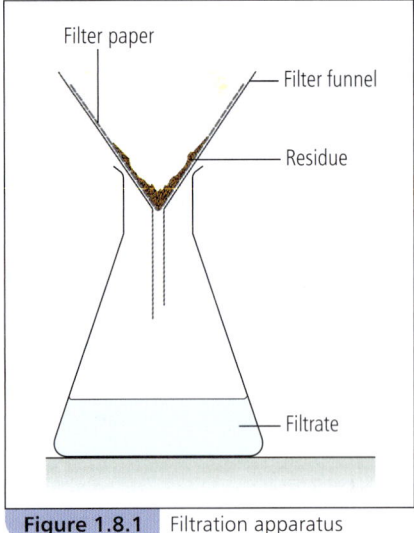

Figure 1.8.1 Filtration apparatus

We can use filtration to separate a solid from a solution

Decanting and centrifugation

Decanting is simply pouring off the solution. It is suitable for solids that have very heavy particles – for example, to separate sand from water.

A **centrifuge** is a machine which spins test tubes around and around at very high speeds. The spinning pulls the solid to the bottom of the tube. You can then decant the liquid from the solid. To get rid of any solution between the particles of solid, you can break up the solid, wash the solid with water and centrifuge again.

Crystallisation

Crystallisation is used to obtain a crystalline solid from a solution. The solution is gently heated in an evaporating basin to concentrate it. The solvent, usually water, is evaporated until the crystallisation point is reached. You can tell when this is by placing a drop of solution onto a cold tile from time to time. If crystals form quickly, the crystallisation point has been reached. The concentrated solution is then left to cool.

STUDY TIP

When describing crystallisation, writing just 'heat the solution' is not as complete as writing 'evaporate off some of the water and then leave to cool'.

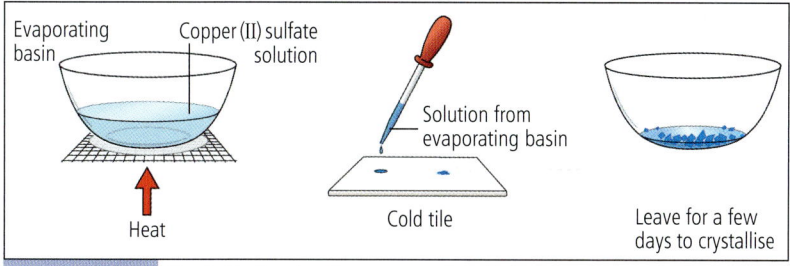

Figure 1.8.2 Crystallisation involves evaporation to the point of crystallisation. Crystals form after the solution is cooled.

DID YOU KNOW?
The largest diamond in the world was the Cullinan diamond. When cutting the crystal Joseph Asscher had a doctor and nurse present because he was so worried about his nerves if he made a mistake.

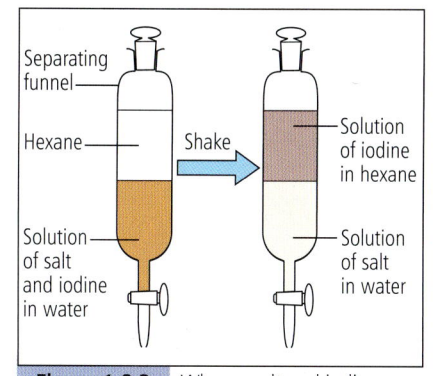

Crystals eventually form at the bottom of the evaporating basin. These can then be filtered off and dried between pieces of filter paper.

If two dissolved substances have different solubilities at different temperatures, fractional crystallisation can be used to separate them. A warm concentrated solution containing the two solutes is cooled. The solute with the lower solubility forms crystals. The solute with the higher solubility remains in solution. The more insoluble solute can be removed by filtration.

Solvent extraction

Solvent extraction can be used to separate two solutes dissolved in a solvent. This is especially useful if one of the solutes is volatile (evaporates readily). A second solvent is used to extract one of the solids from the first solvent. The second solvent must not mix with the first – we say it is immiscible. For example, in Figure 1.8.3 we have a solution of iodine and salt dissolved in water and we want to separate the iodine. We shake the solution of iodine and salt with a solvent called hexane. We do this in a separating funnel. After shaking, the iodine has moved to the hexane layer. The salt will remain in the water layer.

Figure 1.8.3 When a salt and iodine solution in water is shaken with hexane, the iodine moves to the hexane layer because the iodine is more soluble in hexane than in water. The salt is less soluble in hexane than in water.

SUMMARY QUESTIONS

1 Suggest suitable methods to:
   a separate crushed chalk from a mixture of chalk and water
   b get magnesium chloride powder from a solution of magnesium chloride in water
   c get crystals of calcium chloride from a solution of calcium chloride.

2 Copy and complete using the words below:

   **centrifugation   filtrate   mixture   residue
   solid   trapped**

   An insoluble _____ can be separated from a liquid by filtration or _____. When a _____ of a solid and a liquid is filtered, the solid _____ on the filter paper is called the _____ and the liquid which passes through is called the _____.

3 Describe how you can get pure, dry crystals of sodium chloride from a solution of sodium chloride in water.

KEY POINTS

1 Solids can be separated from solutions by filtration, decanting or centrifugation.

2 Crystals are formed when a solution of a crystalline solid is partly evaporated then allowed to cool.

3 Solvent extraction can be used to separate two solids dissolved in a liquid.

1.9 More about purification

LEARNING OUTCOMES

- Describe and explain distillation including the use of a fractionating column
- Suggest suitable purification techniques given information about the substances involved

Simple distillation

Simple distillation is used to obtain a solvent from a solution.

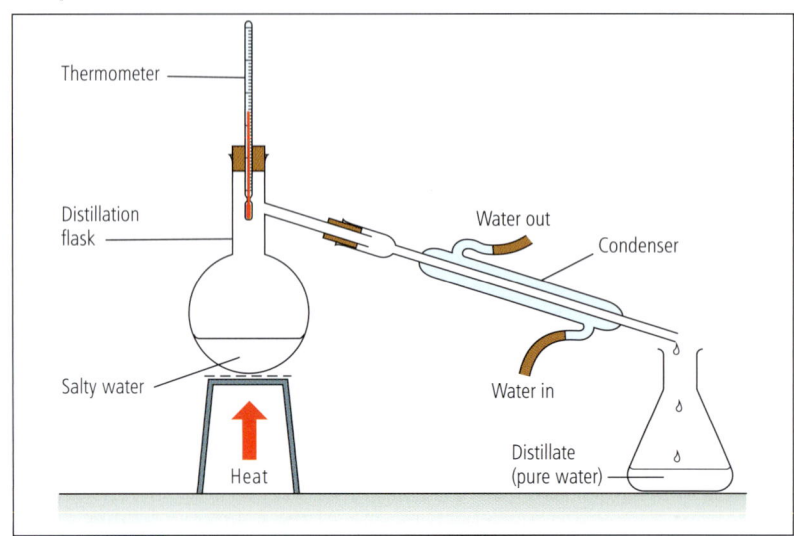

Figure 1.9.1 Simple distillation separates water from dissolved salt. The water is collected in the conical flask.

When a solution of salt in water is heated, the water boils and escapes as steam, leaving the salt behind as a solid. The water is volatile – it has a relatively low boiling point and easily changes to the gaseous state. The steam turns back into water in the condenser. This is because the condenser is cold. The temperature here has fallen below the boiling point of water. The salt remains in the distillation flask because it has a high boiling point. You can see that distillation is a combination of two processes: evaporation and condensation.

Simple distillation can be used to get drinking water from salt water on a large scale. Many countries in the Middle East have built huge, simple distillation plants to get their drinking water.

Fractional distillation

Fractional distillation is used to separate a mixture of liquids with different boiling points. Liquids which are miscible (they can mix with each other) can be separated by this method. The method is very important for the separation of petroleum fractions (see Topic 17.4) and separating ethanol from water (see Topic 20.3). Fractional distillation uses a tall column in which continuous evaporation and condensation of the liquid mixture occurs. There is a range of temperatures in the column – higher at the bottom and lower at the top. When vaporised, the more volatile compounds in the liquid (those with lower boiling points) move further up the column than the less volatile compounds. Eventually, the most volatile compound reaches the condenser, where it changes to a liquid. This is collected as the distillate. Less volatile compounds move more slowly up the column and may eventually reach the condenser. They condense one at a time as **fractions** in order of increasing boiling points.

Jasmine oil is extracted by adding water to jasmine flowers and distilling off the oil

DID YOU KNOW?

In Bhutan, steam distillation is used to make flavouring and perfume from lemon grass.

STUDY TIP

When choosing a method to purify a mixture, think about the states and solubilities of the substances in the mixture.

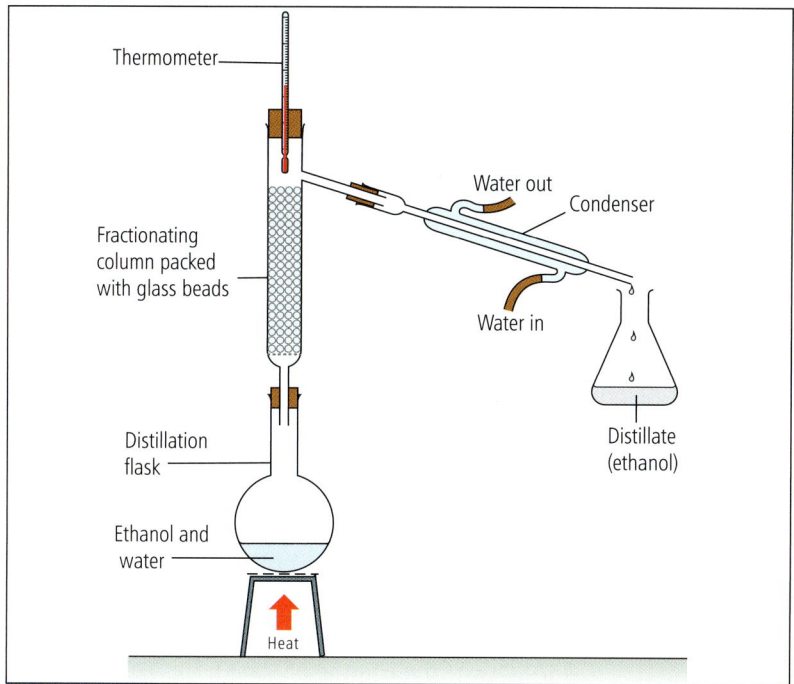

Figure 1.9.2 Fractional distillation is used to separate a mixture of liquids. The more volatile liquid collects in the conical flask.

Which method of purification?

To choose the best method for purifying a mixture you must have a clear idea of how each method works. You may have to use a combination of methods to separate the mixture you want from the unwanted substances. It is often useful to know about the solubility of the substances you are dealing with. Here is an example:

How can you separate a mixture of sand and salt to obtain pure dry samples of sand and salt crystals?

These are both solids, so you can not separate them out by a single method. But salt is soluble in water and sand is not.

- Add water and stir to dissolve the salt.
- Separate the sand from the salt solution by filtration – the sand is the residue and the salt solution is the filtrate.
- Use the process of crystallisation to form the salt crystals.
- Rinse the sand with distilled water.
- Dry the sand and salt crystals separately on filter papers.

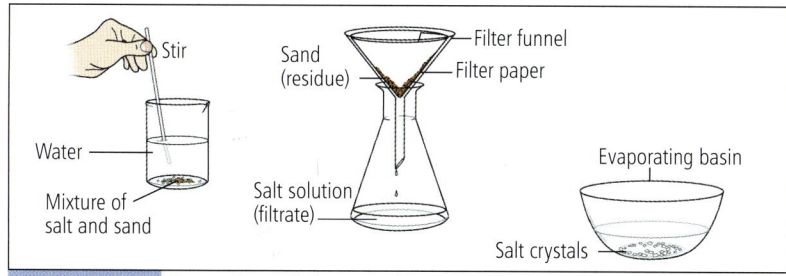

Figure 1.9.3 Separating salt from sand

KEY POINTS

1. Simple distillation is used to separate water from a dissolved salt.
2. Fractional distillation is used to separate more volatile liquids from less volatile liquids.
3. Purification of a mixture often involves a combination of methods.

SUMMARY QUESTIONS

1. Copy and complete the paragraph using these words:

   **boiling   column
   distillation   lower
   temperature   volatile**

   Fractional _____ separates more volatile liquids from less _____ liquids. The more volatile compounds have _____ boiling points. They move further up the distillation _____ Each compound condenses when the _____ in the column falls below its _____ point.

2. Which method or methods can be used to separate the following mixtures?
   a The two volatile liquids ethanol and octanol.
   b A mixture of solid copper(II) sulfate and sand.
   c Water from an aqueous solution of copper(II) sulfate.
   d A mixture of amino acids of similar solubility.

19

SUMMARY QUESTIONS

1. Give definitions of:
   (a) diffusion
   (b) evaporation
   (c) condensation
   (d) distillation.

2. Match each of the words on the left with two of the statements on the right.

solid	particles close together
	particles move everywhere
liquid	particles far apart
	can flow but has a definite surface
gas	has a definite shape
	particles only vibrate

3. Match each piece of apparatus on the left with the phrase on the right that describes it.

volumetric flask	can be used to measure out approximately 20 cm³ of a solution
measuring cylinder	can be used to mix two solutions
burette	can be used to make up 500 cm³ of a solution accurately
beaker	used to change a vapour back to a liquid
condenser	can be used to measure out 18.5 cm³ of a liquid accurately

4. Draw a flow diagram to show how you can separate a mixture of coloured dyes by chromatography.

5. Use ideas about particles to explain why:
   (a) a balloon gets bigger when you blow into it
   (b) solids have a fixed shape
   (c) you can't squeeze a sealed syringe full of water.

1. Which one of these pieces of apparatus is best used for measuring out 13.5 cm³ of acid accurately?

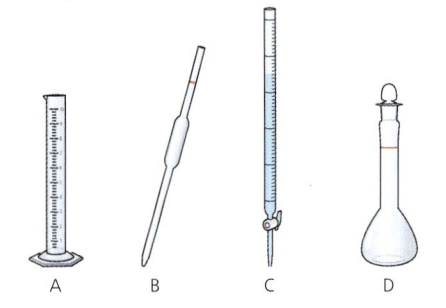

   A B C D

(Paper 1)

PRACTICE QUESTIONS

2. Which one of these methods is used to separate a mixture of liquids?
   A filtration   B crystallisation
   C evaporation   D distillation

   (Paper 1)

3. A crystal of a water-soluble red dye was placed in a beaker of water.
   (a) Describe what you would see (i) after 10 minutes (ii) after several days. [2]
   (b) Describe the arrangement and motion of the particles in:
      (i) the crystal [2]
      (ii) the water. [2]
   (c) Describe how to get dry crystals of the dye from a solution of the dye in water. [3]

   (Paper 3)

4. (a) Name the change of state from:
      (i)   solid to liquid
      (ii) liquid to gas
      (iii) gas to liquid. [3]
   (b) Which two of these changes of state occur when energy is absorbed? Explain your answer using ideas about forces between particles. [2]
   (c) Describe the arrangement and motion of the particles in a gas. [2]

   (Paper 3)

5. Chromatography can be used to separate a mixture of coloured dyes.
   (a) Draw a diagram to show the apparatus used to carry out chromatography. [3]
   (b) Three different mixtures of dyes A, B and C were spotted onto a piece of chromatography paper.

      Two pure dyes, D and E, were also spotted onto the same piece of paper.

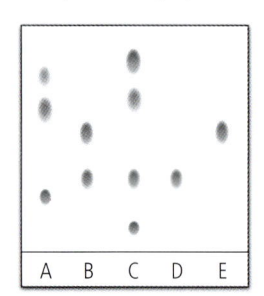

      A B C D E

(i) Where was dye A placed at the start of the experiment? *[1]*

(ii) Which mixture contained the greatest number of different dyes? *[1]*

(iii) Which dye contains both the pure dyes D and E? *[1]*

(iv) Which dye mixture contains neither dye D nor dye E? *[1]*

(Paper 3)

6 A student set up the apparatus shown below.

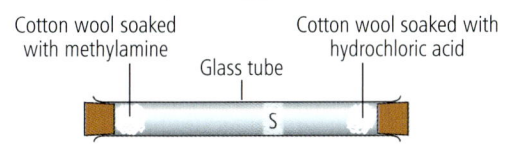

Cotton wool soaked with methylamine — Glass tube — Cotton wool soaked with hydrochloric acid — S

Methylamine and hydrochloric acid give off vapours which react with each other to form a white ring at point S in the tube.

(a) Use ideas about moving particles to explain:
(i) the process of evaporation from the cotton wool *[2]*
(ii) how the particles of methylamine and hydrogen chloride move along the tube. *[2]*

(b) Explain the position of the white ring, S, in the tube. *[2]*

(c) An ammonia molecule has about half the mass of a methylamine molecule. If ammonia was used in this experiment in place of methylamine, predict:
(i) the position of the white ring *[1]*
(ii) why the white ring would be in this position. *[2]*

(Paper 4)

7 Chromatography was used to separate a mixture of amino acids using ethanol as a solvent. The result is shown in the diagram below.

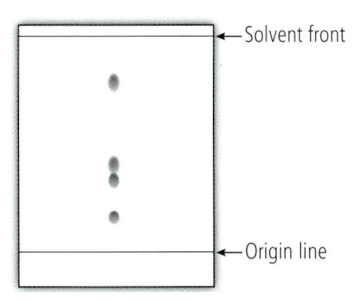

(a) How many different amino acids were there in the mixture? *[1]*

(b) Copy the diagram and mark with an 'M' the amino acid which is the most soluble in ethanol. *[1]*

(c) Calculate the R_f value of the amino acid 'M' using the diagram shown. *[1]*

(d) Some amino acids are not easily separated by paper chromatography using ethanol. Suggest how you could separate these amino acids. *[1]*

(e) Amino acids are colourless compounds. Describe how the spots of the amino acids can be made visible. *[2]*

(Paper 4)

8 Tiny particles of graphite suspended in water show random irregular movement.

(a) Name this type of movement. *[1]*

(b) Explain, using the kinetic particle theory, the reason for this random movement. *[5]*

(Paper 4)

9 The diagram shows a porous pot, which allows the passage of gases through tiny holes in the wall.

(a) Describe how this apparatus can be used to show that hydrogen has a lower molecular mass than carbon dioxide. *[3]*

(b) Explain your answer to part (a) using the kinetic particle theory. *[5]*

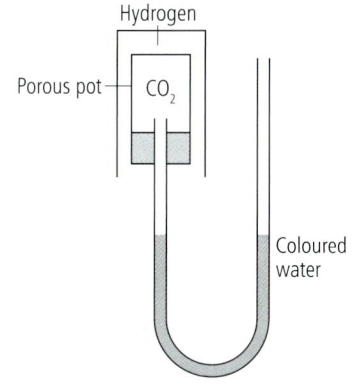

(Paper 4)

2 Atoms, elements and compounds

2.1 Inside the atom

LEARNING OUTCOMES

- State the relative mass and charge of a proton, neutron and electron
- Define proton number and nucleon number
- Use proton number and the simple structure of atoms to explain the basis of the Periodic Table with reference to the elements of proton number 1 to 20

Every substance in our world is made up of atoms. An **atom** is the smallest uncharged particle that can take part in a chemical change. Yet even atoms themselves are made up of still smaller particles called subatomic particles.

At the centre of an atom is a tiny **nucleus**. The nucleus is made up of **protons** and **neutrons**. The general name given to protons and neutrons is **nucleons** – because they are found in the nucleus. The number of protons plus neutrons in the nucleus of an atom is called the **nucleon number** or **mass number**.

Outside the nucleus, the electrons whizz around in areas which are called **electron shells** or **energy levels**. The shells are a certain distance from the nucleus. There can be several electron shells further and further away from the nucleus. Each of these can hold a certain number of electrons.

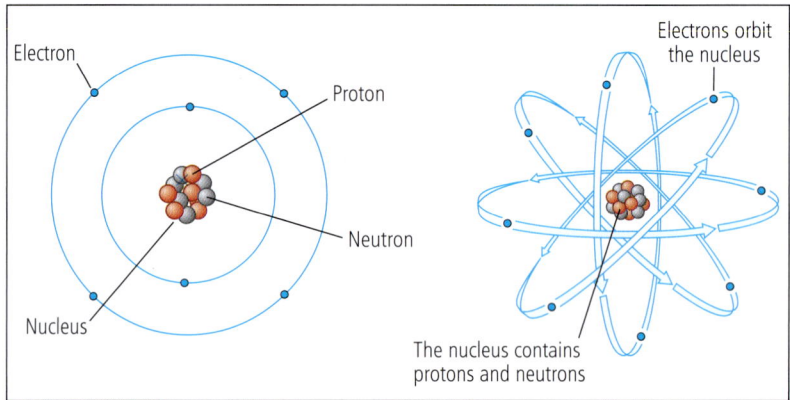

Figure 2.1.1 Two different pictures of an atom. We can draw different models of the atom but it is always difficult to show the position of the electrons.

The three subatomic particles have different charges and properties. These are shown in the table.

subatomic particle	symbol	relative mass	relative charge
proton	p	1	+1
neutron	n	1	no charge
electron	e	0.00054	−1

You can see that nearly all the mass of the atom is in the nucleus: electrons weigh hardly anything. The protons have a positive (+) charge and the electrons have a negative (−) charge. An atom has no overall charge as it is electrically neutral. This is because the number of positive protons is equal to the number of negative electrons.

DID YOU KNOW?

If an atom were the size of a lake 100 metres in diameter, the nucleus would be the size of a large grain of sand. And nearly all the mass of the atom is in the nucleus!

The importance of proton number

The number of protons in the nucleus of an atom is called the **proton number** or **atomic number**. Every atom of the same element has the same number of protons. Hydrogen has one proton in its nucleus, so its proton number or atomic number is 1. Sodium has 11 protons, so its proton number is 11. In the **Periodic Table** of elements, the elements are arranged in order of their proton number.

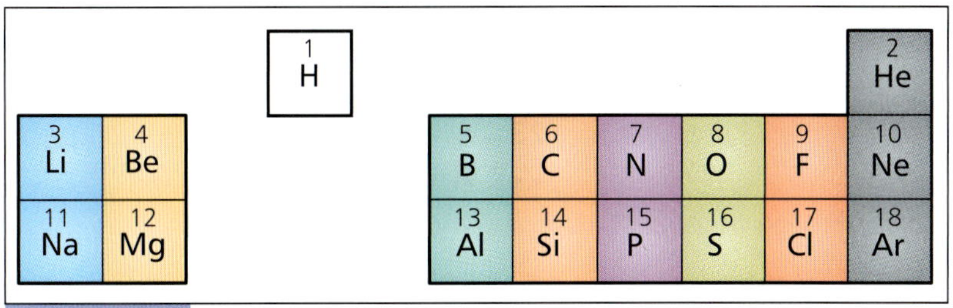

Figure 2.1.2 In the Periodic Table the elements are arranged in order of their proton (atomic) numbers

You read the Periodic Table like a book, line by line starting from the top left. Notice that the proton numbers are put above the symbol for each element. The chemical properties of the elements depend on the electrons. But the number of electrons in an atom is the same as the number of protons. That is why the proton number is important.

STUDY TIP

You will always be given a Periodic Table in Papers 1–4. You can use your Periodic Table to find out the number of protons in an atom. You can also use it to calculate the number of neutrons (see Topic 2.2).

SUMMARY QUESTIONS

1 Copy and complete using the words below:

**electrons   equal   neutrons   no
nucleons   positive   protons   shells**

The nucleus of an atom contains _____ and _____ The general name for protons and neutrons is _____ Protons have a _____ charge but neutrons have _____ charge. Outside the nucleus are the _____, which are negatively charged. The electrons are arranged in _____ or energy levels. The number of protons in an atom is _____ to the number of electrons.

2 An atom of the element neon has 10 protons and 11 neutrons. How many nucleons does it contain?

3 By referring to subatomic particles, explain why an atom of nitrogen is electrically neutral.

KEY POINTS

1 The subatomic particles in atoms are protons, neutrons and electrons.

2 A proton has a positive charge, an electron has a negative charge and a neutron has no charge. The protons and neutrons make up the vast majority of the mass of an atom.

3 Atoms are neutral because the number of positive protons equals the number of negative electrons.

4 Atoms are arranged in the Periodic Table in order of their proton number.

2.2 Isotopes

LEARNING OUTCOMES

- Define *isotope*
- State the two types of isotopes as being radioactive and non-radioactive
- State one medical and one industrial use of isotopes
- Understand that isotopes have the same properties because they have the same number of electrons in their outer shell

Supplement

How many neutrons?

Every element has its own proton number: it has a particular number of protons. The number of neutrons plus protons in the nucleus is the **nucleon number** or **mass number**. We show the nucleon number and proton number of an atom like this:

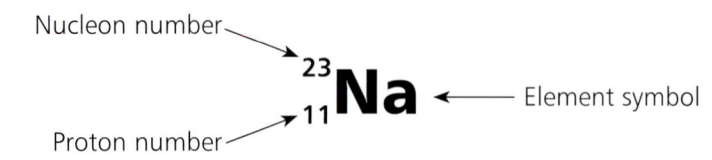

$$^{23}_{11}\text{Na}$$

Nucleon number — 23
Proton number — 11
Element symbol — Na

We can use the nucleon number and proton number to work out the number of neutrons in an atom:

Number of neutrons = nucleon number − proton number *or*

Number of neutrons = mass number − atomic number

For example: sodium has 23 nucleons and 11 protons.

So the number of neutrons in a sodium atom is 23 − 11 = 12 neutrons.

What are isotopes?

Atoms of the same element always have the same proton number. However, in many elements some of the atoms have different numbers of neutrons than others. Atoms of the same element which have the same proton number, but a different nucleon number are called **isotopes**. For example, there are three isotopes of hydrogen:

Hydrogen atoms have 1 proton

Deuterium atoms have 1 proton and 1 neutron

Tritium atoms have 1 proton and 2 neutrons

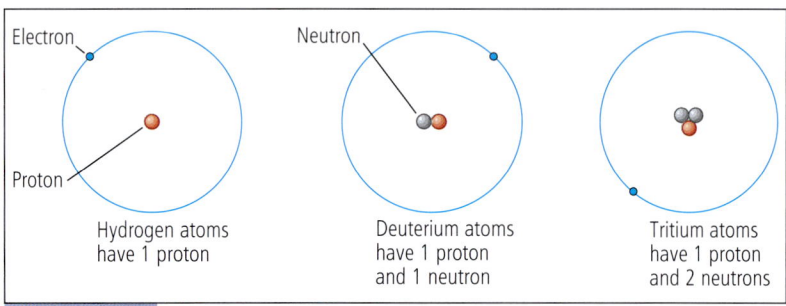

Figure 2.2.1 The three isotopes of hydrogen

Many elements are mixtures of isotopes. For example, in naturally occurring chlorine molecules about ¾ of the atoms have 18 neutrons and about ¼ have 20 neutrons.

isotope	$^{35}_{17}\text{Cl}$	$^{37}_{17}\text{Cl}$
number of neutrons	18	20
percentage of this isotope in naturally occurring chlorine	75.5%	24.5%

STUDY TIP

You do not need to know the details about radioactivity or about α-, β- or γ-radiation. Don't try to remember lots of uses for radioisotopes – just remember one medical and one industrial use.

Isotopes are properly written showing the number of protons and nucleons. However, you will often see isotopes written with just the number of nucleons following the name of the element, for example, chlorine-35. Different isotopes of the same element have the same chemical properties but may have slightly different physical properties.

Some isotopes such as hydrogen-3 (tritium) are **radioactive**, others such as chlorine-35 are non-radioactive.

If an isotope is radioactive, its nucleus is unstable and so it breaks down over a period of time (it decays). As it decays, the nucleus gives out tiny particles or rays. We call these isotopes radioisotopes.

The radiation given out by radioisotopes can be harmful. It can kill cells in your body.

> **Supplement**
>
> The number of electrons in the outer shell of an element determines its chemical properties (see Topic 2.3). The chemical properties of isotopes are the same because they have the same number of electrons in their outer shell. However, some physical properties of isotopes such as density may be slightly different.

Uses of isotopes

Although radioisotopes can be harmful, they have many uses in medicine, industry and for military purposes. They can also be used to date ancient remains. In addition, some isotopes are useful in the home and workplace. For example, the radioisotope americium-241 is used in smoke detectors.

One of the most important *medical uses* of radioisotopes is in cancer treatment. The radiation given out by the radioisotope is used to kill cancer cells. The radiation has to be controlled carefully so that it hits only the cancer cells and not the healthy cells around the tumour. An isotope of cobalt is often used for this. Other medical uses include treatment of overactive thyroid glands and locating tumours in the body. Radioisotopes are also used to sterilise medical equipment.

There are many *industrial uses* for radioisotopes. They can be used to check for leaks in oil and gas pipelines: a radioisotope is added to the oil or gas in the pipeline and a radiation detector called a Geiger counter will give a higher reading over the area of the leak.

> **DID YOU KNOW?**
>
> The thickness of paper can be controlled during its manufacture using a radioisotope. The paper passes between the radioactive source and a Geiger counter. The Geiger counter is connected to a system which controls the pressure on the rollers which roll the paper.

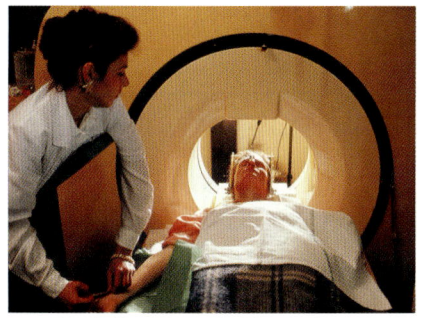

A patient being injected with a radioisotope to locate a possible tumour

SUMMARY QUESTIONS

1 Copy and complete using the words below:

   **different   element   industrial   isotopes
   nucleon   neutrons**

   The _____ number of an atom is the total number of _____ and protons present. Atoms of an _____ with _____ numbers of neutrons are called _____ Radioisotopes have many different medical and _____ uses.

2 What do you understand by the term *radioisotope*?

3 State the number of neutrons in each of these isotopes:

   a $^{235}_{92}U$ b $^{14}_{6}C$ c $^{58}_{26}Fe$

KEY POINTS

1 The number of neutrons in an atom is the nucleon number minus the proton number.

2 Isotopes are atoms with the same number of protons but different numbers of neutrons.

3 Isotopes can be radioactive or non-radioactive.

4 Isotopes have the same chemical properties because they have the same number of electrons in their outer shell. *(Supplement)*

2.3 Electronic structure and the Periodic Table

LEARNING OUTCOMES

- Describe the build up of electrons in shells
- Explain the significance of the noble gas electronic structures and of the outer shell electrons

Electron shells

Electrons move so rapidly that we cannot tell where they are at any point in time. So we have to make a simple model which fits the experimental evidence. In this model we imagine the electrons existing somewhere in spherical regions at certain distances from the nucleus. It is difficult to draw spheres one inside another so we draw the electrons as if they exist in circular orbits around the nucleus. These orbits are called **shells** or **energy levels**.

The first shell is nearest to the nucleus and can hold a maximum of two electrons. The next shell out can hold a maximum of eight electrons. When the third shell has eight electrons, the fourth shell starts filling up.

The arrangement of the electrons in shells is called the **electronic structure** or **electron configuration**. In a sodium atom, there are two electrons in the first shell, eight electrons in the second and one in the third (see Figure 2.3.1). We have a shorthand way of writing this:

We write the electron configuration of sodium as 2,8,1.

Arranging the electrons

The horizontal rows in the Periodic Table are called **periods**. As we move across a period, each element has one more electron to fit into its outer shell. The electrons fill up the shells one by one, starting from the shell nearest to the nucleus. When one shell is full, the electrons go into the next shell out from the nucleus.

Hydrogen has one electron. This electron is in its first shell. Helium has two electrons, so these electrons are also in its first shell. Lithium has three electrons. Two of lithium's electrons are in its first shell; the third goes into the second shell because the first shell can hold only two electrons. In this way the electrons are 'put into their shells'. Figure 2.3.2 shows the electronic structures and electronic configurations of the first 20 elements.

A vertical column in the Periodic Table is called a **group**. You can see that the atoms of elements in the same group have the same number of electrons in their outer shell. These are called the **valency electrons** or outer shell electrons.

The chemical properties of an element depend on its number of valence electrons. This is why elements within some groups of the Periodic Table have similar chemical properties. You can also see that the number of electrons in the outer shell is the same as the group number. So atoms of Group I elements all have one electron in their outer shells, atoms of Group II elements have two and so on.

The noble gas structure

When compounds are formed by combining elements, the electrons in the outer shell are transferred or shared. Group 0 elements,

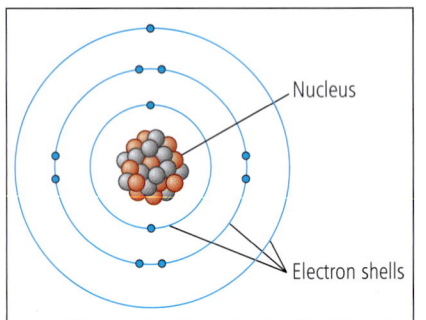

Figure 2.3.1 It is difficult to show where the electrons are. So we use a model like this one to help us.

We know about the electron arrangement in atoms thanks to the work of the Danish physicist Niels Bohr

DID YOU KNOW?

The colours of fireworks are caused by the movement of electrons between different electron shells.

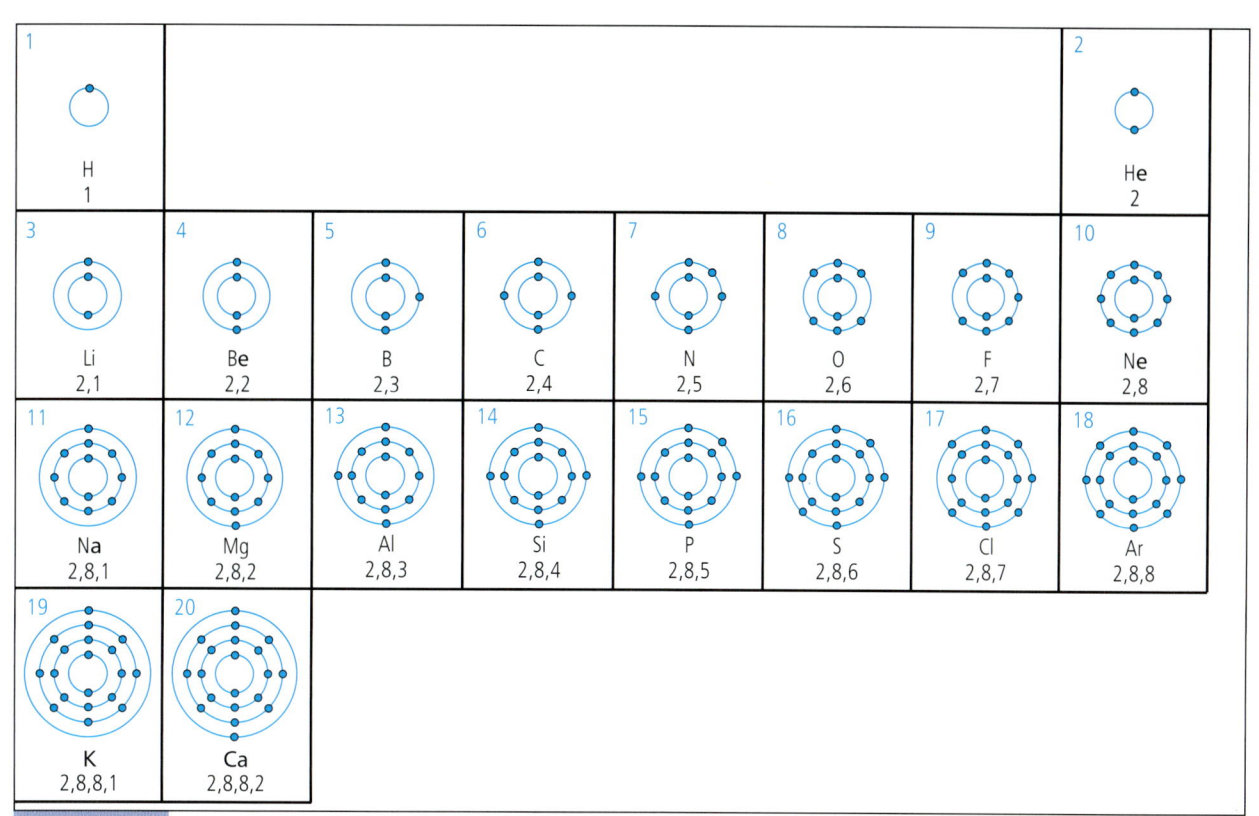

Figure 2.3.2 The electronic structure of the first 20 elements

the **noble gases**, have a full outer shell of electrons. This type of electronic structure is very stable – the atoms cannot gain or lose electrons very easily. This stability of a full outer shell of electrons makes the noble gases very unreactive. They do not combine with other elements to form compounds.

When atoms of other elements combine, the compounds formed have the noble gas structure of full electron shells around each atom.

STUDY TIP

Make sure that you can draw the electronic structure of the first 20 elements in rings containing electrons. The electronic structure of sodium, for example, is 2, 8, 1.

SUMMARY QUESTIONS

1 Copy and complete using the words below:

eight   levels   maximum   shells   two   valency

The electrons in an atom are arranged in _____ or energy _____ around the nucleus. A maximum of _____ electrons can fit into the first shell and a _____ of _____ electrons can fit into the second shell. The outer shell electrons are called the _____ electrons.

2 How many electrons do atoms of each of the following have in their outer shell?

   a Group VI elements   b Group II elements
   c Group 0 elements

3 Write the electronic structure in shorthand form for the following elements:

   a carbon   b potassium   c chlorine   d aluminium

KEY POINTS

1 Electrons in atoms are arranged in shells or energy levels.

2 The number of outer shell electrons in an atom determines the chemical properties of that element.

3 Atoms of elements in the same group in the Periodic Table have the same number of outer shell electrons.

2.4 Elements, compounds and mixtures

LEARNING OUTCOMES

- Describe the differences between elements, compounds and mixtures

Elements

Every substance around us is made up of atoms. There are just over 100 different types of atom. A substance made up of only one type of atom is called an **element**. Elements cannot be broken down into anything simpler by chemical reactions.

Each element is given a symbol. For example, fluorine has the symbol F and the symbol for zinc is Zn (see Topic 4.1). Elements have particular properties which distinguish them from other elements. For example, silver is a shiny solid, chlorine is a green gas and bromine is a reddish-brown liquid.

The atoms of elements generally combine to form molecules or giant structures of atoms (see Topic 3.5). They rarely exist on their own. The link joining the atoms is called a **chemical bond**.

Compounds

The atoms of different elements can join together to form compounds. A **compound** is a substance made up of two or more different types of atom joined together by **chemical bonds**. Millions of compounds are known to us and many new ones are being discovered as you are reading this!

A compound always has a fixed amount of each element in it. Water always has two hydrogen atoms for every one oxygen atom. Copper(II) oxide always contains 80% copper and 20% oxygen by mass. There are two types of compound:

- molecular compounds where atoms are bonded together. Water is an example.
- ionic compounds where many ions (charged atoms) are joined together. For example, sodium chloride or 'salt'.

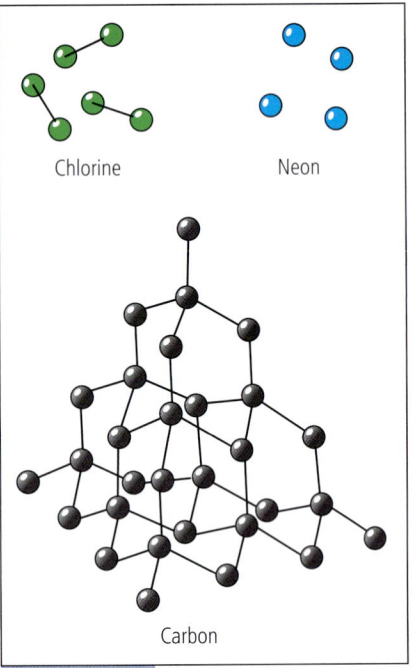

| Figure 2.4.1 | The structures of some elements |

Bromine is an element. It has only one type of atom.

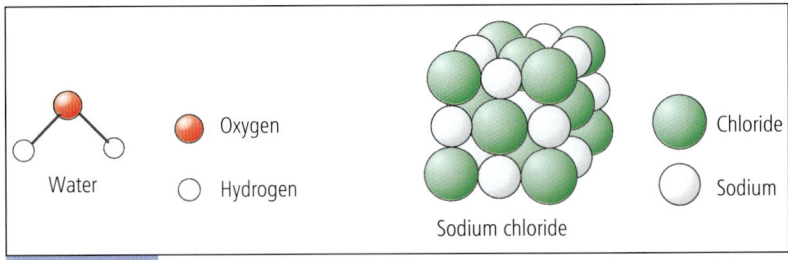

| Figure 2.4.2 | Water is a molecular compound of hydrogen and oxygen. Salt is an ionic compound of sodium and chloride ions. |

Compounds have very different properties from the elements from which they are made. For example, a compound of sodium and chlorine looks completely different from the elements chlorine and sodium. The elements in a compound cannot be separated by physical means, although it might be possible to separate them by chemical means.

DEMONSTRATION

Making sodium chloride from its elements

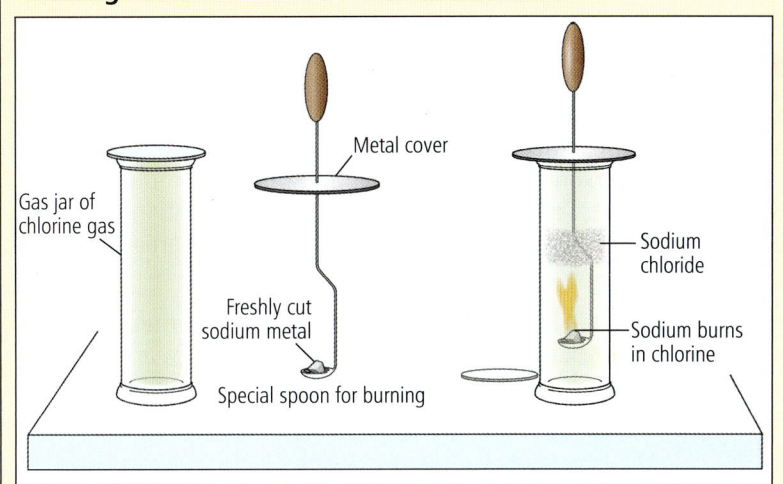

Figure 2.4.3 Burning sodium in chlorine

Sodium is a reactive, silvery metal. Chlorine is a poisonous green gas. The sodium is melted on a special spoon and placed in a gas jar of chlorine. The sodium burns in the chlorine. A white powder is left on the side of the gas jar. This is sodium chloride – a white solid that we can put on our food!

Mixtures

A **mixture** contains two or more elements or compounds that are *not* chemically bonded together. A mixture does not have a fixed amount of each element or compound in it. For example, a mixture of iron and sulfur could contain various amounts of iron and sulfur. Since they are not bonded together, the elements or compounds in a mixture still have their characteristic properties.

We can separate the substances in a mixture by one of the physical methods described in Topics 1.8 and 1.9. For example, we can separate a mixture of sand and salt by dissolving the salt in water, filtering off the sand then evaporating the water to leave salt.

SUMMARY QUESTIONS

1 Copy and complete using the words below:

atoms   bonds   compound   ions

When two or more _____ join together, a _____ is formed. The atoms or _____ in a compound are held together by _____.

2 Make a list of the differences between a compound and a mixture.

3 Some grey iron powder was heated in reddish-brown bromine vapour. A yellowish-green powder was formed. Is the yellowish-green powder an element, a compound or a mixture? Explain your answer.

STUDY TIP

You should learn the definitions of elements, compounds and mixtures.

DID YOU KNOW?

Not all elements are found naturally. Some are made artificially and can exist only for a fraction of a second. An atom of rutherfordium-260 exists only for 0.3 seconds before breaking down.

KEY POINTS

1 Elements contain only one type of atom.

2 A compound is a substance containing two or more different types of atoms bonded together.

3 Mixtures do not have a fixed composition. The substances in a mixture can be separated by physical means.

2.5 Metals and non-metals

LEARNING OUTCOMES

- Describe the differences between metals and non-metals

Elements can be metallic or non-metallic. Can you tell which are metals and which are non-metals?

STUDY TIP

When stating the difference between a metal and a non-metal it is best to select conductivity, malleability or ductility as properties. These have fewer exceptions to the general rules.

Metals and non-metals

Most of the elements in the Periodic Table are **metals**, for example iron, sodium, tin and aluminium are metals. The rest are **non-metals**. Carbon, sulfur, chlorine and helium are non-metals. Although metals have many properties in common, there is a wide variation in these properties. Non-metals usually have the opposite properties to metals, but there is an even wider variation in their properties.

The best way to tell if a substance is a metal or a non-metal is to look at some of its **physical properties**. Physical properties are those such as density, melting point and **electrical conductivity**.

PRACTICAL

Comparing electrical conductivity

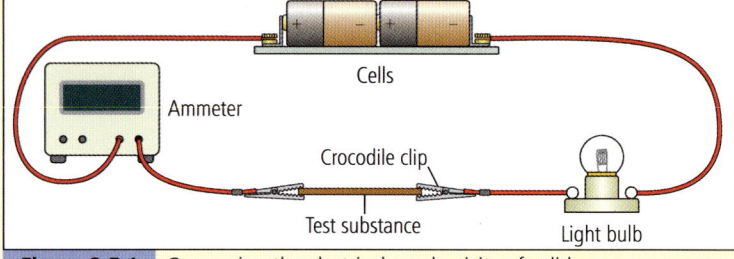

Figure 2.5.1 Comparing the electrical conductivity of solids

1. Set up the circuit with a known substance between the crocodile clips and see if the bulb lights. If it does, the substance is an electrical conductor.
2. Repeat the experiment with different metals and non-metals.

Many physical properties do not change however much substance you use. For example, the density and melting point are the same however much substance you use. In chemistry, the **density** of a substance is measured in grams per centimetre cubed (g/cm^3):

$$\text{density (in } g/cm^3) = \frac{\text{mass (in g)}}{\text{volume (in cm}^3)}$$

The table opposite shows the six best physical differences between a metal and non-metal. There are some exceptions, but not too many.

There are also some physical properties that are less useful for telling the difference between metals and non-metals:

- Many metals have high densities. Exceptions: Group I metals and some others like gallium have low densities.
- Most non-metals have low densities.
- Many metals have high melting and boiling points. Exceptions: Group I metals and mercury have low melting and boiling points.
- Many non-metals have low melting and boiling points. Exceptions: carbon and silicon.

- Many metals are hard (they cannot be scratched easily) and strong. Exceptions: Group I metals, mercury, gallium.
- Many non-metals are soft and can be scratched easily. Exception: diamond.

physical property	metal	non-metal
conductor of electricity	conducts	does not conduct (exception: graphite)
conductor of heat	conducts	does not conduct (exception: graphite)
malleable – can be beaten into a different shape with a hammer	malleable	not malleable. Non-metals are brittle: they break easily when hit
ductile – can be drawn out into wires	ductile	not ductile. Non-metals are brittle: they break easily when a pulling force is applied
lustrous – has a shiny surface when polished	lustrous	dull surface (exceptions: graphite and iodine have shiny surfaces)
sonorous – makes a ringing sound when hit with a hard object	sonorous	not sonorous. Non-metals make a dull sound when hit with a hard object

We can also use some **chemical properties** to tell the difference between metals and non-metals. Chemical properties are to do with chemical reactions and energy changes during the reactions. Here are some differences:

- Many metal oxides are basic.
- Many non-metal oxides are acidic.
- Many metals react with acids to give off hydrogen gas.
- Most non-metals do not react with acids.
- When they react, metals form positive ions by losing electrons.
- When they react with metals, non-metals form negative ions by gaining electrons. (Exception: hydrogen can form positive ions.)

The last two of these points are the best chemical distinction between a metal and a non-metal (see Topic 3.1).

DID YOU KNOW?

Alloys are mixtures of metals with other metals or non-metals. Early civilisations in the Middle East made alloys of copper and tin (bronze) over 5000 years ago.

SUMMARY QUESTIONS

1 Copy and complete using the words below:

conductors   ductile   graphite   heat   low   opposite

Metals are good _____ of electricity and _____ and are malleable and _____. Non-metals usually have properties _____ to those of metals. There are some exceptions. _____ is a non-metal that conducts electricity. Mercury is a metal that has a _____ boiling point.

2 State the names of:
   a two metals that are soft
   b a non-metal that conducts electricity.

3 List four physical properties that help you to tell the difference between a metal and a non-metal.

KEY POINTS

1 Metals are good conductors of electricity and heat, and are malleable and ductile.

2 Non-metals are poor conductors of electricity and heat, and are brittle.

3 Common exceptions: graphite is a non-metal that conducts electricity; most metals have high melting and boiling points, but the Group I metals and mercury have low melting and boiling points.

SUMMARY QUESTIONS

1. For each of the following sub-atomic particles describe **(i)** its position in the atom **(ii)** its relative mass and **(iii)** its charge:
   - **(a)** proton
   - **(b)** neutron
   - **(c)** electron

2. Copy and complete the table to show the number of protons, neutrons and electrons in the isotopes shown.

atom	number of protons	number of neutrons	number of electrons
$^{16}_{8}O$			
$^{207}_{82}Pb$			
$^{1}_{1}H$			
$^{37}_{17}Cl$			

3. Give definitions of the following terms:
   - **(a)** nucleon number
   - **(b)** isotopes
   - **(c)** mixture
   - **(d)** compound

4. Draw a diagram of a boron atom to show the correct number and positions of the electrons, protons and neutrons. Use the Periodic Table to help you.

5. Match each word on the left with the correct description on the right.

isotopes	a substance containing two or more different elements chemically bonded together
element	atoms of the same element with different numbers of neutrons
compound	a positively charged particle in the nucleus
proton	a substance containing only one type of atom

6. Copy each of the following physical properties then write 'metal' or 'non-metal' after each one.
   - **(a)** conducts heat
   - **(b)** is brittle
   - **(c)** is an insulator
   - **(d)** has a dull surface
   - **(e)** gives a ringing sound when hit
   - **(f)** conducts electricity

PRACTICE QUESTIONS

1. Which one of these statements about an atom of $^{14}_{6}C$ is correct?
   - **A** It has 14 electrons and 6 protons.
   - **B** It has 8 protons and 6 neutrons.
   - **C** It has 6 electrons and 8 neutrons.
   - **D** It has 6 protons and 14 neutrons.

   (Paper 1)

2. Which one of the following statements is true?
   - **A** All metals make a dull sound when hit.
   - **B** All non-metals are ductile.
   - **C** Mercury is a solid at room temperature.
   - **D** Graphite is a non-metal that conducts electricity.

   (Paper 1)

3. Which one of the following statements about isotopes are true?
   - **A** The isotope $^{1}_{1}H$ has the same number of neutrons as the isotope $^{2}_{1}H$.
   - **B** The isotope $^{1}_{1}H$ has different chemical properties to the isotope $^{2}_{1}H$.
   - **C** An isotope of hydrogen $^{1}_{1}H$ has only one neutron.
   - **D** The isotope $^{1}_{1}H$ has the same chemical properties as the isotope $^{2}_{1}H$.

   (Paper 2)

4. Chlorine has two isotopes, $^{35}_{17}Cl$ and $^{37}_{17}Cl$.
   - **(a)** What do you understand by the term *isotope*? [1]
   - **(b)** Write down the number of neutrons in each of these isotopes of chlorine. [2]
   - **(c)** Draw and label a diagram of an atom of chlorine to show:
     - the nucleus
     - the electron shells, with the correct number of electrons in each shell. [3]
   - **(d)** A chloride ion has eight electrons in its outer shell. What is the importance of this electronic configuration? [1]

   (Paper 3)

5 Iron is a grey magnetic metal that reacts with hydrochloric acid to produce hydrogen. Sulfur is a yellow non-metal that does not react with acids.
Iron(II) sulfide is a brownish-black compound. It is a solid that is not magnetic when pure. It reacts with hydrochloric acid to produce smelly hydrogen sulfide gas.

(a) What do you understand by the term *compound*? [1]

(b) Describe two differences between a mixture of iron and sulfur and a compound of iron and sulfur. [4]

(c) (i) Give the electronic configuration of sulfur. [1]

(ii) How does this electronic configuration show that sulfur is in Group VI of the Periodic Table? [1]

(iii) What determines the position of an element in the Periodic Table? [1]

(d) State three differences between the properties of iron and sulfur that have not been described in this queston.

(Paper 3) [6]

6 Isotopes can be classed as radioactive or non-radioactive.

(a) What do you understand by the term *radioactive*? [1]

(b) (i) State one medical use of a radioactive isotope. [1]

(ii) State one industrial use of a radioactive isotope. [1]

(c) Deuterium, $^{2}_{1}H$, is a non-radioactive isotope.

(i) Draw a diagram to show the complete atomic structure of deuterium. Label your diagram. [4]

(ii) How does an atom of deuterium differ from a normal hydrogen atom? [1]

(Paper 3)

7 Sodium is a soft, shiny metal with a low melting point.
Chlorine is a poisonous green gas.
Sodium reacts with chlorine to form sodium chloride, a white, crystalline solid with a high melting point.

(a) (i) Describe three properties shown by most metals. [3]

(ii) Describe two properties of sodium that make it an unusual metal. [2]

(b) Give two reasons why chlorine is classed as a non-metal. [2]

(c) Use the information above to explain why sodium chloride is a compound of sodium and chlorine rather than a mixture of sodium and chlorine. [2]

(d) State the electronic configuration of:
(i) a sodium atom [1]
(ii) a chlorine atom. [1]

(Paper 3)

8 The electronic structure of the atoms of five elements, A, B, C, D and E, are shown below:

A 2,8,2   B 2,8,6   C 2,1
D 2,8   E 2,6

(a) Which one of these elements is in Group II of the Periodic Table? Give a reason for your answer. [1]

(b) Which two of these elements are in the same group in the Periodic Table? [1]

(c) Which one of these elements has the highest proton number? [1]

(d) Element E is oxygen. An isotope of oxygen has eight neutrons.

(i) What do you understand by the term *isotope*? [1]

(ii) State the nucleon number of this isotope of oxygen. [1]

(e) Element C is lithium. Lithium is a metal. Describe three differences between the physical properties of lithium and oxygen. [3]

(Paper 3)

3 Structure and bonding

3.1 Ionic bonding

LEARNING OUTCOMES

- Describe the formation of ions by electron loss or gain
- Describe the formation of ionic bonds between elements from Groups I and VII
- Describe the formation of ionic bonds between metallic and non-metallic elements

Supplement

These crystals of sodium chloride are made up of millions of ions

DID YOU KNOW?

In just one gram of salt there are about 10 000 000 000 000 000 000 000 ions of sodium and the same number of chloride ions.

STUDY TIP

When drawing the electronic structure for an ion, make sure that the charge of the ion is shown at the top right-hand corner just outside the square brackets. Do NOT put the charge in the nucleus.

How are ions formed?

An **ion** is an electrically charged particle. Ions are formed when atoms lose or gain one or more electrons. We saw in Topic 2.3 that an atom with a full outer shell of electrons is stable and unreactive – it has the noble gas structure. Most atoms do not have a full outer shell of electrons so tend to react. One way of gaining a full outer shell of electrons is by completely transferring outer shell electrons from one atom to another. The diagram shows how this is done in sodium chloride.

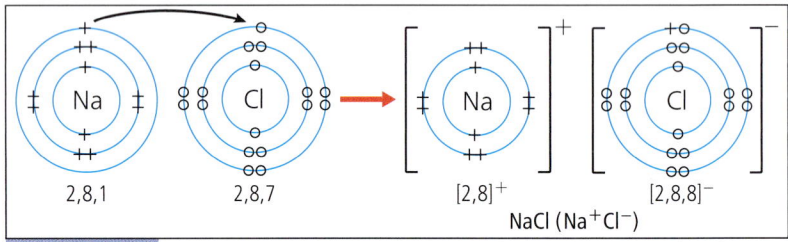

Figure 3.1.1 A sodium ion and a chloride ion are formed by the transfer of an electron from a sodium atom to a chlorine atom

You will notice that in the diagrams above the electrons have been **paired** up. There are four pairs of electrons in a shell of eight electrons. Pairing up the electrons like this helps to keep track of what happens when bonds are formed.

The sodium ion has one positive charge because it has 11 protons (+) in its nucleus but only 10 electrons (−). The chloride ion has one negative charge because it has 17 protons in its nucleus and 18 electrons.

The charge on the ion is written at the top right-hand side. Square brackets are used to show that the charge on the ion is spread evenly all over the ion. This sort of diagram is called a **dot-and-cross diagram**. This does not mean that the electron transferred is any different from the others. It shows where the electrons have come from. It is just a record-keeping exercise.

We can write similar dot-and-cross diagrams for compounds of other Group I and Group VII elements.

Forming a stable structure

Look again at Figure 3.1.1. You will notice that the electron is transferred from a metal atom to a non-metal atom. This is always true when ions are formed. You will also notice that the outer shells of both atoms have become complete. The sodium ion has the electronic structure $[2,8]^+$ and the chloride ion has the structure $[2,8,8]^-$. So both ions have a complete outer shell and have the noble gas structure of eight outer electrons.

The sodium ion has the same electronic structure as neon and the chloride ion has the same electronic structure as argon. The noble gas structure makes the ions stable. This full outer shell of 8 electrons is often called a stable **octet** of electrons. Remember, however, that the first electron shell has a maximum of two electrons, so the stable electronic structure for a lithium ion will be $[2]^+$.

The attraction between the positive ions and the negative ions is the **ionic bond**.

Ions with multiple charges

We can draw dot-and-cross diagrams to show the electron transfer in more complex cases. When we draw dot-and-cross diagrams we usually only show the outer shells because these are the ones that take part in the reaction.

This diagram shows how the ionic structure for magnesium oxide is obtained.

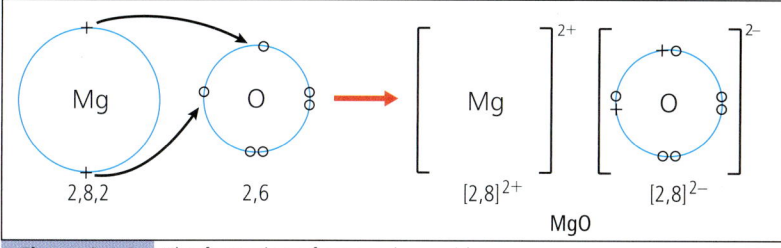

Figure 3.1.2 The formation of magnesium oxide

A magnesium atom has two electrons in its outer shell. An oxygen atom has six electrons in its outer shell. The two magnesium electrons are transferred to the oxygen atom to complete its outer shell. By doing so, both the magnesium and oxide ions have an electronic structure which is the same as the noble gas neon. So a stable electronic structure has been formed.

Drawing the ionic structure for calcium chloride needs a little more thought: the calcium atom has two electrons in its outer shell but each chlorine atom needs only one of these to get a stable octet of electrons. So two chlorine atoms are needed in the reaction. Each of these gains one electron.

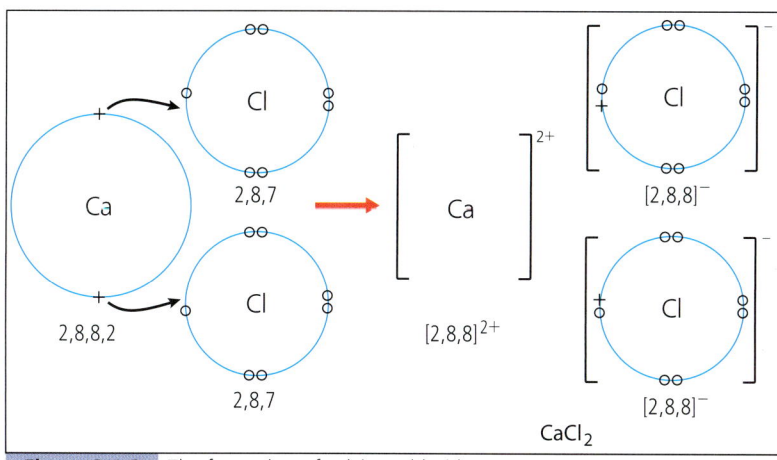

Figure 3.1.3 The formation of calcium chloride

KEY POINTS

1. When elements in Group I and VII react, the atom of the Group I element loses the electron from its outer shell. This can be transferred to the outer shell of the atom of a Group VII element.

2. Ions have the noble gas electronic structure with a complete outer shell of eight electrons (or two for lithium).

3. Dot-and-cross diagrams can be drawn to show the electronic structure of ions.

4. The attraction between oppositely charged ions is called ionic bonding.

SUMMARY QUESTIONS

1. Copy and complete using the words below:

   chlorine electron complete ions noble outer transferred

   When sodium reacts with chlorine the _____ from the _____ shell of sodium is _____ to the outer shell of the _____ atom. Sodium and chloride _____ are formed, which both have the electronic structure of a _____ gas with a _____ outer shell of electrons.

2. Write the electronic structure in numbers for the following ions:
   a Mg^{2+}
   b Na^+
   c F^-

3. Draw dot-and-cross diagrams to show the electronic structure of the ions in:
   a calcium oxide
   b aluminium chloride
   c potassium fluoride.

3.2 Covalent bonding (1): simple molecules

LEARNING OUTCOMES

- Describe the formation of single covalent bonds in H_2, Cl_2, H_2O, CH_4, NH_3 and HCl as the sharing of electrons leading to the noble gas configuration

How are covalent bonds formed?

In the formation of ions when a metal and non-metal react together, the electrons are transferred from the outer electron shell of the metal to the outer shell of the non-metal. When two non-metal atoms react they share a pair of electrons. This is called **covalent bonding**. In a hydrogen molecule both hydrogen atoms share one electron with each other to form a pair. This pair of **shared** electrons is a single **covalent bond**.

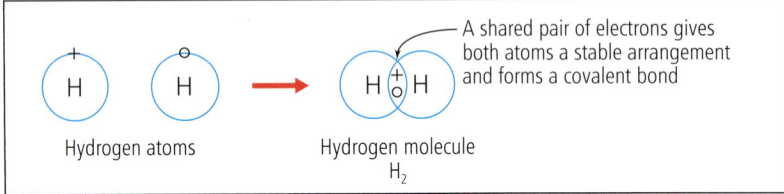

Figure 3.2.1 Atoms of hydrogen share electrons to form a single covalent bond

The electronic structure of simple molecules

We can draw the electronic structure of simple molecules containing a single covalent bond by pairing up the electrons in a similar way. In each case we aim to pair up the electrons from each atom so that there is a stable octet of eight electrons around each atom. Remember though that hydrogen atoms will be stable with two electrons around them. Because it is only the outer shell electrons that are involved in bonding, we do not generally worry about drawing the inner shells of electrons.

When two atoms form covalent bonds we use a dot for the electrons from one of the atoms and a cross for the electrons from the other as we did in dot-and-cross diagrams of ionic bonding. This is simply so we can see where the electrons have come from. You will see from the examples below that not all the electrons are used in bonding. The pairs of electrons that are not used in bonding are called **lone pairs**.

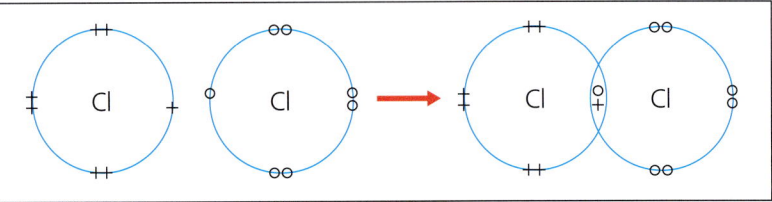

Figure 3.2.2 A chlorine molecule: the two chlorine atoms overlap and share a pair of electrons (one electron from each atom) in the covalent bond. Each atom now has eight electrons in its outer shell so has a stable electronic structure.

STUDY TIP

When drawing dot-and-cross diagrams remember to pair up the bonding electrons in the overlap area between the atoms. Don't put them outside the area where the atoms join.

DID YOU KNOW?

In 1704, the English scientist Sir Isaac Newton suggested that the particles of matter were fixed together by forces.

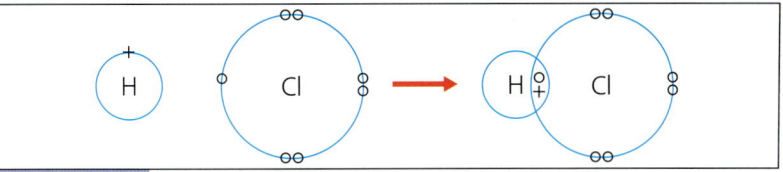

Figure 3.2.3 A hydrogen chloride molecule: hydrogen has one electron in its outer shell and chlorine has seven. One pair of electrons is shared, giving the hydrogen two electrons in its outer shell and chlorine eight.

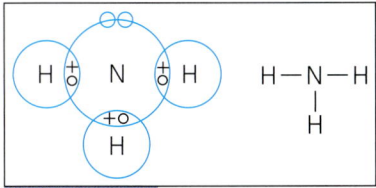

Figure 3.2.4 An ammonia molecule

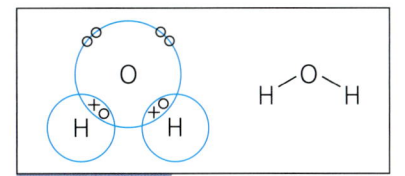

Figure 3.2.5 Water (H_2O): each hydrogen atom shares a pair of electrons. Each can form one bond. Oxygen has six electrons in its outer shell so can form two bonds with hydrogen atoms. We can show a single covalent bond by a line.

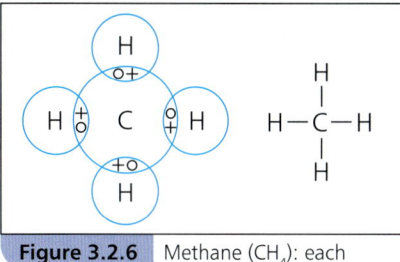

Figure 3.2.6 Methane (CH_4): each hydrogen atom shares a pair of electrons. Each can form one bond. Carbon has four electrons in its outer shell so can form four bonds with hydrogen atoms.

SUMMARY QUESTIONS

1 Copy and complete using the words below:

   electrons   line   non-metal   single

   Covalent bonds are formed when _____ atoms combine. A _____ covalent bond is a pair of _____ shared between two atoms. A single covalent bond is shown by a single _____ between the atoms.

2 Draw dot-and-cross diagrams for the following molecules. Show only the electrons in the outer shells.
   a A fluorine molecule (Fluorine is in the same group of the Periodic Table as chlorine.)
   b A hydrogen bromide molecule (Bromine is in the same group of the Periodic Table as chlorine.)
   c A hydrogen sulfide molecule (Sulfur is in the same group of the Periodic Table as oxygen.)

3 How many bonds are formed when the following atoms combine?
   a Hydrogen and oxygen in water
   b Hydrogen and chlorine in hydrogen chloride
   c Hydrogen and carbon in methane

KEY POINTS

1 A covalent bond is formed when atoms share a pair of electrons.

2 When atoms combine to form covalent bonds each atom has eight electrons in its outer shell, except for hydrogen which has two.

3 A single covalent bond is shown by a line between the atoms. For example, Cl—Cl.

3.3 Covalent bonding (2): more complex molecules

LEARNING OUTCOMES

- Describe the electron arrangement in more complex covalent molecules such as N_2, C_2H_4, CH_3OH and CO_2

Living things contain thousands of different molecules. The atoms in each of these molecules are held together by covalent bonds.

DID YOU KNOW?

It was not until 1916 that an explanation of the covalent bond was given by the American chemist Gilbert Lewis. The modern theory of covalent bonding suggests that clouds of electrons from each atom overlap to form the bond. The greater the overlap, the stronger the bond.

Molecules with three or more types of atoms

Many molecules contain three or more different atoms. For example, a molecule of methanol has four hydrogen atoms, one carbon atom and one oxygen atom. In order to draw a dot-and-cross diagram for more complex molecules it is useful to have some idea of the structure of the molecules. In methanol we know that a carbon atom has four electrons in its outer shell. These electrons must pair up with four others. They cannot pair up with all four hydrogen atoms otherwise we will have methane! One of the electrons must pair up with an oxygen atom, so the structure of methanol must be:

$$\begin{array}{c} H \\ | \\ H-C-O-H \\ | \\ H \end{array}$$

We can now draw a dot-and-cross diagram for methanol. We have three types of atom. So we need to choose another symbol other than a dot or a cross for the electrons from the third type of atom. We could use a circle, a triangle or a square. Using the same ideas as before, we pair up the electrons so that there are eight electrons in the outer shell of the carbon and oxygen atoms, and two electrons in the outer shell of the hydrogen atoms.

+ Electrons from hydrogen
• Electrons from carbon
o Electrons from oxygen

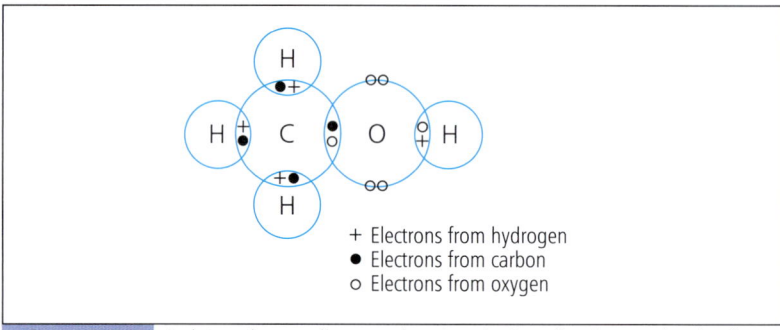

Figure 3.3.1 A dot-and-cross diagram showing the bonding in a molecule of methanol

Compounds with double and triple bonds

In order to form a stable octet of electrons, some atoms combine to form double or triple bonds. A double bond is shown as a double line. For example, in oxygen the **double bond** is shown as O=O. In nitrogen the **triple bond** is shown as N≡N. Some examples are:

An oxygen molecule: each oxygen atom has six electrons in its outer shell. Each atom needs to gain two electrons to complete its outer electron shell. So two pairs of electrons are shared and two covalent bonds are formed.

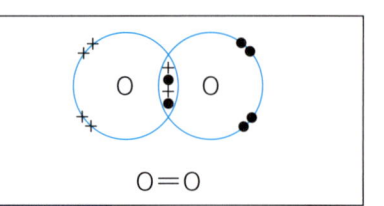

Figure 3.3.2 An oxygen molecule

A nitrogen molecule: each nitrogen atom has five electrons in its outer shell. Each atom needs to gain three electrons to complete its outer electron shell. So three pairs of electrons are shared and three covalent bonds are formed.

Carbon dioxide: each oxygen atom has six electrons in its outer shell. Each of these atoms needs to gain two electrons to complete its outer electron shell. The carbon atom has four electrons in its outer shell so it needs to gain four electrons to complete its outer shell. So four electrons (two pairs) are shared between the carbon atom and the oxygen atoms. The structure of carbon dioxide can therefore be shown as O=C=O.

Ethene: each hydrogen atom has one electron in its outer shell. Each of these atoms needs to gain one electron to complete its outer shell. The two carbon atoms have four electrons in their outer shells so each needs to gain four electrons. Four bonds are formed between the hydrogen atoms and the carbon atoms, two with each carbon. This leaves two electrons free on each carbon atom to share with each other as a double bond.

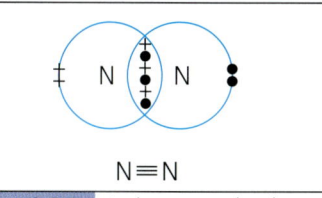

Figure 3.3.3 A nitrogen molecule

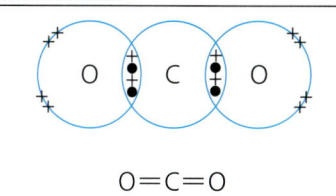

Figure 3.3.4 A carbon dioxide molecule

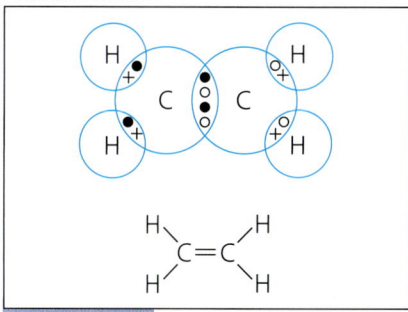

Figure 3.3.5 An ethene molecule

DID YOU KNOW?

VALENCY

We have seen that the outer shell electrons are important in pairing up electrons in covalent compounds or being involved in electron transfer in ionic compounds. The **valency** of an atom can be defined as the number of electrons lost or gained to form a complete electron shell in a covalent or ionic compound. So the valency of metals is the same as the number of electrons in their outer shell. For example, the valency of magnesium is two. For non-metals the valency is the number of electrons needed to make a complete outer shell. So for nitrogen this is three and for oxygen it is two.

It is often said that valency is the **combining power** of an atom. For example, hydrogen has a combining power of one and oxygen has a combining power of two. So in water, two hydrogen atoms combine with one oxygen atom.

STUDY TIP

When drawing the electronic structure of compounds with double and triple bonds, make sure that you draw the atoms large enough so that all the bonding electrons can fit into the overlap area of the atoms.

SUMMARY QUESTIONS

1 Copy and complete using the words below:

nitrogen   oxygen   three   triple   two

A double covalent bond is formed when _____ pairs of electrons are shared between two atoms. A _____ covalent bond is formed when _____ pairs of electrons are shared between two atoms. An _____ molecule has a double bond and a _____ molecule has a triple bond.

2 Draw dot-and-cross diagrams for:
   a   chloroform, $CHCl_3$ b   ethene, C_2H_4 c   ammonia, NH_3.

KEY POINTS

1 A double covalent bond is formed when two pairs of electrons are shared between two atoms.

2 A triple covalent bond is formed when three pairs of electrons are shared between two atoms.

3.4 Ionic or covalent?

LEARNING OUTCOMES

- Describe the differences in volatility, solubility and electrical conductivity between ionic and covalent compounds
- Explain the differences in melting and boiling points of ionic and covalent compounds
- Describe the lattice structure of ionic compounds

More about ionic bonding

In ionic bonding the attractive forces between the positive and negative ions result in the formation of a giant ionic structure. The ions are regularly arranged in the crystal. This arrangement carries on in three dimensions throughout the crystal. This three-dimensional network is called a crystal **lattice** (see Figure 3.4.1). The electrostatic attractive forces between the ions act in all directions and are very strong. It takes a lot of energy to overcome these forces. This is why giant ionic structures have high melting points and boiling points.

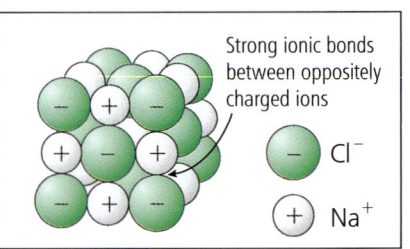

Figure 3.4.1 A giant ionic lattice of sodium and chloride ions

More about molecules

The covalent bonds within molecules are very strong. However, the forces between the separate molecules are weak. These weak, attractive forces are called **intermolecular forces**.

Because the forces between molecules are weak it requires only a little energy to overcome these forces and get the molecules to move away from each other. This is why molecular substances have low melting points and boiling points.

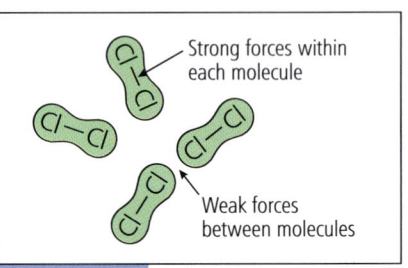

Figure 3.4.2 The forces between chlorine molecules are weak, but within the molecules the forces (covalent bonds) are strong

DID YOU KNOW?

Solid carbon dioxide changes directly to a gas at −78 °C without forming a liquid. This is called sublimation. Solid carbon dioxide ('dry ice') is used in theatres and films to create 'smoke' or 'cloud' effects.

PRACTICAL

Is this compound ionic or covalent?

We can tell if a compound is likely to be ionic or covalent by testing how easily it melts, its electrical conductivity and its solubility in water.

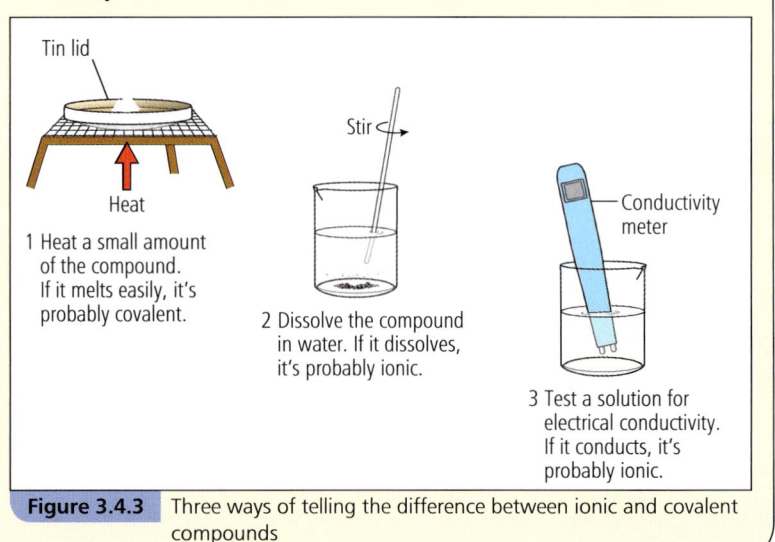

1 Heat a small amount of the compound. If it melts easily, it's probably covalent.

2 Dissolve the compound in water. If it dissolves, it's probably ionic.

3 Test a solution for electrical conductivity. If it conducts, it's probably ionic.

Figure 3.4.3 Three ways of telling the difference between ionic and covalent compounds

- Ionic compounds have high melting and boiling points because of the strong attractive forces between the ions in the giant ionic lattice.
- Covalent compounds have low melting and boiling points because the intermolecular attractive forces are weak. There are some exceptions: giant covalent structures such as silicon dioxide have very high melting points (see Topic 3.5).
- Ionic compounds are soluble in water but insoluble in organic solvents. They are soluble in water because the water molecules are able to separate the ions from one another and keep the ions in solution.
- Covalent compounds are insoluble in water. There are some exceptions: sugar and some amino acids are water-soluble.
- Ionic compounds conduct electricity only when molten or when dissolved in water. This is because the ions are mobile. A solid ionic compound does not conduct electricity because the ions are packed together and are not free to move around (they can vibrate only in their fixed positions in the lattice).

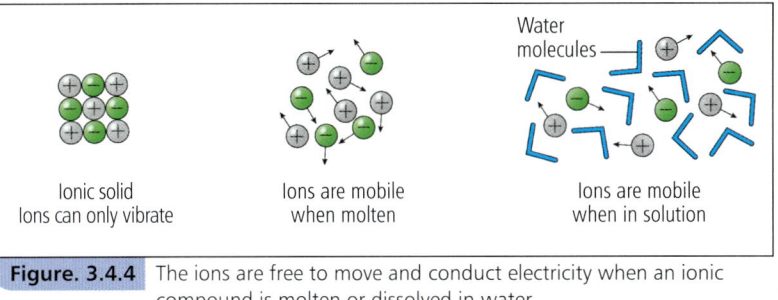

Testing an element for electrical conductivity

Ionic solid — Ions can only vibrate
Ions are mobile when molten
Ions are mobile when in solution

Figure 3.4.4 The ions are free to move and conduct electricity when an ionic compound is molten or dissolved in water

- Covalent compounds do not conduct electricity because they have no ions. They only have uncharged molecules. Some covalent compounds react with water to form ions. Hydrogen chloride reacts with water to form hydrochloric acid which splits up into ions. But this is not just dissolving in water – it is reacting.

STUDY TIP

Remember that compounds of metals with non-metals are likely to be ionic. Compounds of non-metals with other non-metals are covalent.

SUMMARY QUESTIONS

1. Make a list of the differences between the structures and properties of ionic and covalent compounds.
2. Explain why molten sodium chloride conducts electricity but solid sodium chloride does not.
3. Copy and complete the paragraph using these words:

   electrostatic   energy   forces   high   ionic   negative

   In giant _____ structures the _____ attractive _____ between the positive and _____ ions are very strong. Ionic compounds have very _____ melting points and boiling points because it takes a lot of _____ to overcome these forces.

KEY POINTS

1. In a giant ionic structure the attractive forces between the ions are very strong.
2. The attractive forces between molecules are weak.
3. We can tell if a compound is ionic or covalent by testing how easily it melts, its solubility in water and its electrical conductivity when dissolved in water.

3.5 Giant covalent structures

LEARNING OUTCOMES

- Describe the giant covalent structures of graphite and diamond
- Relate the structures of graphite and diamond to their uses
- Describe the macromolecular structure of silicon(IV) oxide
- Describe the similarity in properties between diamond and silicon(IV) oxide, related to their structures

Supplement

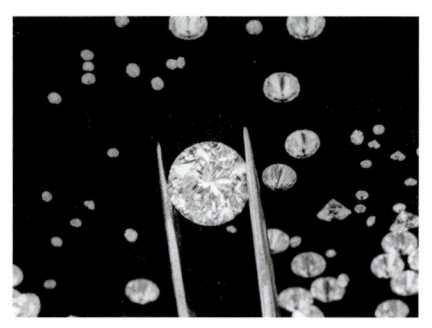

Giant covalent structures

Many covalently bonded substances are small molecules. In some covalent structures, however, there is a network of covalent bonds throughout the whole structure. We call these structures **giant covalent structures**. They are sometimes described as **macromolecules**, but this term may give you a misleading impression of their properties.

Giant covalent structures have a rigid three-dimensional network of strong covalent bonds throughout the crystal. It takes a lot of energy to break these bonds. So, unlike simple molecules, giant covalent structures have very high melting and boiling points. Giant covalent structures can be elements, such as carbon in the form of graphite and diamond, or compounds such as silicon(IV) oxide (silicon dioxide).

Diamond and graphite

Diamond and graphite are two of the several forms of carbon that exist. Different forms of the same element are called **allotropes**.

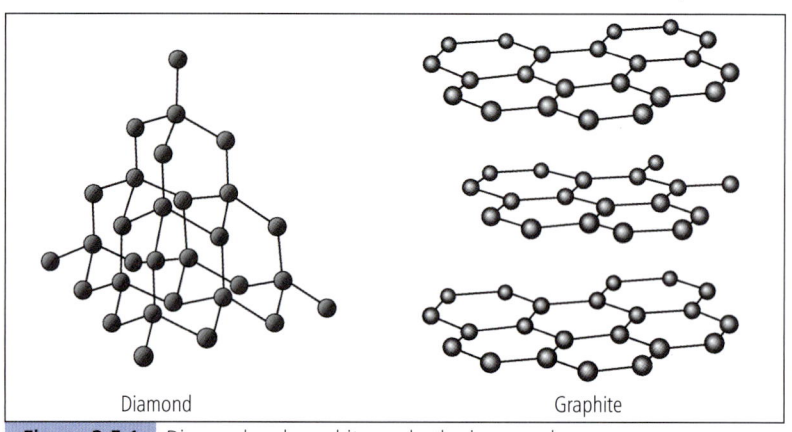

Diamonds owe their hardness and long-lasting nature to the way the carbon atoms are arranged

Figure 3.5.1 Diamond and graphite are both giant covalent structures

In diamond, a carbon atom forms four covalent bonds with other carbon atoms. The carbon atoms link together to form a giant lattice. As well as having high melting and boiling points, diamond is very hard. This means that it can't be scratched easily. That is why diamond is used for cutting and drilling metals and glass. Diamond forms colourless glittering crystals. These are much in demand for jewellery.

Graphite is a black shiny solid. Its carbon atoms are arranged in layers. Each carbon atom is joined to three other carbon atoms. These are arranged in hexagons – six-sided rings. The strong covalent bonding within the layers means that a lot of energy is needed to break the bonds. So the melting and boiling points of graphite are very high.

However, the bonding between the layers in graphite is weak. This means that the layers can slide over each other if a force is applied. This is why graphite has a slippery feel and can be easily scratched.

DID YOU KNOW?

Artificial diamonds for drill tips are made by heating other forms of carbon at very high temperatures and pressures.

Because of this weak bonding, the layers of graphite can flake off easily. That is why graphite is used as a lubricant and in pencil 'leads'. An unusual property of graphite is that it conducts electricity.

Why does graphite conduct electricity?

Each carbon atom in graphite is bonded to three other carbon atoms. But carbon has four valence electrons. So what has happened to the other valence electron from each carbon atom? The answer is that these electrons move around along the layers. They are called **delocalised electrons** because they do not belong to any particular atom. Graphite conducts electricity because the delocalised electrons can drift along the layers when a voltage is applied. Graphite is used to make carbon electrodes for electrolysis (see Topic 6.1). Diamond does not conduct electricity because all its valence electrons are used in bonding – it has no mobile electrons.

Silicon(IV) oxide

Silicon(IV) oxide is found in several forms. Its common name is silicon dioxide. Sand is largely silicon(IV) oxide. The silicon(IV) oxide found in quartz has a structure similar to diamond. Each silicon atom is bonded to four oxygen atoms but each oxygen atom is bonded to only two silicon atoms. This accounts for the formula SiO_2. Silicon(IV) oxide has a similar structure to diamond. They also have similar properties. Silicon(IV) oxide forms very hard colourless crystals and has a high melting point and a high boiling point. This is because it takes a lot of energy to break the strong covalent bonds. Silicon(IV) oxide does not conduct electricity because it has no delocalised electrons.

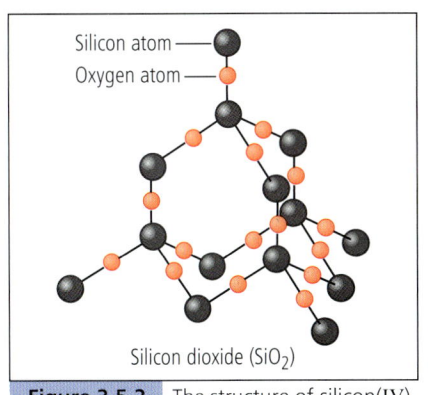

Figure 3.5.2 The structure of silicon(IV) oxide

> **STUDY TIP**
>
> When explaining why graphite conducts electricity, make sure that you state that electrons _in the layers_ can move along. Stating that 'the electrons move' suggests the electrons in the covalent bonds can move through the structure as well.

SUMMARY QUESTIONS

1 Copy and complete using the words below:

   covalent   energy   hexagons   layered   melting   weak

   Graphite has a _____ structure with the carbon atoms arranged in _____ The carbon atoms in the layers have strong _____ bonding. The bonding between the layers is _____. Graphite has a high _____ point because it takes a lot of _____ to break the strong covalent bonds.

2 Explain why graphite is used as a lubricant.

3 Explain why graphite conducts electricity but diamond does not.

4 Describe the similarities and differences in the structures of diamond and silicon dioxide.

KEY POINTS

1 Diamond and graphite have giant covalent structures.

2 Graphite can conduct electricity because of the delocalised electrons along its layers.

3 Silicon(IV) oxide has a similar structure to diamond.

3.6 Metallic bonding

LEARNING OUTCOMES

- Describe metallic bonding as a lattice of positive ions in a 'sea of electrons' and use this to describe the electrical conductivity and malleability of metals

What is metallic bonding?

Metals form a third type of giant structure. The metal atoms are packed closely together in a regular arrangement. Because they are so close together, the valence electrons tend to move away from their atoms. A 'sea' of free, delocalised electrons is formed surrounding a lattice of positively charged metal ions. The positively charged ions are held together by their strong attraction to the mobile electrons which move in between the ions. This is **metallic bonding**. The electrostatic attraction between the electrons and the metal ions acts in all directions.

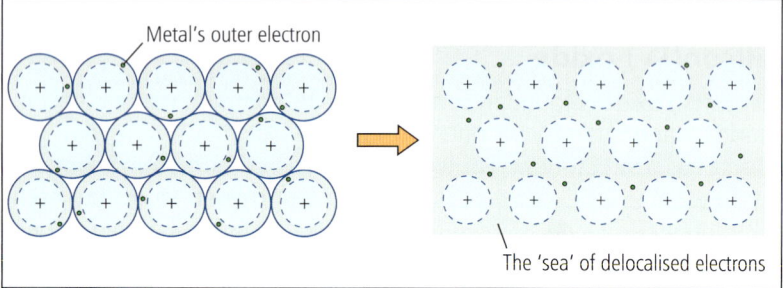

Figure 3.6.1 A metallic structure consists of positively charged metal ions surrounded by a 'sea' of delocalised electrons

Drawing copper out into wires depends on being able to make the layers of metal atoms slide easily over each other

PRACTICAL

Modelling metallic structure

1. First fill a shallow dish with water and add some detergent.

2. Then blow bubbles onto the surface of the water by moving a gas syringe backwards and forwards. Keep doing this so you have a single layer of lots of bubbles.

3. Each bubble represents a metal atom. You will see that the bubbles are regularly arranged. But in some places there are 'grain boundaries' where the directions of the layers change. The structure of metals is rather like this with irregularly shaped grains formed within the metal.

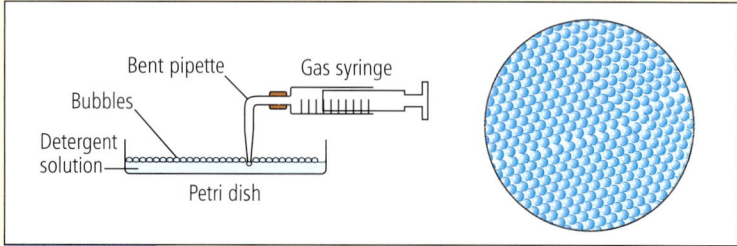

Figure 3.6.2 A bubble raft models the structure of metals

The properties of metals can be explained using this model of metallic structure.

- Most metals have high melting points and boiling points:

It takes a lot of energy to weaken the strong forces of attraction between the metal ions and the delocalised electrons in the lattice. These attractive forces can be overcome only when the temperature is high.

- Metals are good conductors of heat and electricity:

When a voltage is applied, the delocalised electrons move through the metal lattice towards the positive pole of the cell or power pack. Conduction of heat is due to vibrations of the atoms passing on the energy from one atom to the next. The metallic structure allows the atoms to vibrate more freely compared with a covalent structure. The delocalised electrons can also carry energy through the metallic structure quickly. However, the electrons will not move as readily through the lattice if the atoms are vibrating faster. That is why the electrical conductivity of a metal decreases with increasing temperature.

- Metals are malleable and ductile:

The positive ions in a metal are arranged regularly in layers. When a force is applied, the layers can slide over each other. In a metallic bond the attractive forces between the metal ions and the electrons act in any direction. So when the layers slide new bonds can easily form. This leaves the metal with a different shape.

STUDY TIP

It is a common error to think that conduction in metals is due to moving ions. Remember that it is only the delocalised electrons which move. The positive ions remain fixed in position within the giant lattice.

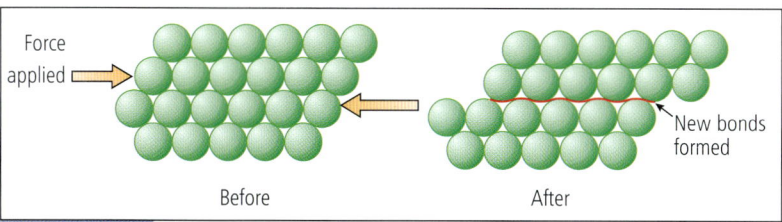

Figure 3.6.3 When a force is applied to a metal the layers slide but soon form new bonds

SUMMARY QUESTIONS

1 Copy and complete using the words below:

attraction   delocalised   energy   high melting   mobile   positive

In metallic bonding the _____ metal ions are held together by a sea of _____ electrons. The strong electrostatic _____ between the ions and the _____ electrons gives many metals their strength. Because it takes a lot of _____ to overcome these strong electrostatic forces the _____ and boiling points of most metals is _____.

2 Explain why metals conduct electricity.

3 When you pull a metal wire with enough force it stretches. Draw diagrams to show what happens to the layers of metal atoms when you pull a metal wire.

KEY POINTS

1 The 'atoms' in metals are arranged in regular layers.

2 The metallic structure consists of positive metal ions held together by a sea of delocalised electrons.

3 Metals conduct electricity because the delocalised electrons can move freely through the structure.

SUMMARY QUESTIONS

1 Match each type of structure and bonding on the left with a description on the right.

simple molecular — the solid conducts electricity

giant ionic — has a high melting point but doesn't conduct electricity

giant covalent — has a high melting point and conducts electricity when it dissolves in water

metallic — has a low melting point and does not conduct electricity

2 State whether the following pairs of elements form ionic or covalent compounds when they combine:
   (a) sodium and chlorine
   (b) hydrogen and oxygen
   (c) hydrogen and chlorine
   (d) lithium and chlorine
   (e) sodium and iodine

3 Copy and complete using words from the list:

**eight   electrons   molecules   outer
pair   shares   shell   stable**

A covalent bond is a _____ of shared electrons. When chlorine atoms combine to form chlorine _____, each chlorine atom _____ a pair of _____ to form a covalent bond. Each chorine atom now has an _____ electron _____ with _____ electrons. This makes the chlorine molecule _____.

4 The table shows some properties of different substances.

substance	melting point / °C	conductivity of solid	conductivity of liquid	solubility in water
A	−56	does not conduct	does not conduct	insoluble
B	610	does not conduct	conducts	soluble
C	−70	does not conduct	does not conduct	insoluble
D	2310	conducts	conducts	insoluble
E	680	does not conduct	conducts	soluble

Classify each of these substances as metals, giant ionic structures or simple molecular structures.

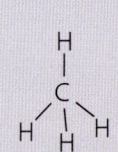

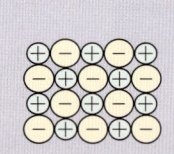

PRACTICE QUESTIONS

1 Which one of these statements about the outer shell electrons in sodium chloride is true?
   A Sodium shares two electrons with chlorine.
   B Sodium ions have eight electrons in their outer shells.
   C Chloride ions have seven electrons in their outer shells.
   D Sodium ions have one electron in their outer shells.

(Paper 1)

2 Which of the following attributes about the outer shell electrons of the sulfur atom in hydrogen sulfide, H_2S, is correct?
   A Two lone pairs and one bonding pair.
   B One lone pair and two bonding pairs.
   C One lone pair and three bonding pairs.
   D Two lone pairs and two bonding pairs.

(Paper 2)

5 Draw dot-and-cross diagrams for:
   (a) sodium chloride
   (b) ammonia (NH_3)
   (c) hydrogen.

6 Name the type of structures represented in each of these diagrams:

3 Potassium chloride, KCl, has a giant ionic structure but methane, CH_4, is a simple molecule.
  (a) State three differences between the physical properties of potassium chloride and methane. *[3]*
  (b) Draw a dot-and-cross diagram to show the electronic structure of a potassium ion and a chloride ion. *[2]*
  (c) Draw a dot-and-cross diagram to show the structure of methane. *[1]*

(Paper 3)

4 The diagram shows the structures of diamond and graphite.

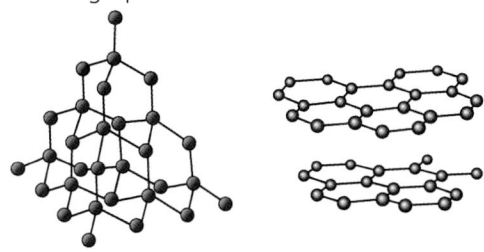

Diamond   Graphite

  (a) Describe two ways in which these structures are similar. *[2]*
  (b) Describe two differences between these structures. *[2]*
  (c) State the name given to these types of structure. *[1]*
  (d) Explain why graphite is used as a lubricant by referring to its structure. *[2]*
  (e) State one use of diamond. *[2]*

(Paper 3)

5 Hydrogen chloride is a covalent compound. It is a colourless gas.
  (a) Draw a dot-and-cross diagram for hydrogen chloride, showing all the electron shells. *[3]*
  (b) Hydrogen chloride reacts with water to form a solution that contains hydrogen ions and chloride ions.
    (i) In what way does hydrogen chloride differ in its properties from other simple covalent compounds such as carbon monoxide and methane? *[1]*
    (ii) State one property of hydrogen chloride that demonstrates that it is a covalent compound. *[1]*
    (iii) Draw the structure of a hydrogen ion and a chloride ion. *[3]*

(Paper 3)

6 The structures of carbon dioxide, diamond and silicon dioxide are shown below.

$O=C=O$
Carbon dioxide   Diamond   Silicon dioxide (SiO_2)

  (a) (i) In what state does carbon dioxide exist at room temperature? Explain your answer. *[2]*
    (ii) In what state does diamond exist at room temperature? Explain your answer. *[2]*
  (b) Draw a dot-and-cross diagram for carbon dioxide, showing only the outer electron shells. *[2]*
  (c) (i) Describe two ways in which the structure of diamond and silicon dioxide are similar. *[2]*
    (ii) Describe one difference in the structure of silicon dioxide and diamond. *[1]*
  (d) State the formula for silicon dioxide. *[1]*
  (e) Describe two properties of silicon dioxide that are similar to diamond. Suggest how each property is related to the structure of silicon dioxide. *[4]*

(Paper 4)

7 Zinc is a metal and sulfur is a non-metal.
  (a) State three differences between the physical properties of zinc and sulfur. *[3]*
  (b) Draw a diagram to show the metallic bonding in zinc. *[3]*
  (c) Sulfur is a non-metal which forms rings containing eight atoms.

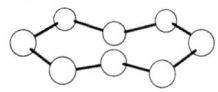

  Draw a diagram to show the electronic structure of a sulfur molecule. Show only the outer electron shells. *[3]*

(Paper 4)

4 Formulae and equations

4.1 Chemical formulae

LEARNING OUTCOMES

- Know how to write symbols for chemical elements and formulae of simple compounds
- Deduce the formula of a simple compound from the relative number of atoms present

DID YOU KNOW?

John Dalton produced a list of symbols for the atoms in 1808. Each symbol was placed in a circle which represented an atom.

STUDY TIP

When writing symbols containing two letters, make sure that the second letter is in lower case. Cl is correct for chlorine. CL is wrong.

Chemical symbols

Different languages may have different names for a particular element. For example, in Italian, nitrogen is called *azoto* and in German it is called *Stickstoff*. So it is useful to have a set of symbols for the elements that is recognised throughout the world.

Every element has its own **chemical symbol**. For example, sulfur is S, oxygen is O and bromine is Br. Many symbols are not as obvious. This is because they come from Latin, Greek or Arabic words. For example, iron is Fe and potassium is K. If you do not know the symbol for an element, you can look it up in the Periodic Table. Notice that the first letter in a symbol is always a capital letter. If there is a second letter, it is always small (lower case).

John Dalton used different symbols for the elements from those we use today

Formulae

We can work out the **formula** of a compound by knowing the valencies or combining powers of the elements it contains. You can usually work out the valencies from the group in the Periodic Table to which the element belongs.

Group	I	II	III	IV	V	VI	VII	VIII
Valency	1	2	3	4	3	2	1	0

You can see that the valencies of the non-metals on the right-hand side of the Periodic Table in Groups V to VIII are found by taking the group number away from 8. So nitrogen in Group V has a valency of = 3 (8 − 5). The valencies are the combining powers of the elements. So how do we get the formula from the valencies? Let's look at some examples:

What is the formula of magnesium chloride?

Magnesium is in Group II so it has a valency of 2.

Chlorine is in Group VII so it has a valency of 1.

You can use the following method to find the formula:

- Write down the symbols in the same order as in the name of the compound.

- Write down the valencies underneath each element.
- Swap the numbers around.
- Where possible, cancel down the numbers to get the smallest whole number ratio when necessary.

	Example 1	Example 2	Example 3
What is the formula of:	magnesium chloride	aluminium oxide	calcium oxide
Valencies	Mg Cl 2 1	Al O 3 2	Ca O 2 2
Formula	$MgCl_2$	Al_2O_3	CaO

There are three points to bear in mind when writing formulae:

- You cannot predict the valency of the transition elements (see Topic 12.5).
- The formulae of some compounds containing only non-metals are not simplified by cancelling down.
  For example, hydrogen peroxide is H_2O_2.
- Some non-metals form oxides with different formulae.
  For example, CO is carbon monoxide and CO_2 is carbon dioxide; SO_2 is sulfur dioxide and SO_3 is sulfur trioxide.

Simple rules for naming compounds

Naming compounds of two elements

If a compound contains a metal and a non-metal, the metal is put first and the ending of the non-metal changes to 'ide'. For example, the compound of chlorine and magnesium is named magnesium chlor<u>ide</u> and the compound of calcium and oxygen is calcium ox<u>ide</u>.

If a compound contains two non-metals, naming is a bit more complicated. If the compound contains hydrogen, this comes first. If not, the non-metal with the lower group number comes first. For example, nitrogen dioxide – nitrogen is in Group V so it comes before oxygen, which is in Group VI. If both non-metals are in the same group then the one further down the group comes first. For example: sulfur dioxide – sulfur is lower than oxygen in Group VI so it comes first.

Some compounds which have been known for a long time are called by their common names. For example, H_2O is water and NH_3 is ammonia.

Naming compounds with three elements

You should be able to recognise the following groups which contain oxygen as well as another element.

OH	hydroxide	Example: sodium hydroxide, NaOH
NO_3	nitrate	Example: magnesium nitrate, $Mg(NO_3)_2$
SO_4	sulfate	Example: calcium sulfate, $CaSO_4$
CO_3	carbonate	Example: sodium carbonate, Na_2CO_3

> **STUDY TIP**
>
> Take care when writing the second atom in a formula. Co2 is not acceptable for carbon dioxide and neither is H²o for water. The symbol for oxygen is always a capital O.

KEY POINTS

1. Each chemical element has a symbol.
2. When naming a compound containing two elements the second name often changes to -ide.
3. The formula of a simple compound can be worked out from the valency of the elements present.

SUMMARY QUESTIONS

1. Write the name of the compounds formed by combining:
   a bromine and sodium
   b magnesium and oxygen
   c hydrogen and iodine.
2. Work out the formulae of:
   a magnesium bromide
   b potassium chloride
   c barium oxide
   d magnesium nitride
   e gallium oxide.
3. Name the following compounds:
   a $MgBr_2$
   b $AlCl_3$
   c K_2SO_4
   d KNO_3

4.2 Working out the formula

LEARNING OUTCOMES

- Deduce the formula of a simple compound from the relative number of atoms present
- Deduce the formula of a simple compound from a diagrammatic representation
- Determine the formula of an ionic compound from the charges of the ions present

DID YOU KNOW?

Every chemical that has been written about has been given an identification number called the CAS number. In January 2008, there were 33 565 050 chemicals with CAS numbers.

STUDY TIP

When writing the formula of an ionic compound from a diagram of its structure, make sure that you write the formula as the simplest ratio. For example, write $CaBr_2$, not Ca_8Br_{16}.

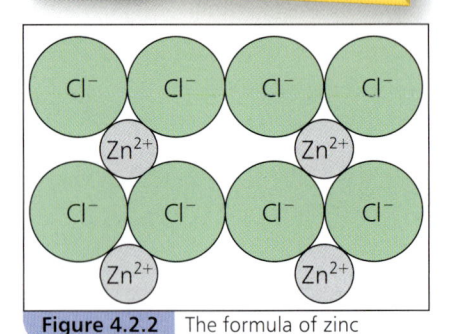

Figure 4.2.2 The formula of zinc chloride is $ZnCl_2$

Working out the formula from diagrams

If we are given a picture of a molecule showing all atoms and bonds we can easily work out its formula. We do this simply by counting the number and type of each atom. We then use the rules for naming compounds to write the name.

Oxygen (2 O atoms) formula O_2

Ammonia (1 N atom and 3 H atoms) formula NH_3

Phosphorus(V) chloride (1 P atom and 5 Cl atoms) formula PCl_5

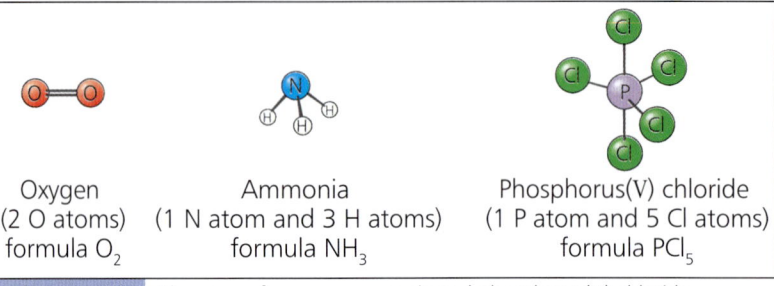

Figure 4.2.1 Diagrams of oxygen, ammonia and phosphorus(V) chloride

For some molecules, especially those containing carbon and hydrogen, we can write two types of formula:

- **Molecular formula**: this shows the number of each type of atom present.
- **Empirical formula**: this shows the simplest whole number ratio of the different atoms present.

For example, for the compound ethane:

The full **structural formula** is:

$$H-\underset{\underset{H}{|}}{\overset{\overset{H}{|}}{C}}-\underset{\underset{H}{|}}{\overset{\overset{H}{|}}{C}}-H$$

The molecular formula is C_2H_6.

The empirical formula is CH_3.

If you are just asked to work out the formula of a compound, you should write the molecular formula.

Working out the formula of an ionic compound

You can also work out the formula of an ionic compound from a diagram by:

- counting up the number of positive ions
- counting up the number of negative ions
- cancelling the numbers down to find the simplest whole number ratio.

For example, the diagram in Figure 4.2.2 of zinc chloride shows four positive ions and eight negative ions. So Zn = 4 and Cl = 8. If we divide each by four, we get the simplest ratio which is Zn = 1 and Cl = 2. So the formula is $ZnCl_2$.

In an ionic compound the number of positive charges is balanced out by the number of negative charges. The total charge is zero. We can work out the formula for a compound if we know the charges on the ions. Figure 4.2.3 shows how the charges on some common ions are related to the position of the element in the Periodic Table.

You can see that for metal ions the positive charge on the ion is the same as the group number. For non-metal ions the negative charge is 8 minus the group number.

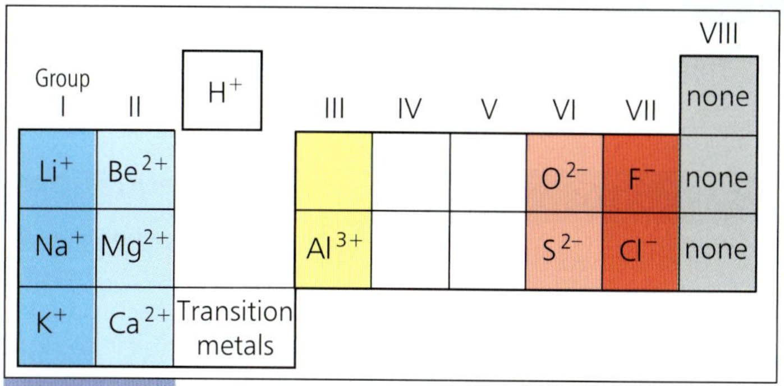

Figure 4.2.3 Ions formed by some elements of the Periodic Table

You work out the formula for an ionic compound knowing the charge on each ion in a similar way to the one you used in Topic 4.1

Some ions contain more than one type of atom. These are called **compound ions**. You need to remember the names and charges of these ions.

NH_4^+	OH^-	NO_3^-	SO_4^{2-}	CO_3^{2-}	HCO_3^-
ammonium ion	hydroxide ion	nitrate ion	sulfate ion	carbonate ion	hydrogencarbonate ion

The formulae of compounds formed from these ions are found in the same way as above by balancing out the charges on the ions. Sometimes you have to use brackets around the compound ions to make sure that the number of atoms is correct. Here are two examples in Figure 4.2.4.

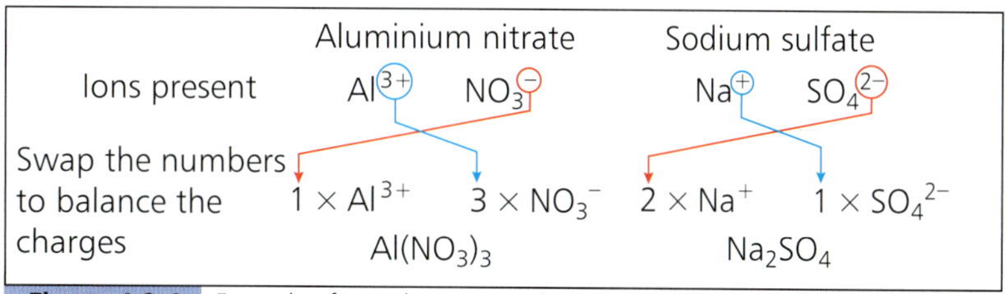

Figure 4.2.4 Formulae from changes on the ions

SUMMARY QUESTIONS

1 Write the molecular formula for each of the following compounds:
  a $N{\equiv}N$
  b $O{=}C{=}O$
  c $\begin{array}{c}H\\ \end{array}C{=}C\begin{array}{c}H\\ \end{array}$ with H, H

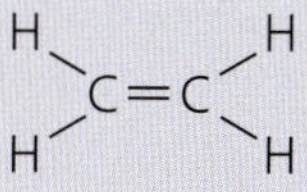

2 Name the following compounds:
  a K_2SO_4
  b $Mg(NO_3)_2$
  c $Cu(OH)_2$

3 Write the formula of each of the following compounds by balancing the charges on the ions.
  a calcium carbonate
  b magnesium nitrate
  c ammonium chloride
  d aluminium sulfate

KEY POINTS

1 The formula of a molecular compound can be worked out from a diagram of the full structural formula by counting the number of each type of atom.

2 The formula of an ionic compound can be found by counting the ions and finding the simplest ratio.

3 The formula of an ionic compound can be worked out using the charges on the ions.

4.3 Chemical equations

LEARNING OUTCOMES

- Construct word equations and simple balanced chemical equations
- Deduce the balanced equation for a chemical reaction, given relevant information

Supplement

Word equations

We show chemical reactions by chemical equations. The simplest type of equation is a **word equation**. This shows the reactants on the left and the products on the right. The arrow shows that the reaction goes from left to right – from reactants to products. For example:

$$\text{magnesium} + \text{oxygen} \longrightarrow \text{magnesium oxide}$$

Any conditions such as heating or adding catalysts are written over the arrow. Heat is not a definite substance so it does not appear as a reactant or product.

$$\text{nitrogen} + \text{hydrogen} \xrightarrow{\text{heat + catalyst + pressure}} \text{ammonia}$$

Word equations are less useful than symbol equations because:

- they do not show the number of molecules that react together
- they can be very long if the name of a chemical is complicated
- the names for different chemicals are not the same in different languages.

Symbol equations

A **symbol equation** is a shorthand way of describing a chemical reaction. In any chemical reaction the bonds in the reactants are broken and the new bonds are formed so the atoms combine differently. Atoms cannot be formed from nothing and cannot be destroyed. So there must be the same number of each type of atom on both sides of the equation. We can balance an equation by counting the number of each type of atom in the reactants and products. If the numbers are equal, the equation is balanced.

This equation is balanced: $\quad Fe + S \longrightarrow FeS$

There is one atom of iron and one atom of sulfur on each side of the equation.

This equation is not balanced: $\quad Mg + O_2 \longrightarrow MgO$

There are two atoms of oxygen on the left but only one on the right.

When balancing an equation you must not change the formula of any of the reactants or products.

In order to balance an equation you need to follow these steps:

Step 1. Write down the formulae for the reactants and products. For example:

$$H_2 + O_2 \longrightarrow H_2O$$

Some gaseous elements exist as **diatomic** molecules. They have two atoms per molecule. You should know that the following molecules are diatomic: hydrogen, H_2; nitrogen, N_2; oxygen, O_2; fluorine, F_2; chlorine, Cl_2; bromine, Br_2; and iodine, I_2.

STUDY TIP

When balancing symbol equations you must not change any of the formulae. Always balance by putting large numbers in front of the formulae. For example, balancing CaO by making it into CaO_2 is wrong. It should be 2CaO.

DID YOU KNOW?

The first diagram showing a type of chemical equation was made by the Frenchman Jean Beguin in 1615.

Step 2. Count up the atoms. You can use coloured dots beneath if it helps you.

$$H_2 + O_2 \longrightarrow H_2O$$

There are two oxygen atoms on the left but only one on the right.

Step 3. Balance the atoms by putting a number in front of one of the reactants or products. The number in front multiplies all the way through the molecule. So $2H_2O$ has (2×2) 4 hydrogen atoms and (2×1) 2 oxygen atoms.

$$H_2 + O_2 \longrightarrow 2H_2O$$

Now the oxygen atoms are balanced. But the hydrogen atoms have become unbalanced.

Step 4. Keep balancing in this way until you get all the atoms balanced.

$$2H_2 + O_2 \longrightarrow 2H_2O$$

The equation is now balanced.

More symbol equations

When brackets are used, the small number at the bottom right of the brackets multiplies through what is in the brackets. So $Mg(NO_3)_2$ has 1 'atom' of magnesium, (1×2) 2 atoms of nitrogen and (2×3) 6 atoms of oxygen. If we wrote $2Mg(NO_3)_2$ we would have twice as many of each of these atoms: 2 magnesium, 4 nitrogen and 12 oxygen.

Some examples are given here:

Example 1: aluminium + water $\longrightarrow$ aluminium + hydrogen
chloride   hydroxide   chloride

$$AlCl_3 + H_2O \longrightarrow Al(OH)_3 + HCl$$

Balance the chlorine atoms $AlCl_3 + H_2O \longrightarrow Al(OH)_3 + 3HCl$

Balance the oxygen atoms $AlCl_3 + 3H_2O \longrightarrow Al(OH)_3 + 3HCl$

Example 2:   calcium + nitric $\longrightarrow$ calcium + water
hydroxide   acid   nitrate

$$Ca(OH)_2 + HNO_3 \longrightarrow Ca(NO_3)_2 + H_2O$$

Balance the nitrate (NO_3) $Ca(OH)_2 + 2HNO_3 \longrightarrow Ca(NO_3)_2 + H_2O$

The hydrogen is not balanced: there are 4 atoms on the left and only 2 on the right.

Balance the hydrogen atoms $Ca(OH)_2 + 2HNO_3 \longrightarrow Ca(NO_3)_2 + 2H_2O$

KEY POINTS

1 There is the same number of each type of atom on each side of a chemical equation.

2 In a chemical reaction, the mass of the products always equals the mass of the reactants.

3 Equations are balanced by writing numbers in front of particular reactants or products.

SUMMARY QUESTIONS

1 Write word equations for the following reactions:

  a $2Na + 2H_2O \longrightarrow 2NaOH + H_2$

  b $Mg + ZnSO_4 \longrightarrow Zn + MgSO_4$

  c $CuO + H_2SO_4 \longrightarrow CuSO_4 + H_2O$

2 Write balanced symbol equations for the following reactions:

  a $Na + Cl_2 \longrightarrow NaCl$

  b $Cl_2 + H_2 \longrightarrow HCl$

  c $Na + O_2 \longrightarrow Na_2O_2$

3 Write balanced symbol equations for the following reactions:

  a zinc + hydrochloric acid $\longrightarrow$ zinc chloride + hydrogen

  b chlorine + potassium bromide $\longrightarrow$ bromine + potassium chloride

  c copper(II) chloride, $CuCl_2$ + sodium hydroxide $\longrightarrow$ copper hydroxide, $Cu(OH)_2$ + sodium chloride

4.4 More about equations

LEARNING OUTCOMES

- Construct symbol equations including state symbols
- Construct balanced ionic equations

Using state symbols

We use special symbols in equations to show if a substance is a solid, liquid, gas or dissolved in water. These are called state symbols. They are:

(s) solid   (l) liquid   (g) gas

(aq) **aqueous** solution – solute dissolved in water

State symbols are written after the formula of each reactant and product:

$$Zn(s) + H_2SO_4(aq) \longrightarrow ZnSO_4(aq) + H_2(g)$$

When water or other liquids are reactants or products they can be liquid or gas according to the conditions used:

$$MgO(s) + 2HCl(aq) \longrightarrow MgCl_2(aq) + H_2O(l)$$

$$CO(g) + H_2O(g) \longrightarrow CO_2(g) + H_2(g)$$

Ionic equations

Many chemical reactions involve ionic compounds. When ionic compounds dissolve in water, the ions separate:

$$MgCl_2(s) + aq \longrightarrow Mg^{2+}(aq) + 2Cl^-(aq)$$

Notice how we write the separate ions. The $2Cl^-$ shows that there are two separate chloride ions in solution.

You can often tell if a substance forms ions. Examples are:

- compounds containing a metal and a non-metal
- acids such as hydrochloric acid, HCl; sulfuric acid, H_2SO_4; and nitric acid, HNO_3
- ammonium compounds – these contain the ammonium ion, NH_4^+.

In many ionic reactions, only some of the ions take part in the reaction. The ones that do not are called spectator ions.

An **ionic equation** is a special form of symbol equation that shows only those ions that react. We can change an ordinary symbol equation into an ionic equation in the following way:

Step 1. Write down the balanced ionic equation with the state symbols.

$$BaCl_2(aq) + Na_2SO_4(aq) \longrightarrow 2NaCl(aq) + BaSO_4(s)$$

Step 2. Write down all the ions present in the equation. You can use the bullet-pointed list above to help you. Any reactant or product that is a solid, a liquid or a gas is not split into ions.

$$Ba^{2+}(aq) + 2Cl^-(aq) + 2Na^+(aq) + SO_4^{2-}(aq) \longrightarrow 2Na^+(aq) + 2Cl^-(aq) + BaSO_4(s)$$

The products of a reaction may have a different state from the reactants

DID YOU KNOW?

In a lecture in 1827, Michael Faraday gave several chemical demonstrations to show the meaning of solids, liquids and gases. This was one of the first times that this had been done.

Step 3. Cross out the ions that are the same on both sides of the equation. These are the **spectator ions**.

$$Ba^{2+}(aq) + 2\cancel{Cl^-}(aq) + 2\cancel{Na^+}(aq) + SO_4^{2-}(aq) \longrightarrow 2\cancel{Na^+}(aq) + 2\cancel{Cl^-}(aq) + BaSO_4(s)$$

Step 4. Write down only the reactants and products that are left to get the ionic equation.

$$Ba^{2+}(aq) + SO_4^{2-}(aq) \longrightarrow BaSO_4(s)$$

Another example:

An aqueous solution of chlorine reacts with an aqueous solution of sodium bromide to form an aqueous solution of sodium chloride and an aqueous solution of bromine. Write an ionic equation for this reaction.

Step 1. $Cl_2(aq) + 2NaBr(aq) \longrightarrow Br_2(aq) + 2NaCl(aq)$

Step 2. $Cl_2(aq) + 2Na^+(aq) + 2Br^-(aq) \longrightarrow Br_2(aq) + 2Na^+(aq) + 2Cl^-(aq)$

The chlorine and bromine are not ionic.

Step 3.

$$Cl_2(aq) + 2\cancel{Na^+}(aq) + 2Br^-(aq) \longrightarrow Br_2(aq) + 2\cancel{Na^+}(aq) + 2Cl^-(aq)$$

Step 4. $Cl_2(aq) + 2Br^-(aq) \longrightarrow Br_2(aq) + 2Cl^-(aq)$

If two solutions are added together and a precipitate (solid) is formed you can simplify the method. All you have to do is to write down the formula of the precipitate as a product and the ions that go to make up the precipitate as the reactants:

symbol equation: $FeCl_3(aq) + 3NaOH(aq) \longrightarrow Fe(OH)_3(s) + 3NaCl(aq)$

ionic equation: $Fe^{3+}(aq) + 3OH^-(aq) \longrightarrow Fe(OH)_3(s)$

In this reaction we know that the iron must be Fe^{3+}, because each of the three chloride ions has one negative charge and there is no overall charge on the iron chloride.

> **STUDY TIP**
>
> When writing ionic equations, first identify the reactants or products that are not ionic. These will be solids, liquids or simple molecules like chlorine. It is only then that you can separate the other compounds into ions.

SUMMARY QUESTIONS

1 Copy and complete the paragraph using these words:

equations   ions   precipitate   reaction   spectator

Ionic _____ show only the _____ that take part in a reaction to form a product which is either not ionic or is a _____. The ions that do not take part in a _____ are called _____ ions.

2 Write ionic equations for the following reactions:
   a $AgNO_3(aq) + KBr(aq) \longrightarrow KNO_3(aq) + AgBr(s)$
   b $CuCl_2(aq) + 2NaOH(aq) \longrightarrow Cu(OH)_2(s) + 2NaCl(aq)$
   c $Br_2(aq) + 2KI(aq) \longrightarrow I_2(aq) + 2KBr(aq)$

KEY POINTS

1 The state symbols (s), (l), (g) or (aq) are added after the formula of each reactant and product.

2 Ionic equations are simplified symbol equations showing only those ions that react and the products of their reaction.

3 In reactions involving ions not all the ions take part in the reaction. Those that are not involved are called spectator ions.

SUMMARY QUESTIONS

1. Write word equations for the reactions of:
   (a) magnesium with chlorine
   (b) potassium with oxygen
   (c) sodium with bromine
   (d) carbon with oxygen.

2. Write the formulae for:
   (a) sodium chloride
   (b) magnesium oxide
   (c) aluminium oxide
   (d) calcium chloride
   (e) aluminium chloride.

3. Match the names on the left with the elements or ions on the right.

oxide	Cl_2
hydrogen	O_2
calcium	O^{2-}
chlorine	H^+
chloride	Ca^{2+}
oxygen	Cl^-

4. Balance these equations.
   (a) $Ca + O_2 \longrightarrow CaO$
   (b) $SO_2 + O_2 \longrightarrow SO_3$
   (c) $Na + Cl_2 \longrightarrow NaCl$
   (d) $C_4H_8 + O_2 \longrightarrow CO_2 + H_2O$
   (e) $Fe_2O_3 + CO \longrightarrow Fe + CO_2$
   (f) $PbO + C \longrightarrow Pb + CO_2$

5. Write word equations for these reactions:
   (a) $Zn + CuSO_4 \longrightarrow ZnSO_4 + Cu$
   (b) $4Fe + 3O_2 \longrightarrow 2Fe_2O_3$
   (c) $NH_4Cl + NaOH \longrightarrow NH_3 + NaCl + H_2O$
   (d) $Cu(NO_3)_2 + Mg \longrightarrow Cu + Mg(NO_3)_2$

6. Change these equations into ionic equations:
   (a) $Cl_2(aq) + 2NaBr(aq) \longrightarrow Br_2(aq) + 2NaCl(aq)$
   (b) $FeCl_2(aq) + 2KOH(aq) \longrightarrow Fe(OH)_2(s) + 2KCl(aq)$
   (c) $Mg(s) + 2HCl(aq) \longrightarrow MgCl_2(aq) + H_2(g)$

PRACTICE QUESTIONS

1. Part of the structure of magnesium chloride is shown below.

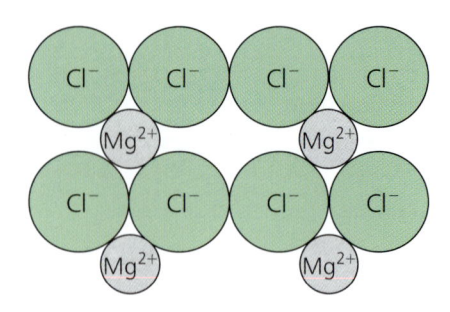

   The correct formula for magnesium chloride is:
   A Mg_2Cl
   B Mg_4Cl_8
   C Mg_8Cl_4
   D $MgCl_2$

   (Paper 1)

2. Gallium is in Group III of the Periodic Table. Oxygen is in Group VI. The formula of gallium oxide is:
   A Ga_2O_3
   B Ga_3O_6
   C Ga_3O_2
   D GaO_2

   (Paper 2)

3. The formulae for some halogens and halogen compounds are shown below.

   NaCl KBr Br_2 Cl_2 I_2 LiBr KI

   (a) Which of these are compounds? [3]
   (b) Write the names of (i) NaCl (ii) LiBr. [2]
   (c) Copy and complete the equation for the following reaction:

   $Cl_2 + \_\_\_ KBr \longrightarrow 2KCl + \_\_\_\_\_$ [2]

   (d) Name the following ions:
   (i) I^- (ii) O^{2-} (iii) Cl^- [3]

   (Paper 3)

4. Magnesium reacts with hydrochloric acid to form magnesium chloride and with sulfuric acid to produce magnesium sulfate.
   (a) Write the formula for
   (i) hydrochloric acid
   (ii) sulfuric acid
   (iii) magnesium sulfate. [3]

(b) Magnesium also reacts with very dilute nitric acid.
  (i) Copy and complete the equation for this reaction
  $Mg + \ldots HNO_3 \longrightarrow Mg(NO_3)_2 + H_2$ [1]
  (ii) Write a word equation for this reaction. [2]
(c) Magnesium reacts with oxygen to form magnesium oxide, MgO. Write a symbol equation for this reaction. [2]

(Paper 3)

5 Calcium carbonate reacts with hydrochloric acid to form calcium chloride, carbon dioxide and water.
  (a) Copy and complete the equation for this reaction.
  $CaCO_3 + \_\_ HCl \longrightarrow CaCl_2 + \_\_ + \_\_$ [3]
  (b) When heated, calcium carbonate decomposes:
  $CaCO_3 \longrightarrow CaO + CO_2$
  Write a word equation for this reaction. [1]
  (c) The 'model equation' below describes the combustion of methane.

  $H-\underset{\underset{H}{|}}{\overset{\overset{H}{|}}{C}}-H + \begin{matrix} O=O \\ O=O \end{matrix} \longrightarrow O=C=O + \begin{matrix} H^{\diagdown O \diagup} H \\ H^{\diagdown O \diagup} H \end{matrix}$

  Write a balanced chemical equation for this reaction. [2]

(Paper 3)

6 This question is about some compounds containing nitrogen.
  (a) Under certain conditions, nitrogen reacts with hydrogen to make ammonia gas, NH_3.
  Write a balanced equation for this reaction, including state symbols. [2]
  (b) When ammonia is bubbled through a solution of hydrochloric acid, a solution of ammonium chloride is formed.
  Write a balanced equation for this reaction including state symbols. [2]

(c) Aqueous ammonia contains ammonium ions, NH_4^+, and hydroxide ions, OH^-. When aqueous ammonia reacts with aqueous copper(II) chloride, $CuCl_2$, the products are a precipitate of copper(II) hydroxide, $Cu(OH)_2$, and aqueous ammonium chloride. Write an ionic equation for this reaction. [2]

(d) Nitric acid reacts with calcium hydroxide to form calcium nitrate and water.
  Write a balanced equation for this reaction. [2]

(Paper 4)

7 This question is about sodium and some compounds of sodium.
  (a) When sodium metal reacts with water the products are hydrogen gas and a solution of sodium hydroxide.
  Write a balanced equation for this reaction, including state symbols. [3]
  (b) An aqueous solution of sodium chloride reacts with an aqueous solution of silver nitrate, $AgNO_3$. The products are a precipitate of silver chloride, AgCl, and a solution of sodium nitrate.
    (i) Write a full symbol equation for this reaction, including state symbols. [3]
    (ii) Convert the full symbol equation to an ionic equation. [2]

(Paper 4)

8 When iron(III) oxide, Fe_2O_3, and aluminium are heated together at a high temperature, iron and aluminium oxide are formed.
  (a) What is the meaning of the symbol (III) in iron(III) oxide? [1]
  (b) Write a balanced equation for this reaction. [3]
  (c) Iron reacts with hydrochloric acid:
  $Fe(s) + 2HCl(aq) \longrightarrow FeCl_2(aq) + H_2(g)$
  Explain the meaning of (i) (aq) and (ii) (g) in the symbol equation. [2]

(Paper 4)

5 Chemical calculations

5.1 Reacting masses

LEARNING OUTCOMES

- Define relative atomic mass and relative molecular mass
- Do simple chemical calculations involving reacting masses

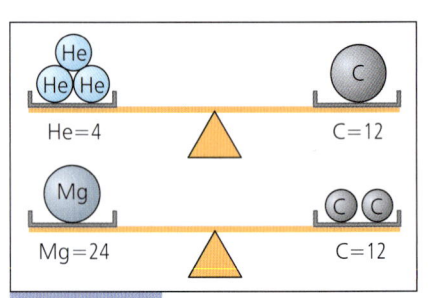

Figure 5.1.1 Three helium atoms have the same mass as one carbon atom. One magnesium atom has the same mass as two carbon atoms.

DID YOU KNOW?

The mass of a single proton is 0.000 000 000 000 000 000 000 001 67 grams!

Relative atomic mass

A chemical equation shows the number and type of atoms in the reactants and products. But this does not take into account the mass of each atom. The atoms of different elements have different masses. So if we want to know what quantities of reactants to use to make a certain amount of product we have to know how heavy one atom is compared with another – their relative mass.

The mass of a single atom is so small that you cannot weigh it using even the most accurate balance. To overcome this problem scientists weigh a lot of atoms and then compare them with the mass of the same number of 'standard' atoms. Atomic scientists have chosen to use as a standard an isotope of carbon called carbon-12.

Scientists have given an atom of carbon-12 a mass of exactly 12 units. The mass of other atoms is then found by comparing their mass with the same number of carbon-12 atoms. The mass found is called the **relative atomic mass**.

So the definition of relative atomic mass is:

The average mass of naturally occurring atoms of an element on a scale where the carbon-12 atom has a mass of exactly 12 units.

The symbol for relative atomic mass is A_r.

The reason why we use the *average* mass of the atoms is to take account of naturally occurring isotopes.

For most work in the school laboratory we do not need to know relative atomic masses very accurately. We can use whole numbers. But there are some exceptions. For example, the A_r of chlorine = 35.5. Relative atomic masses are shown on most Periodic Tables.

We can also use relative atomic masses for the mass of ions such as Cl^-. That is because the electrons which are lost or gained when ions form have hardly any mass compared with the nucleus.

We can use relative atomic masses to see how heavy one atom is compared with another. For example the A_r of sulfur is 32, the A_r of copper is 64 and the A_r of oxygen is 16. This means that one sulfur atom has the same mass as two oxygen atoms. One copper atom has the same mass as two sulfur atoms or four oxygen atoms.

Relative molecular mass

Relative molecular mass is the sum of the relative atomic masses of all the atoms shown in the formula of a molecule. So all we have to do is to add all the relative atomic masses together. The symbol for relative molecular mass is M_r. Here are two examples:

1. Relative molecular mass of methane:

   formula CH_4
   atoms present   1 carbon + 4 hydrogen
   adding A_rs $1 \times A_r$ carbon + $4 \times A_r$ hydrogen
   M_r $(1 \times 12) + (4 \times 1) = 16$

2. Relative formula mass of calcium hydroxide:

   formula $Ca(OH)_2$
   atoms present   1 calcium + 2 oxygen + 2 hydrogen
   adding A_rs $(1 \times A_r$ calcium$) + (2 \times A_r$ oxygen$) + (2 \times A_r$ hydrogen$)$
   M_r $(1 \times 40) + (2 \times 16) + (2 \times 1) = 74$

We follow exactly the same method for ionic compounds but these are not molecules, so we use the term **relative formula mass**. The term 'relative formula mass' is more general and we can apply it to all compounds.

Simple chemical calculations

We can work out how much product we get from a given amount of reactant by simple proportion. We can do this very easily if we already have some information about the mass of product produced by a different amount of reactant.

Example 1: A student obtains 48 g of magnesium sulfate from 9.6 g of magnesium. What mass of magnesium sulfate can the student get from 1.2 g of magnesium?

9.6 g magnesium gives 48 g magnesium sulfate

so   1.2 g magnesium gives $\frac{1.2}{9.6} \times 48 = 6$ g magnesium sulfate

Example 2: In the reaction $Mg + CuSO_4 \longrightarrow MgSO_4 + Cu$
6.4 g of copper are formed from 2.4 g of magnesium. What mass of magnesium is needed to get 32 g of copper?

6.4 g copper is formed from 2.4 g magnesium

so to get 32 g copper requires $\frac{32}{6.4} \times 2.4 = 12$ g magnesium

STUDY TIP

If a formula has brackets, first work out the atomic masses inside the brackets then multiply by the number outside. Finally, add the atomic masses which were not bracketed.

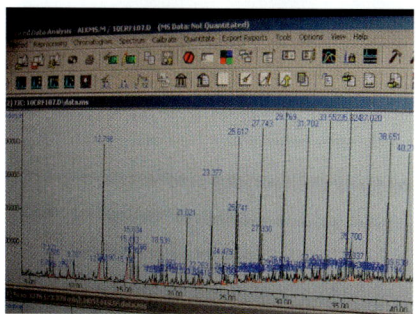

Mass spectrometry is one way of finding the relative molecular mass of a molecule

SUMMARY QUESTIONS

1 Copy and complete using the words below:

adding   atoms   carbon   exactly   formula   relative   scale

The masses of _____ are compared on a _____ where an atom of _____ -12 has a mass of _____ 12 units. The relative _____ mass is found by _____ together the _____ atomic masses in the formula of a molecule or ionic compound.

2 a   How many helium atoms have the same mass as one atom of sulfur?
  b   How many lithium atoms have the same mass as one atom of iron?
  c   How many helium atoms have the same mass as four lithium atoms?

3 Calculate the relative formula mass of:
  a   potassium chloride, KCl   b   calcium nitrate, $Ca(NO_3)_2$.

KEY POINTS

1 The relative atomic masses of atoms are compared on a scale which gives an atom of carbon-12 a mass of exactly 12 units.

2 The relative molecular mass is the sum of the relative atomic masses in the formula of a molecule.

3 We can use simple proportion to calculate the mass of a product formed if we have enough information.

5.2 Chemical calculations

LEARNING OUTCOMES

- Define the mole and the Avogadro constant
- Use the mole in calculations involving stoichiometric reacting masses

The Avogadro constant and the mole

The mass of even 1000 atoms is far too small to weigh on a balance. If we want to calculate the mass of a reactant needed to get a given amount of product we must scale up much more than this. If we have 6×10^{23} (6 with 23 zeros added) atoms, ions or molecules of a substance, we have an amount which is easily weighed. This number of atoms, ions or molecules is called the Avogadro number or the **Avogadro constant**.

The amount of substance with the Avogadro number of particles is called the **mole**.

6×10^{23} atoms of hydrogen, H, have a mass of 1 gram – its $A_r = 1$.

6×10^{23} molecules of oxygen, O_2, have a mass of 32 grams – its $M_r = 32$.

6×10^{23} sodium ions, Na^+, have a mass of 23 grams – its $M_r = 23$.

You can see that one mole of a substance is simply the relative atomic mass or the relative formula mass in grams. We can also use the term **molar mass** for the relative formula mass in grams.

- A mole is the amount of substance that contains 6×10^{23} atoms, ions or molecules
- the Avogadro constant is the number of atoms, ions or molecules in one mole.

It is very important to make clear what types of particles you are referring to. If you just state 'moles of oxygen' it is not clear whether you are thinking about moles of oxygen atoms or moles of oxygen molecules. A mole of oxygen molecules (O_2) contains 6×10^{23} molecules but it contains twice as many atoms – 1.2×10^{24} atoms. This is because there are two atoms in every molecule of oxygen.

Calculations using the mole

You can find the number of moles of atoms, molecules or ions using the following formula:

$$\text{number of moles} = \frac{\text{mass of substance taken}}{\text{mass of one mole of the substance}}$$

← in grams
← formula mass in grams

Example:

How many moles of water are there in 4.5 grams of water?

M_r water = 18

1 mole water = 18 g

So 4.5 g water is $\frac{4.5}{18} = 0.25$ mol

Note that the abbreviation for moles is mol.

In chemical calculations you may need to rearrange the mole equation above. If you rearrange the equation you can see that you can find the mass of a substance in a given number of moles:

The German scientist Wilhelm Ostwald suggested the term 'mol' in 1893, which he made up by abbreviating the German word for molecule (*Molekül*). For his work on catalysis, Ostwald was awarded the Nobel Prize for Chemistry in 1909.

DID YOU KNOW?

Amedeo Avogadro found that equal volumes of gases contained equal numbers of molecules. A mole of any element or compound contains 6.02×10^{23} atoms, ions or molecules. This is called Avogadro's number, and it is used to gauge the results of chemical reactions.

mass of substance (g) = number of moles × mass of 1 mole of the substance (formula mass)

Example:

What is the mass of sodium hydroxide in 0.5 moles of sodium hydroxide? M_r of NaOH = 40

mass = number of moles × mass of 1 mole (formula mass)

= 0.5 × 40 = 20 g NaOH

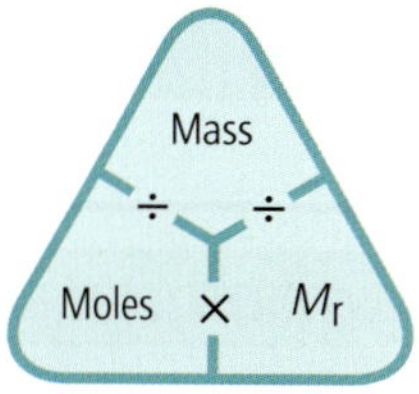

Figure 5.2.1 If you have difficulty rearranging the mole equation this triangle may help you. Cover the quantity you want to find. You will see the correct form of the equation to use.

Reacting masses

We often want to know what mass of one reactant we need to add to another so that they react exactly and there is no waste. To do this we need to know the ratio in which the reactants combine. The ratio of the various reactants and products in an equation is called the **stoichiometry** of the equation. In the equation:

$$2Mg + O_2 \longrightarrow 2MgO$$

two moles of magnesium atoms react with one mole of oxygen molecules to form two moles of magnesium oxide.

We can use relative formula masses to work out the minimum mass of oxygen needed to react completely with a given mass of magnesium. Remember that one mole is the relative formula mass in grams.

Example:

Calculate the mass of oxygen needed to react with 12 g of magnesium. A_r values: Mg = 24; O = 16

using formula masses and stoichiometry

$$2Mg \; + \; O_2 \; \longrightarrow \; 2MgO$$
$$2 \times 24\,g \quad 2 \times 16\,g$$

48 g magnesium reacts with 32 g of oxygen.

So 12 g magnesium will react with $\frac{12}{48} \times 32 = 8\,g$ of oxygen.

> **STUDY TIP**
>
> When doing calculations, put the relative formula masses or moles below the appropriate reactants or products in the symbol equation so that you can see which reactants or products are relevant. Be sure to take the stoichiometry of the equation into account.

SUMMARY QUESTIONS

1 Copy and complete using the words below:

amount   equal   formula   grams   mole   molecules

The mole is the _____ of substance that contains 6 × 10²³ atoms, _____ or ions. The Avogadro constant is the number of atoms, molecules or ions in one _____ of a substance. A mole of a compound has a mass _____ to the relative _____ mass of the compound in _____.

2 Calculate the number of moles of:
  a sodium hydroxide in 8 g of sodium hydroxide, NaOH
  b water in 5.4 g of water
  c aluminium oxide in 12.75 g of aluminium oxide, Al_2O_3.

A_r values: Na = 23, O = 16, H = 1, Al = 27.

KEY POINTS

1 One mole of a substance contains 6 × 10²³ atoms, molecules or ions.

2 The Avogadro constant is the number of atoms, molecules or ions in one mole of substance.

3 We can use chemical equations and relative formula masses to calculate the mass of product obtained from a given amount of reactant.

5.3 How much product?

LEARNING OUTCOMES
- Calculate the theoretical yield of product obtained from a given amount of reactant
- Apply the concept of limiting reactants

Limiting reactants

When we carry out a reaction we sometimes use an excess of one of the reactants. The reactant that is not in excess is called the **limiting reactant** or **limiting reagent**. The reaction stops when the limiting reactant is used up. You can calculate which reactant is limiting by calculating which reactant has the lower number of moles. You must, however, take into account the stoichiometry of the reaction. If you just work out the number of moles from the masses given, it does not take into account the fact that one of the reactants might be used up twice as fast as the other.

Figure 5.3.1 shows how the particles of calcium carbonate and hydrochloric acid decrease in number as the reaction proceeds. The equation for the reaction is:

$$CaCO_3 + 2HCl \longrightarrow CaCl_2 + CO_2 + H_2O$$

Calcium carbonate ×
Hydrochloric acid ○   Reaction proceeds ⟶

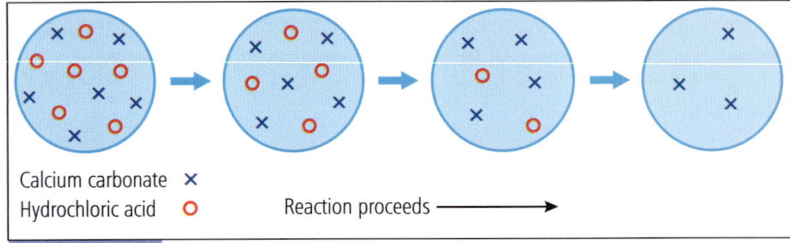

Figure 5.3.1 Hydrochloric acid is the limiting reactant because it gets used up before the calcium carbonate

The equation shows that for each mole of calcium carbonate which reacts, two moles of hydrochloric acid are converted to products. So, in this case, although we had the same number of particles of calcium carbonate and hydrochloric acid to start with, the hydrochloric acid gets used up quicker and runs out first. So hydrochloric acid is the limiting reactant.

The example below shows how to work out problems involving limiting reagents.

Example:

1.2 g of magnesium is reacted with a solution containing 2.74 g of hydrochloric acid. Which is the limiting reactant?

$$Mg + 2HCl \longrightarrow MgCl_2 + H_2$$

A_r values: Mg = 24, Cl = 35.5, H = 1.

So M_r [HCl] = 35.5 + 1 = 36.5

number of moles of magnesium = $\frac{1.2}{24}$ = 0.05 mol

number of moles of hydrochloric acid = $\frac{2.74}{36.5}$ = 0.075 mol

STUDY TIP

The limiting reactant is the reactant that is NOT in excess. It has the smaller number of moles. Be careful though – you must also take into account the ratio in which the reactants combine.

Bleach is used in some swimming pools to control and kill harmful bacteria. Getting the quantities right involves some careful calculation! Adding too much bleach can cause burns to the skin and eyes.

DID YOU KNOW?

Stoichiometry is derived from the Greek words meaning 'element measure'.

From the equation: 1 mol of magnesium reacts with 2 mol of HCl

so to react completely, 0.05 mol magnesium will need to react with $2 \times 0.05 = 0.1$ mol of HCl.

But we have only 0.075 mol of HCl, so the HCl is the limiting reactant.

From this type of calculation, you can also find out by how much one reactant is in excess. In the example all the hydrochloric acid was used up.

Hydrochloric acid used up = 0.075 mol

From the equation: 2 mol hydrochloric acid reacts with 1 mol of magnesium.

So 0.075 mol hydrochloric acid will react with $\frac{0.075}{2}$

= 0.0375 mol magnesium

Therefore

Excess magnesium = moles of magnesium at start − moles of magnesium reacted

= 0.05 − 0.0375 = 0.0125 mol

How much product?

We often need to know how much product we can get from a given amount of reactant. There are two ways of doing this: we can use relative formula masses and simple proportion or we can work out the number of moles of reactant and product. These two methods are shown for the following question:

Calculate the mass of water formed when 4 g of methane is completely burned in oxygen.

A_r values: C = 12, H = 1, O = 16

Method 1: using relative formula masses

$$CH_4 + 2O_2 \longrightarrow CO_2 + 2H_2O$$

M_rs   12 + (4 × 1)   2 × [(2 × 1) + 16]

16 g methane   will give   36 g water

4 g methane   will give $\frac{4}{16} \times 36 = 9$ g water

Method 2: using moles

$$CH_4 + 2O_2 \longrightarrow CO_2 + 2H_2O$$

moles of methane = $\frac{4}{16}$ = 0.25 mol

mole ratio of methane : water = 1 : 2

So moles of water formed = 2 × 0.25 = 0.5 mol

mass of water in g = moles × M_r of water

So mass of water = 0.5 × 18 = 9 g

KEY POINTS

1 The limiting reactant is the one that is <u>not</u> in excess.

2 We can work out which reagent is limiting by comparing the number of moles of each reactant, taking into account the stoichiometry of the equation.

3 We can use the chemical equation and relative formula masses to calculate the mass of product formed from a given amount of reactant.

SUMMARY QUESTIONS

1 Calculate the mass of copper(II) sulfate formed when 2.5 g of copper(II) oxide reacts with excess sulfuric acid. The equation for the reaction is:

$$CuO + H_2SO_4 \longrightarrow CuSO_4 + H_2O$$

A_r: Cu = 63.5; O = 16; S = 32

2 In the reaction:

$$Mg + 2CH_3CO_2H \longrightarrow (CH_3CO_2)_2Mg + H_2$$

2.4 g of magnesium is reacted with 6 g of ethanoic acid, CH_3CO_2H.

Which reagent is in excess?

A_r: Mg = 24; C = 12; H = 1; O = 16

5.4 Percentages and volumes

LEARNING OUTCOMES

- Calculate the percentage by mass of an element present in a compound
- Do calculations involving reacting volumes of gases

DID YOU KNOW?

It takes about 16 000 000 moles of helium gas to fill a modern airship.

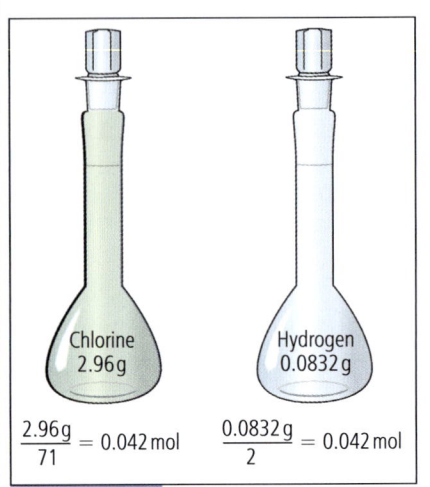

$\dfrac{2.96\,g}{71} = 0.042\,mol \qquad \dfrac{0.0832\,g}{2} = 0.042\,mol$

Figure 5.4.1 There are the same number of moles in each flask

Percentage by mass

We can use formulae and relative molecular masses to work out the **percentage by mass** of a particular element present in a compound. It is useful to know this, as it enables us to compare the amount of nitrogen in different fertilisers or to work out how much metal we can obtain from a metal ore. To find the percentage by mass of an element in a compound we use the relative atomic mass of the element and the relative formula mass of the compound.

$$\% \text{ by mass} = \dfrac{\text{mass of a particular element in one mole of a compound}}{\text{mass of one mole of compound}} \times 100$$

Since the relative formula mass is equal to the molar mass in grams, we can rewrite this as:

$$\% \text{ by mass} = \dfrac{\text{sum of the relative atomic masses of a particular element in a compound}}{\text{relative formula mass of compound}} \times 100$$

Example: Calculate the percentage by mass of iron in 1 mole of iron oxide, Fe_2O_3.

A_r: Fe = 56; O = 16

There are two moles of iron in every mole of iron oxide. So mass of iron = 2 × 56 = 112.

The relative formula mass of Fe_2O_3 is (2 × 56) + (3 × 16) = 160.

So % by mass of iron in Fe_2O_3 = $\dfrac{112}{160} \times 100 = 70\%$

Gas volume calculations

Look at the two flasks in Figure 5.4.1. Each has a volume of 1 dm³. They are each filled with a different gas and the volume of the gas is measured accurately.

Each flask contains the same number of moles. This means that the same volume of gas has the same number of moles.

PRACTICAL

What is the volume of one mole of gas?

1. Put 0.1 g of magnesium in the flask and add excess hydrochloric acid.
2. Record the maximum amount of gas produced. It is 100 cm³. The equation hydrogen is produced when one mole of magnesium reacts:

$Mg + 2HCl \longrightarrow MgCl_2 + H_2$

Moles of magnesium used = 0.1/24 = 0.004 166 mol = moles of hydrogen

If 100 cm³ hydrogen = 0.004 166 mol, then the volume of one mole = 100 × 1/0.004 166 = 24 000 cm³.

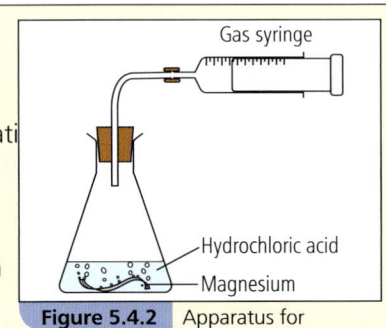

Figure 5.4.2 Apparatus for following the progress of a reaction where a gas is given off

At room temperature and pressure the volume of one mole of any gas is 24 dm³ or 24 000 cm³. This is called the **molar gas volume**. Room temperature and pressure (r.t.p.) is 20 °C and 1 atmosphere pressure.

We can use the fact that one mole of gas occupies 24 dm³ to do chemical calculations for reactions where gases are produced.

volume of gas (in dm³) = number of moles of gas × 24

Example 1:

Calculate the volume of 0.2 moles of carbon dioxide at r.t.p.

1 mol occupies 24 dm³. So 0.2 mol occupies 0.2 × 24 = 4.8 dm³.

Example 2:

Calculate the mass of carbon dioxide present in 60 cm³ of carbon dioxide. M_r [CO_2] = 44

moles of CO_2 = $\dfrac{\text{volume of gas (in dm}^3\text{)}}{24}$ = $\dfrac{0.06}{24}$ = 0.0025 moles

mass of CO_2 = moles CO_2 × M_r = 0.0025 × 44
= 0.11 g of carbon dioxide

Example 3:

Calculate the volume of carbon dioxide which is produced when 2.8 g of butene burns in excess air.

$C_4H_8(g) + 6O_2(g) \longrightarrow 4CO_2(g) + 4H_2O(l)$

M_r [butene] = 56 moles of butene = 2.8/56 = 0.05 moles

From the equation, 1 mole of butene produces 4 moles of carbon dioxide.

So 0.05 moles of butene will produce 0.05 × 4 = 0.2 moles of carbon dioxide.

0.2 moles of carbon dioxide gas will occupy 0.2 × 24 dm³ = 4.8 dm³.

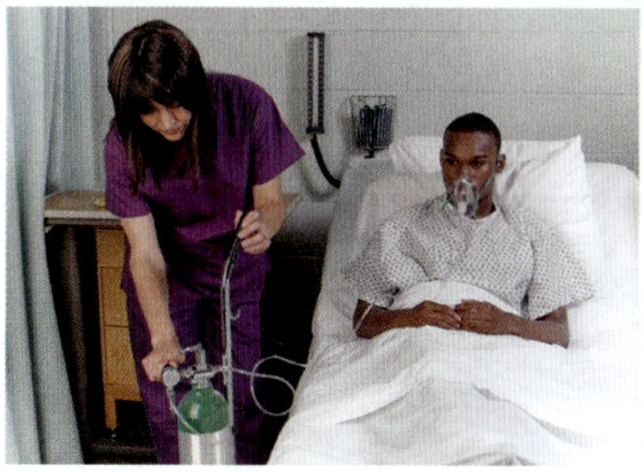

Doctors need to know about gas volumes when using equipment to help patients to breathe

STUDY TIP

When working out gas volumes, first find the number of moles and then multiply this by 24. The volume is then in dm³. Remember that the molar gas volume is given at the bottom of your Periodic Table.

SUMMARY QUESTIONS

1 Copy and complete using the words below:

any   mole   pressure   room   volume

The _____ of one _____ of a gas occupies 24 dm³ at _____ temperature and _____. This is the same for _____ gas.

2 Calculate the percentage by mass of nitrogen in ammonium nitrate, NH_4NO_3.

3 a Calculate the volume occupied by 11 g of carbon dioxide gas at r.t.p.

  b What mass of nitrogen gas is present in 1200 cm³ of nitrogen at r.t.p.?

KEY POINTS

1 The percentage by mass of an element in a compound can be found using the relative atomic masses and relative formula masses.

2 For reactions involving gases, the molar gas volume (24 dm³ at r.t.p) can be used to calculate reacting masses.

5.5 Yield and purity

LEARNING OUTCOMES
- Calculate percentage yield
- Calculate percentage purity

Percentage yield

When you carry out a chemical reaction in the laboratory not all the reactants are changed to the products you want. This is because there may be other reactions happening at the same time or the reaction doesn't go to completion. The word **yield** describes how much of a particular product you can get from the reactants in a chemical reaction. It is more useful to talk about percentage yield:

$$\% \text{ yield} = \frac{\text{actual yield}}{\text{predicted yield}} \times 100$$

The **actual yield** is the amount of product we get in a reaction. We don't know what this will be until we have weighed the product. The **predicted yield** is found by using relative formula masses together with the equation for the reaction to calculate the maximum amount of product we could get from a given amount of reactants. This is sometimes called the theoretical yield, because we have calculated it.

Example:

A student reacts 9 g of aluminium powder with excess chlorine. The mass of aluminium chloride produced is 35.6 g. Calculate the percentage yield. A_r: Al = 27, O = 16.

The actual yield is 35.6 g.

The predicted yield is calculated using formula masses and the equation:

$$2Al + 3Cl_2 \longrightarrow 2AlCl_3$$

1 mole of aluminium produces 1 mole of aluminium chloride.

Using formula masses: 27 g of Al produces 27 + (3 × 35.5)
= 133.5 g of $AlCl_3$

So theoretically 9 g Al produces $9 \times \dfrac{133.5}{27}$ = 44.5 g of $AlCl_3$

So % yield = $\dfrac{\text{actual yield}}{\text{predicted yield}} \times 100 = \dfrac{35.6}{44.5} \times 100 = 80\%$

STUDY TIP
Always show your working in calculations if a question is worth more than one mark. If you make an error at the start – for example, use an incorrect molar mass – you can still demonstrate how to do the rest of the calculation.

PRACTICAL

How much copper(II) sulfate can we get from malachite?

1. Put a known amount of crushed malachite (copper carbonate ore) in a beaker. This is treated as shown.

Figure 5.5.1 Extracting copper(II) sulfate from malachite

- Heat with excess sulfuric acid
- Filter the mixture. The filtrate is a solution of copper(II) sulfate
- Evaporate the water and weigh the copper(II) sulfate

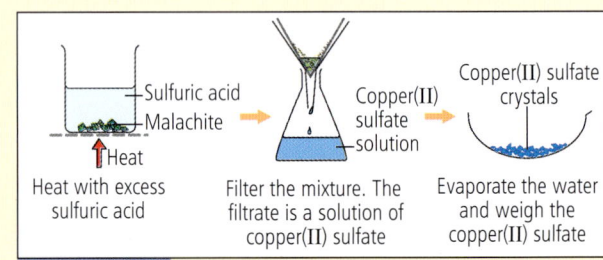

$$\% \text{ yield} = \frac{\text{actual mass of copper(II) sulfate}}{\text{predicted mass of copper(II) sulfate}} \times 100$$

Percentage purity

We said in Topic 1.7 that when a chemical company makes medical drugs or food additives the products have to be very pure. However, the product may still contain very small amounts of impurities mixed with it. These impurities may be small amounts of unreacted starting material, other products formed or additional products caused by unwanted reactions. You can work out **percentage** purity in a similar way to percentage yield:

$$\% \text{ purity} = \frac{\text{mass of pure product}}{\text{mass of impure product}} \times 100$$

Example:

A chemist made 60 g of aspirin. Chemical analysis showed that this sample contained 58.5 g of pure aspirin and 1.5 g of impurities. Calculate the percentage purity of this aspirin sample.

$$\% \text{ purity} = \frac{\text{mass of pure product}}{\text{mass of impure product}} \times 100$$

$$= \frac{58.5}{60} \times 100$$

$$= 97.5\% \text{ pure}$$

The impurities make up (100−97.5)% = 2.5% of the product.

Silicon chips used for computers have to be very pure

DID YOU KNOW?

The silicon chips used in computers have to be so pure that more than 600 steps are needed to purify the quartz used to make them.

SUMMARY QUESTIONS

1 Copy and complete using the words below:

   actual   equation   masses   predicted   product   theoretical

   The amount of _____ made in a chemical reaction is called the _____ yield. We can use relative formula _____ and the stoichiometry of the _____ to calculate the _____ yield. The predicted yield can also be called the _____ yield.

2 When 11.50 g of sodium reacts with excess chlorine, 22.3 g of sodium chloride is made.

   $2Na + Cl_2 \longrightarrow 2NaCl$

   A_r: Na = 23, Cl = 35.5

   Calculate:
   a the theoretical yield of sodium chloride
   b the % yield.

3 A 26.5 g sample of impure paracetamol contains 24.5 g of pure paracetamol. Calculate the % purity of this paracetamol.

KEY POINTS

1 The percentage yield in a chemical reaction is found by comparing the actual yield with the predicted yield.

2 The predicted yield is found using relative atomic masses and the stoichiometry of the equation.

3 Percentage purity is given by
$$\frac{\text{mass of pure product}}{\text{mass of impure product}} \times 100$$

5.6 More chemical calculations

LEARNING OUTCOMES
- Calculate empirical formulae
- Calculate molecular formulae

Finding the empirical formula

In Topic 5.4 we worked out the percentage by mass of an element in a compound. We can also do this the other way around to find the formula of a compound. If we know the masses of each element that combine to form a compound, we can work out its formula. This formula is called the **empirical formula**.

The empirical formula shows the simplest whole number ratio of the elements present in the compound. For example, the empirical formula for hydrogen peroxide, H_2O_2, is HO. The formula of an ionic compound is always its empirical formula.

A tiny difference in the amount of iron in the ore might not seem very much, but when millions of tonnes of iron ore are extracted and processed each year, it all adds up!

PRACTICAL

Finding the formula of magnesium oxide

1. Burn a weighed amount of magnesium (0.6 g) in a crucible. The magnesium reacts with the oxygen in the air to form magnesium oxide.

2. When the apparatus has cooled, weigh the magnesium oxide formed. In this experiment it was 1.0 g.

mass of oxygen in magnesium oxide = mass of magnesium oxide formed − mass of magnesium
= 1.0 − 0.6 = 0.4 g

We can calculate the empirical formula from these results knowing $A_r[Mg] = 24$ and $A_r[O] = 16$:

$$\text{moles of magnesium atoms} = \frac{0.6}{24} = 0.025 \text{ mol}$$

$$\text{moles of oxygen atoms} = \frac{0.4}{16} = 0.025 \text{ mol}$$

The numbers of moles of magnesium and oxygen atoms are equal, so the empirical formula of magnesium oxide is MgO.

Worked example 1:

Analysis of a compound of tin (Sn) and chlorine (Cl) showed that the tin chloride contained 29.75 g of tin and 35.5 g of chlorine. Calculate the empirical formula of tin chloride. $A_r[Sn] = 119$; $A_r[Cl] = 35.5$

	Sn	Cl
Step 1: note the mass of each element	29.75 g	35.5 g
Step 2: divide by the relative atomic masses	$\frac{29.75}{119} = 0.25$ mol	$\frac{35.5}{35.5} = 1.0$ mol
Step 3: divide each by the lowest number of moles	$\frac{0.25}{0.25} = 1$	$\frac{1.0}{0.25} = 4$
Step 4: write the formula	$SnCl_4$	

You can also do calculations based on the percentage by mass of the elements in the compound. This works in the same way except you are dealing with percentages.

Worked example 2:

A compound of carbon and hydrogen contains 80% carbon and 20% hydrogen by mass. Calculate the empirical formula of this compound. $A_r[C] = 12$; $A_r[H] = 1$

	C	H
Step 1: divide %s by A_r	$\frac{80}{12} = 6.67$	$\frac{20}{1} = 20$
Step 2: divide by lowest	$\frac{6.67}{6.67} = 1$	$\frac{20}{6.67} = 2.99$

Notice that 2.99 is very near 3, so the ratio is 1 : 3 and the empirical formula is CH_3.

Finding the molecular formula

The **molecular formula** shows the actual number of atoms in a molecule. For example, ethane has the empirical formula CH_3 but its molecular formula is C_2H_6. To find the molecular formula we need to know:

- the relative formula mass of the compound – this is usually found by using an instrument called a mass spectrometer
- the empirical formula of the compound.

Worked example:

A compound has the empirical formula CH_2. Its relative formula mass is 84. Calculate its molecular formula. $A_r[C] = 12$; $A_r[H] = 1$

Step 1: find the empirical formula mass $12 + (2 \times 1) = 14$

Step 2: divide relative formula mass by empirical formula mass $\frac{84}{14} = 6$

Step 3: multiply the empirical formula by the number calculated in Step 2 $6 \times CH_2 = C_6H_{12}$

> **STUDY TIP**
>
> When calculating empirical formulae, make sure that between steps 1 and 2 you don't round up the figures. This often leads to errors.

> **DID YOU KNOW?**
>
> Modern equipment for the analysis of carbon, hydrogen and nitrogen allows the empirical formula for a compound to be worked out using less than 0.001 gram of material.

SUMMARY QUESTIONS

1 Copy and complete using the words below:

   atomic   dividing   empirical   lowest   mass   ratio

   The _____ formula shows the simplest whole number _____ of atoms in a compound. It is found by _____ the _____ of each element by its relative _____ mass and then dividing by the _____ number to get the ratio of atoms.

2 A compound contains 92.3% carbon and 7.7% hydrogen. Calculate its empirical formula. $A_r[C] = 12$; $A_r[H] = 1$

3 Calculate the molecular formula of an oxide of phosphorus which has an empirical formula P_2O_5 and a relative molecular mass of 284. $A_r[O] = 16$; $A_r[P] = 31$

KEY POINTS

1 An empirical formula shows the simplest whole number ratio of atoms in a compound.

2 The empirical formula can be found using the mass of the elements present and their relative atomic masses.

3 A molecular formula shows the actual number of atoms present in a molecule of a compound.

5.7 Titrations

LEARNING OUTCOMES

- Know the units of concentration in solutions
- Understand how to calculate the concentration of a solution

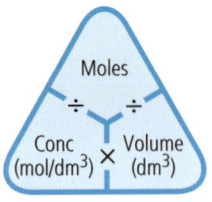

Figure 5.7.1 Cover the moles to show the equation for finding moles from concentration and volume

STUDY TIP

Mole calculations involving concentrations are easier if you change cm³ to dm³ and then use the formula concentration = number of moles ÷ volume of solution in dm³.

Titrations give valuable practice in calculating solution concentration

Solution concentration

The **concentration** of a solution is the amount of solute dissolved in 1 dm³ of solution. The units of concentration are therefore mol/dm³.

$$\text{concentration (mol/dm}^3\text{)} = \frac{\text{number of moles of solute}}{\text{volume of solution (dm}^3\text{)}}$$

There are two points to note about this equation:

- You may be given the amount of solute dissolved in grams. You need to convert this into moles.
- You may be given the volume in cm³ (dm³ = cm³/1000).

You will also need to be able to rearrange the equation so that you can calculate moles when given concentration and volume:

$$\text{number of moles of solute} = \text{concentration (mol/dm}^3\text{)} \times \text{volume of solution (dm}^3\text{)}$$

Two worked examples of calculations involving concentration are shown below:

Example 1:

Calculate the concentration in mol/dm³ of a solution of sodium hydroxide, NaOH, containing 4 g of sodium hydroxide in 50 cm³ of solution. A_r [Na] = 23, A_r [O] = 16, A_r [H] = 1

Step 1: change grams to moles $\frac{4}{(23 + 16 + 1)} = 0.1$ mol NaOH

Step 2: change cm³ to dm³   50/1000   = 0.05 dm³

Step 3: concentration (mol/dm³) =

$$\frac{\text{number of moles of solute}}{\text{volume of solution (dm}^3\text{)}} = \frac{0.1}{0.05} = 2 \text{ mol/dm}^3 \text{ NaOH}$$

Example 2:

Calculate the mass of potassium hydroxide in 20 cm³ of a solution of concentration 0.4 mol/dm³. M_r [KOH] = 56

Step 1: change cm³ to dm³   20/1000 = 0.02 dm³

Step 2: moles of solute = concentration (mol/dm³) × volume of solution (dm³)

= 0.4 × 0.02 = 0.008 moles KOH

Step 3: change moles to grams   mass = moles × M_r [KOH]
= 0.008 × 56 = 0.448 g

Titrations

We can find the concentration of alkali needed to completely react with an acid using a procedure called a **titration**. In a titration, the concentration of one of the solutions is known accurately.

We put this solution into the burette. We put a measured volume of the other solution into the titration flask and add an indicator. The indicator will change colour when the reaction is complete.

We then run the solution from the burette into the flask until the indicator changes colour. We can then use the burette reading together with the known concentration of the solution in the burette to find the concentration of the solution in the flask.

Worked example:

You put 25 cm³ of potassium hydroxide of unknown concentration into a flask with some indicator.

The indicator changes colour when 30 cm³ of hydrochloric acid of concentration 0.2 mol/dm³ has been added from the burette. Calculate the concentration of the potassium hydroxide in mol/dm³.

$$KOH + HCl \longrightarrow KCl + H_2O$$

Step 1: calculate the moles of hydrochloric acid used

moles = concentration × volume (in dm³)

$$= 0.2 \times \frac{30}{1000} = 0.006 \text{ mol HCl}$$

Step 2: look at the equation to find the ratio of KOH to HCl; in this case it is 1 : 1.

So moles HCl = moles KOH = 0.006 mol KOH

Step 3: change moles of KOH to concentration in mol/dm³:

$$\text{concentration (mol/dm}^3) = \frac{\text{number of moles of solute}}{\text{volume of solution (dm}^3)}$$

$$= \frac{0.006}{0.025} = 0.24 \text{ mol/dm}^3 \text{ KOH}$$

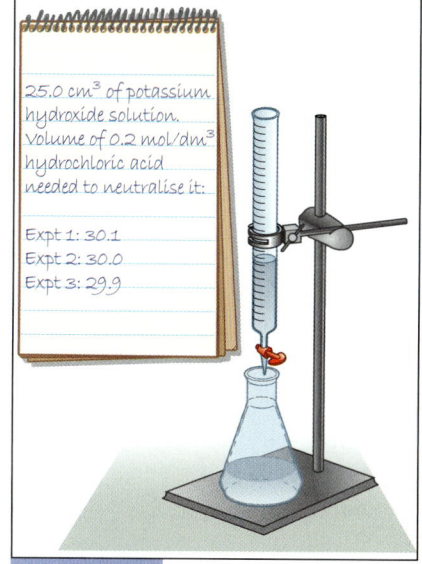

25.0 cm³ of potassium hydroxide solution. Volume of 0.2 mol/dm³ hydrochloric acid needed to neutralise it:

Expt 1: 30.1
Expt 2: 30.0
Expt 3: 29.9

Figure 5.7.2 Titration apparatus

DID YOU KNOW?

Quality control testing for the amount of chlorine in bleach is still regularly carried out by titration.

SUMMARY QUESTIONS

1 Copy and complete using the words below:

   alkali   concentration   moles   titration   volume

   We can find an unknown solution _____ of an alkali by carrying out a _____ with an acid. The first step in the calculation is to use the concentration and _____ of the acid to find the _____ of acid used. We can then find the concentration of the _____ in mol/dm³.

2 The equation shows the reaction of sulfuric acid with sodium hydroxide:

   $$H_2SO_4 + 2NaOH \longrightarrow Na_2SO_4 + 2H_2O$$

   25 cm³ of sulfuric acid of concentration 0.2 mol/dm³ reacted with exactly 10 cm³ of sodium hydroxide. Calculate:

   a   the number of moles of sulfuric acid present
   b   the number of moles of sodium hydroxide reacting
   c   the concentration of the sodium hydroxide in mol/dm³.

KEY POINTS

1 Solution concentration is calculated as number of moles dissolved in 1 dm³.

2 The mass of solute dissolved in a solution can be found if the concentration and volume of a solution are known as well as the relative formula mass.

3 We can use titration to calculate an unknown solution concentration.

SUMMARY QUESTIONS

1 Calculate the relative formula mass of:
   (a) $CaBr_2$
   (b) $NaNO_3$
   (c) Al_2O_3
   (d) Na_2SO_4
   (e) $Al_2(SO_4)_3$
   (f) PCl_5

2 Copy and complete using the words below:

   **actual   amount   formula   limiting
   molecular   product   relative
   theoretical   yield**

   The relative _____ mass is found by adding the _____ atomic masses together. For ionic substances, we use the phrase relative _____ mass. The amount of _____ made in a reaction depends on the _____ reagent. The percentage _____ can be found by comparing the _____ yield with the _____ yield calculated from the _____ of limiting reactant present.

3 Match the words on the left with the phrases on the right.

Avogadro number	the reagent that is not in excess
molar gas volume	the number of specified particles in one mole of those particles
limiting reagent	the sum of the relative atomic masses
relative molecular mass	the volume occupied by one mole of any gas

4 Calculate the number of moles of:
   (a) bromine molecules in 2.4 g of bromine, Br_2
   (b) iron atoms in 9.6 g of iron oxide, Fe_2O_3
   (c) sulfuric acid in 0.49 g of sulfuric acid, H_2SO_4
   (d) chloride ions in 79.17 g magnesium chloride, $MgCl_2$.

5 Calculate the volume in cm^3 at r.t.p. of:
   (a) 0.3 moles of neon
   (b) 1.5 moles of carbon dioxide
   (c) 0.03 moles of ammonia
   (d) 0.0005 moles of nitrogen.

PRACTICE QUESTIONS

1 Iron reacts with hydrochloric acid to form iron(II) chloride, $FeCl_2$, and hydrogen.
   (a) Write a balanced equation for this reaction. [3]
   (b) When 28 g of iron reacts with excess hydrochloric acid, 63.5 g of iron(II) chloride and 1 g of hydrogen are formed.
      (i) Calculate the mass of iron(II) chloride formed from 7 g of iron. [1]
      (ii) Calculate the mass of iron that will be needed to produce 10 g of hydrogen. [1]

(Paper 3)

2 3.2 g of iron is added to 50 cm^3 of 1.0 mol/dm^3 sulfuric acid.

   $$Fe + H_2SO_4 \longrightarrow FeSO_4 + H_2$$

   (a) Calculate the number of moles of sulfuric acid present. [1]
   (b) Show by calculation that iron is in excess. [2]

6 What is the mass of:
   (a) 0.5 moles of hydrogen chloride, HCl
   (b) 0.2 moles of calcium nitrate, $Ca(NO_3)_2$
   (c) 0.015 moles of sodium sulfate, Na_2SO_4
   (d) 3 moles of phosphorus(V) chloride, PCl_5?

7 Calculate the moles of gas in:
   (a) 1680 cm^3 of carbon dioxide
   (b) 960 cm^3 of nitrogen
   (c) 240 cm^3 of neon
   (d) 96 dm^3 of oxygen.

8 Calculate the concentration in mol/dm^3 of the following solutions:
   (a) 0.5 moles of hydrochloric acid in 500 cm^3 of solution
   (b) 0.15 moles of magnesium chloride in 250 cm^3 of solution
   (c) 5.85 g of sodium chloride in 100 cm^3 of solution
   (d) 4.9 g of sulfuric acid in 250 cm^3 of solution

(c) Calculate the mass of iron remaining after the reaction is complete. [2]

(d) Calculate the maximum volume of hydrogen formed in the reaction at r.t.p. [2]

(e) Calculate the theoretical yield of iron(II) sulfate at the end of the reaction. [2]

(f) The actual yield of iron(II) sulfate was 3.61 g. Calculate the percentage yield. [2]

(Paper 4)

3 A solution of potassium hydroxide of unknown concentration was titrated with sulfuric acid.

(a) Describe how to carry out an acid–alkali titration. [4]

(b) Write a balanced equation for the reaction of potassium hydroxide with sulfuric acid. [2]

(c) It required 15 cm³ of 0.05 mol/dm³ sulfuric acid to neutralise 25 cm³ of potassium hydroxide. Calculate:

   (i) the number of moles of sulfuric acid in the titration [1]

   (ii) the concentration of the aqueous potassium hydroxide [2]

   (iii) the mass of potassium hydroxide dissolved in 25 cm³ of solution. [1]

(Paper 4)

4 Compound Z contains 0.96 g of carbon, 0.16 g of hydrogen and 2.84 g of chlorine. No other elements are present.

(a) Calculate the empirical formula of Z. [3]

(b) The relative molecular mass of Z is 99. Calculate the molecular formula of Z. [2]

(c) A different compound of carbon, hydrogen and chlorine was made by reacting 8.1 g of chlorine with 1.6 g of methane.

$$CH_4 + Cl_2 \longrightarrow CH_3Cl + HCl$$

   (i) Show by calculation whether the methane or chlorine is in excess. [2]

   (ii) Calculate the theoretical yield of CH_3Cl. [1]

   (iii) The actual yield of CH_3Cl was 3.79 g. Calculate the percentage yield. [2]

(d) CH_3Cl is a gas. Calculate the volume of 5.05 g of CH_3Cl at r.t.p. [2]

(Paper 4)

5 Calcium carbonate reacts with hydrochloric acid:

$$CaCO_3 + 2HCl \longrightarrow CaCl_2 + CO_2 + H_2O$$

(a) Calculate the number of moles of hydrochloric acid required to react exactly with 5 g of calcium carbonate. [2]

(b) What volume of 0.5 mol/dm³ hydrochloric acid will react exactly with 5 g of calcium carbonate? [2]

(c) Calculate the volume of carbon dioxide produced at r.t.p. when 35 g of calcium carbonate react with excess hydrochloric acid. [3]

(d) A student crystallised the calcium chloride produced in this reaction. The percentage yield was only 90%.

   (i) Give two possible reasons why the yield was not 100%. [2]

   (ii) Describe how to calculate percentage yield. [2]

(Paper 4)

6 Ammonium nitrate, NH_4NO_3, and ammonium sulfate, $(NH_4)_2SO_4$, are used as fertilisers.

(a) (i) Calculate the percentage by mass of nitrogen in ammonium sulfate. [2]

   (ii) Calculate the percentage by mass of nitrogen in ammonium nitrate. [2]

   (iii) Calculate the percentage by mass of nitrogen in a mixture containing 4 parts of ammonium nitrate and 1 part of ammonium sulfate. [2]

(b) Ammonium nitrate can be made by titrating ammonia with nitric acid.

$$NH_3 + HNO_3 \longrightarrow NH_4NO_3$$

50 cm³ of a solution of ammonia was neutralised by 15 cm³ of nitric acid of concentration 2 mol/dm³.

Calculate the concentration of the solution of ammonia

   (i) in mol/dm³ [3]

   (ii) in g/dm³. [1]

(Paper 4)

6 Electricity and chemistry

6.1 Electrolysis

LEARNING OUTCOMES

- Define electrolysis
- Describe the electrode products and the observations made during electrolysis of molten lead(II) bromide
- Predict the products of the electrolysis of a specified binary compound in the molten state

A bank of electrolysis cells being used to extract aluminium

What is electrolysis?

Electrolysis is the breaking down of an ionic compound when molten or in an aqueous solution by the passage of electricity. Electrolysis works for ionic compounds only when they are molten (melted) or when dissolved in water. This is because ionic compounds conduct electricity only when they are in these states.

The electric current causes a chemical reaction that breaks down the ionic compound. We say the compound **decomposes**. The important parts of an electrolysis apparatus are shown in Figure 6.1.1. We call this an electrolysis **cell**.

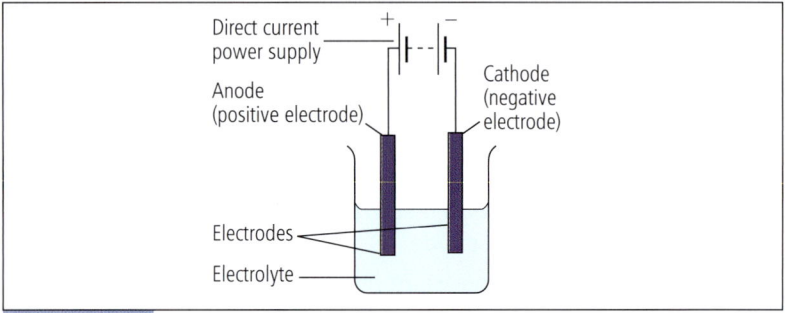

Figure 6.1.1 The key parts of an electrolysis cell

The compound that conducts electricity when molten and breaks down during electrolysis is called the **electrolyte**.

The **electrodes** are rods that carry the electric current to and from the electrolyte. They are normally made from an inert conductor, such as graphite or platinum. This is so that the electrodes do not react with

DEMONSTRATION

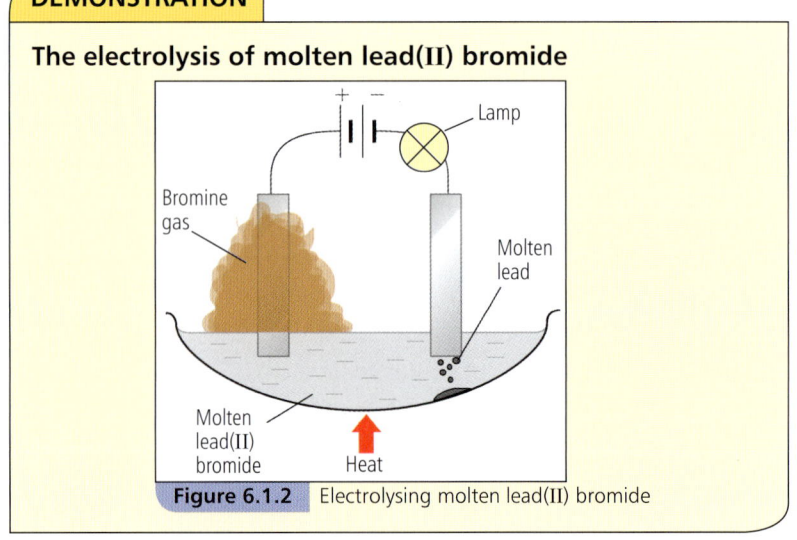

The electrolysis of molten lead(II) bromide

Figure 6.1.2 Electrolysing molten lead(II) bromide

74

This experiment is carried out in a fume cupboard because bromine is toxic and corrosive.

When the lead(II) bromide is heated it begins to melt and the lamp lights. This shows that an electric current is flowing and the ions in the electrolyte are able to move around. The electric current decomposes the lead(II) bromide. Grey beads of molten lead form at the negative electrode. At the positive electrode, reddish-brown bromine gas is formed.

the electrolyte or the products of electrolysis. The positive electrode is called the **anode**. The negative electrode is called the **cathode**.

Predicting the products of electrolysis

The equation for the electrolysis of molten lead(II) bromide shows that it is broken down into its elements, lead and bromine.

$$PbBr_2(l) \longrightarrow Pb(l) + Br_2(g)$$

Does the same happen with other molten ionic compounds? The table gives more examples of electrolysis of molten compounds using inert electrodes.

compound electrolysed	product at the cathode (– electrode)	product at the anode (+ electrode)
aluminium oxide	aluminium	oxygen
copper bromide	copper	bromine
sodium chloride	sodium	chlorine
zinc chloride	zinc	chlorine

You can see a pattern in the table. When the electrolyte decomposes, a metal is formed at the cathode (negative electrode) and a non-metal forms at the anode (positive electrode). So we can easily predict the products of electrolysis of a molten ionic compound.

STUDY TIP

Remember that in an electrolyte, it is the ions that move, not the electrons.

DID YOU KNOW?

Seawater is the source of the 300 000 tonnes of magnesium produced every year by electrolysis.

SUMMARY QUESTIONS

1 Copy and complete using the words below:

anode   cathode   current   electrodes   electrolyte

Electrolysis occurs when an electric _____ passes through a molten _____. The two rods dipping into the electrolyte are called the _____. Metals are deposited at the _____ and non-metals are formed at the _____.

2 What type of compounds can be electrolysed? What conditions are needed for this electrolysis?

3 Predict the products formed at the anode and cathode when the following molten compounds are electrolysed:
 a sodium fluoride
 b copper(II) iodide
 c zinc bromide.

KEY POINTS

1 Electrolysis is the breakdown of an ionic compound (when molten or in aqueous solution) by the passage of electricity.

2 An ionic compound can be electrolysed only when it is molten or in solution in water.

3 When a molten ionic compound is electrolysed, a metal is formed at the negative electrode and a non-metal is formed at the positive electrode.

6.2 More about electrolysis

LEARNING OUTCOMES

- Describe the electrode products and observations made during the electrolysis of concentrated hydrochloric acid and concentrated aqueous sodium chloride
- Predict the products of electrolysis of a specified halide in concentrated aqueous solution
- Describe the manufacture of chlorine, hydrogen and sodium hydroxide from concentrated aqueous sodium chloride

Electrolysing concentrated aqueous sodium chloride

Ionic compounds dissolve in water and conduct electricity. So concentrated solutions of ionic compounds in water should also decompose when they are electrolysed. But do they decompose in the same way as a molten ionic compound?

DEMONSTRATION

Electrolysing a concentrated solution of sodium chloride

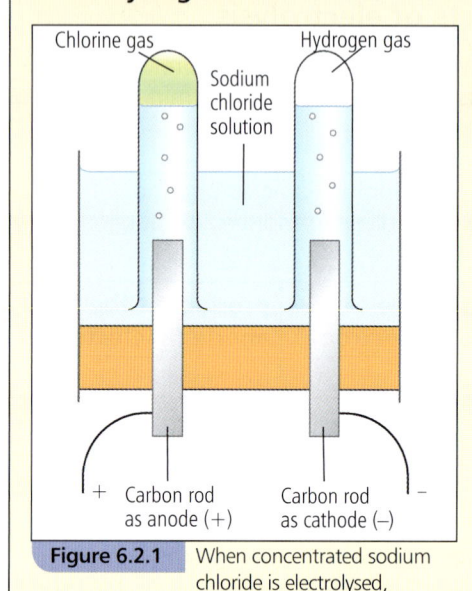

This experiment is done in a fume cupboard because chlorine gas is toxic.

An electric current is passed through the concentrated aqueous solution of sodium chloride. Hydrogen gas is collected at the negative electrode (cathode) and chlorine is collected at the positive electrode (anode).

Figure 6.2.1 When concentrated sodium chloride is electrolysed, chlorine forms at the anode and hydrogen at the cathode.

If we electrolyse concentrated aqueous solutions of ionic metal chlorides we find that chlorine is produced at the positive electrode. This is the same result we get with the molten compounds. However, we do not get a metal at the negative electrode – we get hydrogen instead.

Explaining the results

In concentrated solutions of ionic compounds in water we have a greater variety of ions. Water itself is a weak electrolyte. It has a very low concentration of hydrogen ions, H^+, and hydroxide ions, OH^-.

$$H_2O \rightleftharpoons H^+ + OH^-$$

So in an aqueous solution of sodium chloride we have the following ions: Na^+, H^+, Cl^-, OH^-. So why is hydrogen given off at the cathode instead of sodium?

The answer lies in the ion **discharge series**. This is a particular form of the metal reactivity series (see Topic 13.2). The lower down the series, the more likely it is that the ion will be discharged (changed into an atom or molecule at the electrode). The order of this series is:

STUDY TIP

Remember that when a solution of sodium chloride is electrolysed, hydrogen is formed at the cathode whereas with molten sodium chloride, sodium is formed.

DID YOU KNOW?

Although the electrolysis of brine was discovered about 200 years ago, the large-scale production of chlorine and sodium hydroxide from brine had to wait until the invention of a dynamo which could supply enough electric current.

for positive ions: $\underrightarrow{Na^+ \quad Mg^{2+} \quad Al^{3+} \quad H^+ \quad Cu^{2+}}$
more likely to be discharged

for negative ions: $\underrightarrow{SO_4^{2-} \quad NO_3^- \quad OH^- \quad Cl^- \quad Br^- \quad I^-}$
more likely to be discharged

So when a concentrated aqueous solution of sodium chloride is electrolysed, hydrogen rather than sodium is discharged at the negative electrode. This is because hydrogen is lower in the discharge series.

The electrolysis of concentrated hydrochloric acid

When concentrated hydrochloric acid is electrolysed using inert electrodes, bubbles of hydrogen are given off at the negative electrode. As the electrolysis continues, the concentration of hydrogen ions falls and the acid gets used up. Bubbles of chlorine are formed at the positive electrode since it is low in the discharge series.

Electrolysing brine

Brine is a concentrated aqueous solution of sodium chloride. It is obtained from seawater or from seams of rock salt underground. The electrolysis of brine is used to produce chlorine, hydrogen and sodium hydroxide on a large scale.

Chlorine is used to make solvents, for treating drinking water and for making bleaches. Hydrogen is used for making ammonia, for making margarine and as a fuel. Sodium hydroxide is used for making soap and in the extraction of aluminium.

Brine is electrolysed in a diaphragm cell.

The ions in solution are Na^+, H^+, Cl^- and OH^-.

At the anode: chloride ions lose electrons and are discharged as chlorine gas:

$2Cl^-(aq) \longrightarrow Cl_2(g) + 2e^-$

At the cathode: hydrogen ions accept electrons and hydrogen gas is discharged rather than sodium:

$2H^+(aq) + 2e^- \longrightarrow H_2(g)$

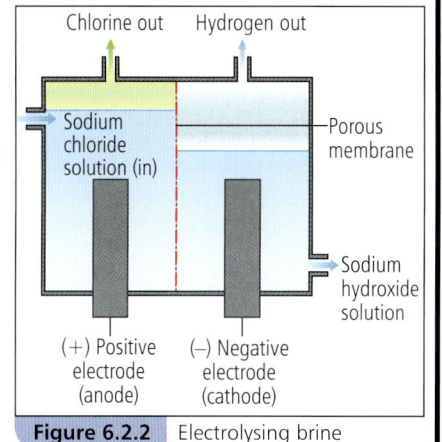

Figure 6.2.2 Electrolysing brine

The ions remaining in solution are Na^+ and OH^-. So an aqueous solution of sodium hydroxide is formed.

Salt is used for the manufacture of chlorine and sodium hydroxide by electrolysis

KEY POINTS

1 When concentrated aqueous solutions of metal ions are electrolysed, hydrogen rather than a metal is formed at the cathode.

2 When concentrated hydrochloric acid is electrolysed, hydrogen is formed at the cathode and chlorine at the anode.

3 When brine is electrolysed in a diaphragm cell chlorine, hydrogen and sodium hydroxide are formed.

SUMMARY QUESTIONS

1 Copy and complete using the words below:

**cathode concentrated
hydrogen hydroxide
sodium**

A _____ aqueous solution of sodium chloride contains _____, chloride, hydrogen and _____ ions. When this solution is electrolysed, _____ rather than sodium is discharged at the _____.

2 State the products formed at the + and − electrodes when the following are electrolysed:
   a molten sodium chloride
   b concentrated aqueous sodium chloride.

3 State the name of the products formed at each electrode when concentrated solutions of the following are electrolysed:
   a copper(II) chloride
   b hydrogen bromide
   c sodium iodide.

6.3 Explaining electrolysis

LEARNING OUTCOMES

- State that metals or hydrogen are formed at the negative electrode and non-metals (other than H_2) are formed at the positive electrode
- Predict the products of electrolysis of a specified halide in dilute aqueous solution
- Construct ionic half-equations for reactions at the cathode
- Describe the transfer of charge during electrolysis

Electrolysis of dilute aqueous solutions

Water can be electrolysed. A little sulfuric acid, which contains hydrogen ions and sulfate ions, is added to the water to improve its conductivity. When water is electrolysed, hydrogen is formed at the negative electrode and oxygen is formed at the positive electrode.

The oxygen comes from the decomposition of the OH^- ions in the water. So oxygen is discharged rather than sulfate (see Topic 6.2). This is because OH^- is lower in the discharge series than sulfate. We can therefore extend our general pattern of electrolysis to state that metals or hydrogen are formed at the negative electrode and halogens or oxygen are formed at the positive electrode.

When dilute aqueous solutions are electrolysed we can usually predict the electrode products from the discharge series. The table shows some examples.

aqueous solution	ions present	product at cathode	product at anode	ions remaining	change to the electrolyte
potassium iodide	K^+, I^-, H^+, OH^-	hydrogen	iodine	K^+, OH^-	becomes alkaline
copper nitrate	Cu^{2+}, NO_3^-, H^+, OH^-	copper	oxygen	H^+, NO_3^-	becomes acidic
dilute sulfuric acid	SO_4^{2-}, H^+, OH^-	hydrogen	oxygen	SO_4^{2-}	water used up

STUDY TIP

Make sure that you know the difference in the products at each electrode when dilute and concentrated aqueous sodium chloride and molten sodium chloride are electrolysed.

From ions to atoms

When an ionic compound is molten, the ions are able to move. When an electric current is applied, the positive ions move towards the cathode which is the negative electrode. Positive ions are called **cations** because they move towards the cathode. Negative ions move towards the anode, which is the positive electrode. Negative ions are therefore called **anions**.

When the ions reach the electrodes they gain or lose electrons. A reaction where electrons are lost is called an **oxidation** reaction. A reaction where electrons are gained is called a **reduction** reaction (see Topic 9.3). When lead(II) bromide is electrolysed using inert electrodes, a reduction reaction occurs at the cathode and an oxidation reaction at the anode.

At the cathode, lead ions in the electrolyte take electrons from the external circuit and become lead atoms. This is a reduction reaction because lead ions gain electrons.

$$Pb^{2+}(aq) + 2e^- \longrightarrow Pb(l)$$

This type of equation, which shows what is happening at only one of the electrodes, is called a **half equation**. Half equations are a form of ionic equation. In this case two electrons are added to balance the 2+ charge on the lead ion.

DID YOU KNOW?

Unwanted hairs on the face can be removed by electrolysis. The electrolyte is the salt solution around the base of the hair. The sodium hydroxide produced during the electrolysis kills the cells at the base of the hair.

At the anode, bromide ions in the electrolyte lose electrons to the anode and become bromine molecules. This is an oxidation reaction because bromide ions lose electrons.

$$2Br^-(aq) \longrightarrow Br_2(g) + 2e^-$$

You will sometimes see this sort of equation written to show the removal of electrons from the negative ion:

$$2Br^-(aq) - 2e^- \longrightarrow Br_2(g)$$

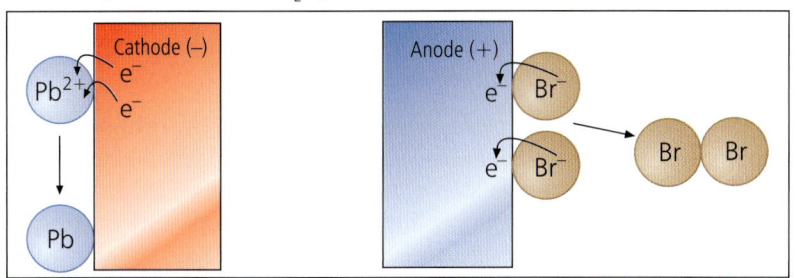

Figure 6.3.1 At the cathode metal ions gain electrons. At the anode non-metal ions lose electrons.

Similar equations can be written for other molten electrolytes. The metal ions gain electrons and the non-metal ions lose electrons.

What happens in aqueous solutions?

In aqueous solutions of metal salts, H^+ and OH^- ions are present as well as the ions from the salt. The less reactive element is discharged at the cathode. So in a solution of sodium chloride, hydrogen ions, rather than sodium ions, gain electrons. Hydrogen gas is formed:

$$2H^+(aq) + 2e^- \longrightarrow H_2(g)$$

When sulfuric or nitric acid or solutions of sulfates or nitrates are electrolysed, oxygen is formed at the anode. The hydroxide ions in the water, rather than the sulfate or nitrate ion, lose electrons. This is because OH^- ions are lower in the discharge series than sulfate or nitrate ions. Oxygen gas is formed:

$$4OH^-(aq) \longrightarrow O_2(g) + 2H_2O(l) + 4e^-$$

The first person to explain electrolysis was Michael Faraday, who worked on this and many other problems in science nearly 200 years ago. His work formed the basis of an understanding of electrolysis that we still use today.

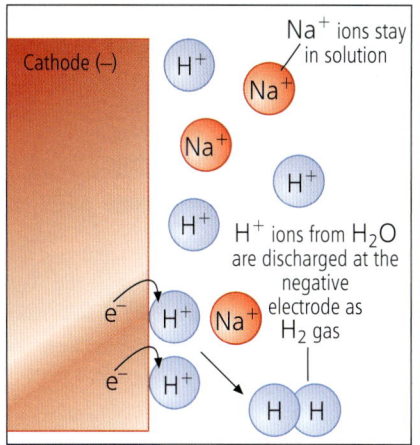

Figure 6.3.2 Hydrogen ions rather than sodium ions are discharged at the cathode

SUMMARY QUESTIONS

1 Copy and complete using the words below:

anode   cathode   gas   hydrogen   lose   oxygen   positive

During electrolysis _____ ions move towards the _____ and negative ions move towards the _____. At the cathode the metal or _____ ions gain electrons and form metal atoms or hydrogen _____. At the anode the non-metal ions _____ electrons and form halogens or _____.

2 Write half equations for the following electrolysis reactions:
   a   lead ions Pb^{2+} reacting to form lead, Pb
   b   hydrogen ions reacting to form hydrogen gas
   c   hydroxide ions reacting to form oxygen and water

3 Explain the differences in the electrolysis products of:
   a   molten sodium chloride
   b   a dilute aqueous solution of sodium chloride.

KEY POINTS

1 In electrolysis, metals or hydrogen are formed at the negative electrode and halogens or oxygen are formed at the positive electrode.

2 When dilute aqueous solutions of acids are electrolysed, oxygen is formed at the anode and hydrogen is formed at the cathode.

3 During electrolysis ions gain electrons at the cathode and lose electrons at the anode.

6.4 Purifying copper

LEARNING OUTCOMES

- Relate the products of electrolysis to the electrolyte and electrodes used
- Describe and explain the refining of copper

Refining copper

Copper is an excellent conductor of electricity. Thousands of tonnes of copper are used every year to make electrical wiring and pieces of electrical equipment. Although copper is easily extracted from copper ore, we need to refine it further. This is done to remove any impurities that will reduce its electrical conductivity.

Copper is purified by electrolysis. We often call this copper **refining**. An impure strip of copper is connected to the positive end of a power supply. This forms the anode. A thin strip of pure copper is connected to the negative end of the power supply. This forms the cathode. The electrolyte is a solution of copper(II) ions, usually copper(II) sulfate solution.

Copper-plated sheets being removed from an electrolysis cell containing copper(II) sulfate electrolyte

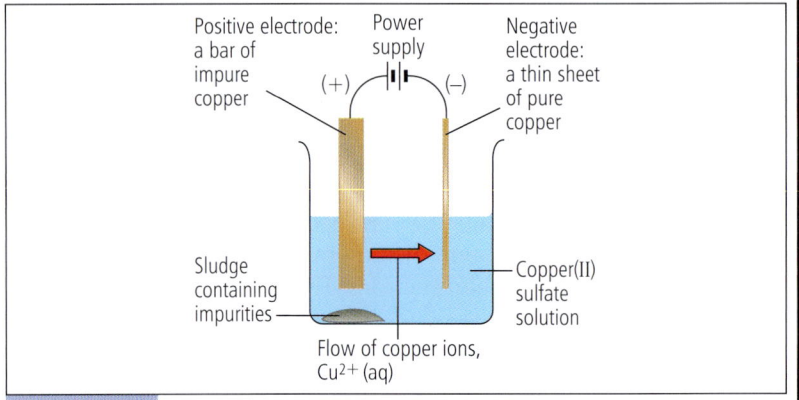

Figure 6.4.1 Copper is refined using electrolysis

At the anode (the positive electrode), copper atoms lose their valency electrons and form copper ions. These go into solution as part of the electrolyte:

$$Cu(s) \longrightarrow Cu^{2+}(aq) + 2e^-$$

At the cathode (the negative electrode), copper ions from the electrolyte gain electrons and form copper atoms. These copper atoms are deposited on the strip of pure copper:

$$Cu^{2+}(aq) + 2e^- \longrightarrow Cu(s)$$

As the electrolysis proceeds the cathode becomes thicker as it gains more and more copper. After a time the cathode of pure copper is removed and replaced by a new one. The anode loses mass and the impurities fall to the bottom of the electrolysis cell as 'anode slime'. Other valuable metals such as gold and platinum can be extracted from this 'anode sludge'.

The overall result of this electrolysis is that pure copper is transferred from the anode to the cathode.

DID YOU KNOW?

Copper ore is extracted from the Collahuasi mine in Chile and immediately refined to produce 70 000 tonnes of copper each year.

Changing the electrodes

We can also electrolyse copper(II) sulfate using inert electrodes (graphite or platinum) or metal electrodes.

There are some differences in the products of electrolysis which are influenced by the type of electrodes we use.

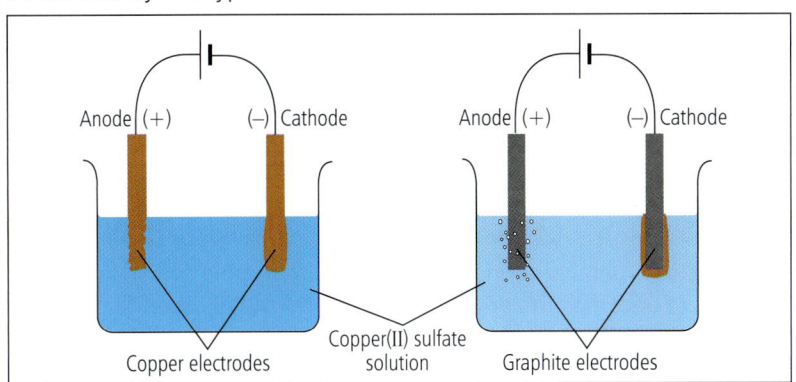

Figure 6.4.2 The products of electrolysis depend on the type of electrodes used

The ions present in an aqueous solution of copper(II) sulfate are $Cu^{2+}(aq)$, $SO_4^{2-}(aq)$, $H^+(aq)$ and $OH^-(aq)$.

Electrolysis with inert electrodes:

- At the anode:

  The anode cannot lose electrons because it is inert. Hydroxide ions rather than sulfate ions are discharged. This is because hydroxide ions are lower in the discharge series. Oxygen gas bubbles off.

  $$4OH^-(aq) \longrightarrow O_2(g) + 2H_2O(l) + 4e^-$$

- At the cathode:

  Copper ions rather than hydrogen ions are discharged because they are lower in the discharge series. Copper metal is deposited.

  $$Cu^{2+}(aq) + 2e^- \longrightarrow Cu(s)$$

- The electrolyte gradually loses its blue colour. This is because the copper ions in solution are turning to copper atoms at the cathode but are not being replaced in the solution at the anode (see below).

Electrolysis with copper electrodes:

- At the anode:

  Because the anode is not inert, it loses electrons and copper ions go into solution. The anode gets smaller.

  $$Cu(s) \longrightarrow Cu^{2+}(aq) + 2e^-$$

- At the cathode:

  Copper ions rather than hydrogen ions are discharged because they are lower in the discharge series.

  $$Cu^{2+}(aq) + 2e^- \longrightarrow Cu(s)$$

- The electrolyte remains the same deep blue colour. This is because the copper ions removed from the solution at the cathode are replaced in solution by copper ions formed at the anode.

STUDY TIP

Remember that in electrolysis the electrodes are usually inert (graphite or platinum). If the anode is not inert, it will react and decrease in size.

KEY POINTS

1. Copper is purified by using an impure copper anode and a pure copper cathode.

2. During electrolysis using inert electrodes the negative ions in the electrolyte lose electrons to the anode.

3. In electrolysis using metal electrodes the metal atoms of the anode lose electrons to form positive ions.

SUMMARY QUESTIONS

1. Copy and complete using the words below:

   **cathode   copper
   electrolysed   impure
   solution**

   When a solution of copper(II) sulfate is _____ using _____ electrodes the copper atoms at the _____ anode go into _____ as copper ions. At the _____ the copper ions turn into copper atoms.

2. Draw a table to summarise the similarities and differences when copper(II) sulfate solution is electrolysed using:
   a   platinum electrodes
   b   copper electrodes.

6.5 Electroplating

LEARNING OUTCOMES

- Describe the electroplating of metals
- Outline the uses of electroplating

Plating articles with metal makes them more attractive

STUDY TIP

With electroplating, think about what is happening at each electrode and any changes in the colour of the electrolyte.

Electroplating metals

Electroplating is used to put a thin layer of one metal on top of another metal. This is acheived by electrolysis.

- We connect the object to be electroplated to the negative pole of the power supply. It becomes the cathode. The object can be anything which is made of metal. For example, we can electroplate a spoon or a small metal statue. The object to be electroplated must be very clean so that the metal which is to cover it does not flake off.
- The plating metal is connected to the positive pole of the power supply. It becomes the anode. Some typical metals used for plating are silver, gold, tin and chromium.
- The electrolyte is a solution of an ionic compound of the plating metal. For example, if you want to plate an object with silver, you can use silver cyanide as the electrolyte.

PRACTICAL

Electroplating with copper

When you pass an electric current, the steel ring gets covered with a thin layer of copper. The copper anode gradually gets smaller as the copper is transferred from the anode to the steel cathode. The colour of the copper(II) sulfate electrolyte does not change because copper ions removed from solution at the cathode are continually being replaced at the copper anode.

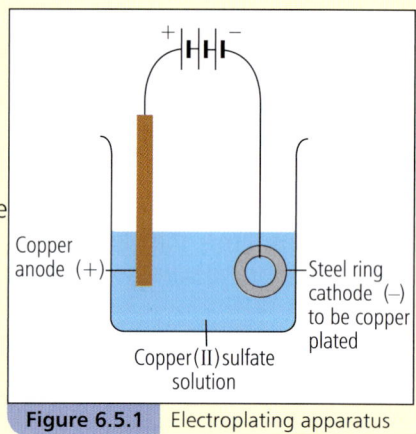

Figure 6.5.1 Electroplating apparatus

How electroplating works

Electroplating works in the same way as metal refining using electrolysis. Figure 6.5.2 shows what happens when an object is electroplated with silver.

At the anode the silver atoms lose electrons. They become silver ions which go into solution.

$$Ag(s) \longrightarrow Ag^+(aq) + e^-$$

The silver ions move to the cathode. At the cathode the silver ions gain electrons to become silver atoms which form a thin layer of silver on the surface of the object to be plated.

$$Ag^+(aq) + e^- \longrightarrow Ag(s)$$

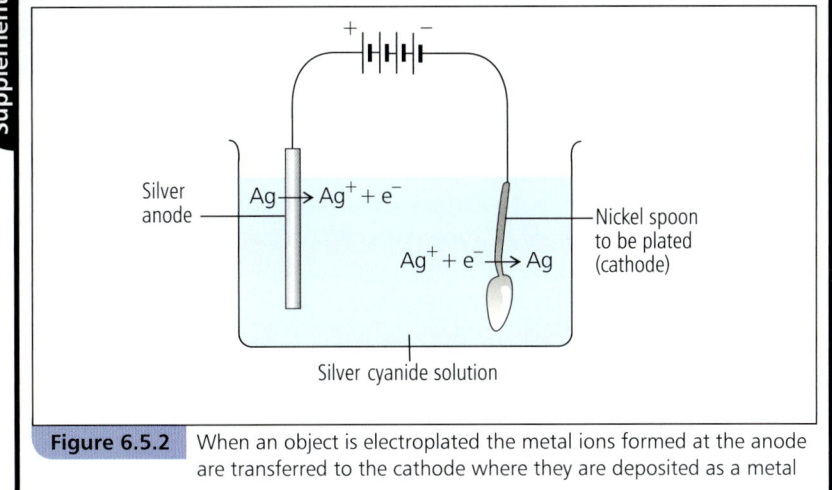

Figure 6.5.2 When an object is electroplated the metal ions formed at the anode are transferred to the cathode where they are deposited as a metal

Uses of electroplating

Copper, chromium, nickel, silver and tin are the most commonly used metals for electroplating articles. Mixtures of metals can also be used for plating. There are two main reasons for electroplating: protection of metals from corrosion and improving their appearance.

Protection of metals from corrosion

Steel cans are electroplated with tin. The layer of tin protects the metal underneath from the air and water so that it does not rust. If the layer of tin is scratched the metal underneath will start to corrode. Chromium is used to plate the metal parts of furniture, bicycle handlebars and the 'trim' on cars. Chromium is very hard, so it does not scratch easily. It also gives a pleasing shiny appearance.

Improving the appearance of metals

Chromium plating gives a very shiny surface to objects that does not go dull. Silver plating is used for jewellery, cutlery and in electronics where it would be too expensive to use solid silver. Gold plating is also used for jewellery and for specialised electronic equipment.

> **DID YOU KNOW?**
> China has more than 1200 companies which specialise in electroplating articles.

> **KEY POINTS**
> 1. An object can be electroplated by making the object to be plated the cathode.
> 2. In electroplating, the metal object to be plated is the cathode and the electrolyte is a solution of a compound of the plating metal.
> 3. When an article is electroplated the ions of the plating metal gain electrons at the cathode and become metal atoms.

> **SUMMARY QUESTIONS**
>
> 1. Copy and complete using the words below:
>
>   **anode   cathode   electrolyte   plate   solution**
>
>   To electroplate an object you make the object to be electroplated the _____. The _____ is the metal that will _____ the surface. The _____ is a _____ of a compound of the plating metal.
>
> 2. A student wants to electroplate a piece of copper with chromium. Draw a labelled diagram of the apparatus that the student could use.
>
> 3. Using half equations explain how silver can be used to electroplate an object.

6.6 Extracting aluminium

LEARNING OUTCOMES

- Describe the extraction of aluminium from bauxite including the role of cryolite and the reactions at the electrodes

Metals such as iron and copper have been in use for thousands of years. People learned how to extract iron from its ores thousands of years ago. This required heating with carbon (see Topic 14.1). Reactive metals such as aluminium, magnesium and sodium cannot easily be extracted by heating with carbon.

It was not until electrolysis was discovered nearly 200 years ago that scientists could begin to work out how to get these reactive metals from their compounds. In fact, it was not until 1886 that the first small drops of liquid aluminium were extracted from aluminium oxide.

Aluminium oxide from aluminium ore

Aluminium is the most abundant metal in the Earth's crust. It is found in the mineral ore bauxite which contains 50–65% aluminium oxide, Al_2O_3. Aluminium oxide is sometimes called alumina. The main impurities in bauxite are oxides of iron, silicon and titanium.

The first step in aluminium extraction is to purify the ore. The ore is first crushed and mixed with sodium hydroxide. The aluminium oxide reacts with the sodium hydroxide and dissolves.

$$Al_2O_3(s) + 2NaOH(aq) \longrightarrow 2NaAlO_2(aq) + H_2O(l)$$
aluminium oxide   sodium aluminate

The impurities are insoluble in sodium hydroxide. These are filtered off. The sodium aluminate undergoes further treatment and is finally heated to make pure aluminium oxide.

Mining bauxite

Extracting aluminium from aluminium oxide

Electrolysis to produce aluminium is carried out in shallow electrolysis cells about 8 metres long and 1 metre deep. In order to carry out electrolysis, the aluminium oxide needs to be molten. Aluminium oxide melts at about 2040 °C. It is difficult to keep the electrolyte at this high temperature for long periods of time. In addition, it is too costly because it needs so much energy – and energy is expensive. Aluminium oxide on its own is a poor conductor of electricity.

The problem is solved by dissolving the aluminium oxide in large amounts of molten cryolite. Cryolite, which is sodium aluminium fluoride, Na_3AlF_6, melts at about 1000 °C. Since the aluminium oxide is dissolved in the cryolite, the melting point of the electrolyte is much lower compared with pure aluminium oxide.

Dissolving the aluminium oxide in cryolite not only saves a lot of energy but also improves the electrical conductivity of the electrolyte. Calcium fluoride, CaF_2 is often added to lower the melting point further. In most cases the melting point of the electrolyte is about 900 °C.

DID YOU KNOW?

The Frenchman Paul Herault and the American Charles Hall both discovered how to extract aluminium by electrolysis. Neither knew about the other's discovery until in 1886 there was a long legal battle to decide who made the discovery first.

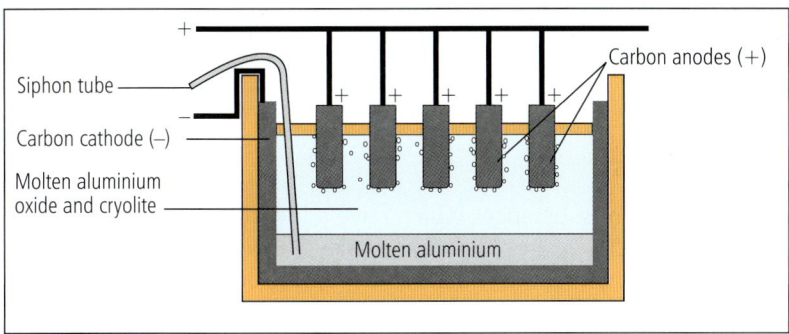

Figure 6.6.1 The electrolytic cell used in the extraction of aluminium

Electrolysis is carried out using graphite electrodes. The overall equation for this electrolysis is:

$$2Al_2O_3 \longrightarrow 4Al + 3O_2$$

The cathode is the carbon lining of the steel electrolysis cell. Several anodes, which can be raised or lowered, dip into the electrolyte. The very high electric current (40 000 amps) used in this electrolysis not only decomposes the aluminium oxide but also keeps the electrolyte molten.

At the cathode, aluminium ions gain electrons and are reduced to aluminium metal. The liquid aluminium metal falls to the bottom of the cell. It is removed from time to time using a siphon tube:

$$Al^{3+} + 3e^- \longrightarrow Al$$

At the anode, the oxide ions lose electrons and are oxidised to oxygen:

$$2O^{2-} \longrightarrow O_2 + 4e^-$$

The oxygen reacts with the hot carbon anodes to form carbon dioxide gas. Because the carbon anodes 'burn away' they need to be replaced from time to time.

> **STUDY TIP**
>
> You do not have to learn the diagram of the cell used to extract aluminium but you should be able to label the different parts. You should also be able to write half equations for the reactions at the electrodes.

SUMMARY QUESTIONS

1 Copy and complete using the words below:

   conductivity   cryolite   dissolved   energy   lower   pure

   Aluminium is extracted by the electrolysis of molten aluminium oxide _____ in _____ This mixture melts at a much _____ temperature than _____ aluminium oxide. Therefore a lot of _____ is saved and the electrical _____ of the electrolyte is improved.

2 Explain why during the electrolysis of aluminium oxide dissolved in cryolite, the carbon anodes have to be replaced from time to time.

3 Balance this half equation by adding electrons to one side of the equation:
   $$4Al^{3+} \longrightarrow 4Al$$

KEY POINTS

1 The electrolytic cell for the extraction of aluminium has carbon anodes and cathode.

2 The electrolyte in the cell is molten aluminium oxide, dissolved in molten cryolite to lower its melting point.

3 During the electrolysis of molten aluminium oxide in cryolite, aluminium forms at the cathode and oxygen is released at the anode.

6.7 Conductors and insulators

LEARNING OUTCOMES

- Describe reasons for the use of steel-cored aluminium in high-voltage electrical cables and copper in electrical wiring
- Explain why plastics and ceramics are used as insulators

Conductors

Conductors are substances which have a low resistance to the passage of electricity. In other words, they allow electricity to flow through them easily. Metals and graphite conduct electricity because they have mobile (free-moving) outer shell electrons in their structures. Most metals are good conductors. Copper, silver and gold are among the best conductors of electricity.

PRACTICAL

Comparing conductors

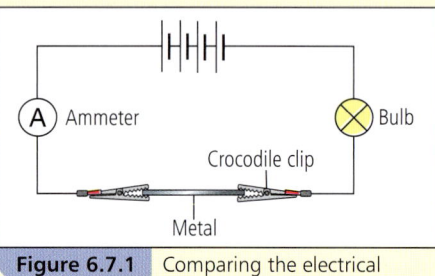

Figure 6.7.1 Comparing the electrical conductivity of solids

The electrical circuit is set up and the ammeter reading is recorded. By placing wires made of different metals in the circuit, we can compare their conductivity. To make it a fair test, we use the same length and thickness of wire for each metal.

Copper is commonly used in electrical wiring and in thicker electricity cables because it is a good conductor of electricity. It is also easily drawn into wires – it is ductile – and it is easily purified by electrolysis.

Thick wires can carry a larger electric current more safely than thin wires. If a wire is very thin, the electrons have to move through a narrow space. The 'friction' produced causes a lot of heat and the wire may even melt. We use very thick wires in the high-voltage power lines used to transfer electricity over long distances as these do not lose as much heat to the air as do a lot of thin wires.

DID YOU KNOW?

More than 17 000 000 tonnes of copper are extracted from copper ores every year.

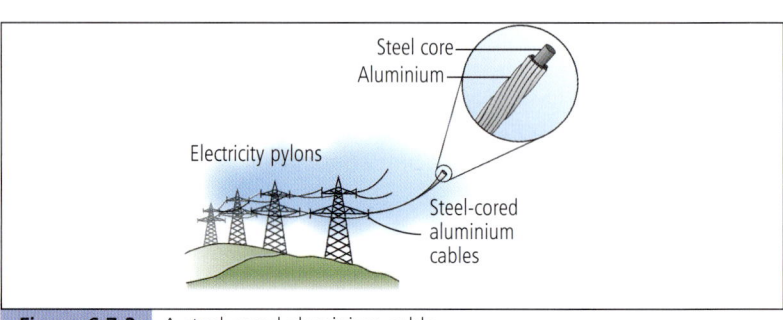

Figure 6.7.2 A steel-cored aluminium cable

High-voltage power lines are sometimes made from aluminium cables with a steel core in the middle. The steel gives the cables additional strength to stop them sagging and breaking.

Aluminium is used in high-voltage power lines because it is a good conductor of electricity. It also has a low density – this is important if very thick wires are suspended in the air over long distances. Aluminium is also resistant to corrosion.

Insulators

An **insulator** is a substance that resists the flow of an electric current – it does not conduct electricity. Insulators do not conduct electricity because they do not have mobile electrons. Examples of insulators include plastics, glass and ceramic materials made by heating clay.

Plastics such as PVC are useful insulators, not only because they do not conduct electricity: they are also flexible and non-biodegradable. This makes them useful for covering electrical wires so that we do not get an electric shock or form short circuits in electrical equipment. Plastics like PVC are less useful as insulators where high electric currents are used. The heat of the electric current can easily melt the plastic. If there is a danger of this, plastics called thermosetting plastics can be used.

Ceramics and glass are useful insulators, not only because they do not conduct electricity: they also have very high melting points so do not melt when high electric currents flow. For this reason they are used in high-voltage electricity towers to keep the wires from touching the metal pylons or from touching each other. Other advantages of ceramics are that they are not affected by water or air and they can be moulded into complex shapes.

Overhead power lines have to be strong as well as good conductors of electricity

STUDY TIP

It is a common mistake to think that the steel core in electricity cables just conducts electricity. It is also there to strengthen the cables.

SUMMARY QUESTIONS

1 Copy and complete using the words below:

   current   electricity   melt   resistance   thicker   thin

   Conductors have a low _____ to the passage of _____. The _____ an electrical wire the greater is the amount of _____ that can flow through it. A _____ wire will heat up when an electric current passes through it. It may even _____.

2 What do you understand by the term *insulator*?

3 Explain why steel-cored aluminium wires are used for high-voltage power cables.

KEY POINTS

1 Steel-cored aluminium cables are used in high-voltage power lines because aluminium is a good conductor and steel strengthens the cable.

2 Insulators such as plastics and ceramics prevent an electric current flowing.

3 Copper is used in electrical wiring because it is a good conductor of electricity.

SUMMARY QUESTIONS

1 Classify the following as either conductors or insulators:
   (a) molten sodium chloride
   (b) iron
   (c) plastic
   (d) wood
   (e) sugar
   (f) a solution of sodium bromide

2 Define these terms:
   (a) electrolyte
   (b) electrode
   (c) ion
   (d) insulator
   (e) electroplating

3 Match each electrolyte on the left with two of the possible products of electrolysis on the right. Each product can be used once, more than once or not at all.

molten lead chloride	chlorine
concentrated aqueous sodium chloride	hydrogen
concentrated hydrochloric acid	lead
dilute aqueous sodium chloride	oxygen
	sodium

4 Draw and label the apparatus used to demonstrate:
   (a) the electrolysis of molten zinc chloride
   (b) how well a solid conducts electricity.

5 Predict the products at each electrode during electrolysis of the following molten salts:
   (a) zinc chloride
   (b) potassium bromide
   (c) lead iodide
   (d) magnesium chloride

6 Copy and complete the paragraph using these words in the list below:

   **accept   anode   aqueous   attracted
   cathode   deposited   electrolyte
   electrons   plated   smaller**

PRACTICE QUESTIONS

1 Which one of these statements about the electrolysis of molten lead(II) bromide is true?
   A Lead is formed at the positive electrode.
   B Hydrogen is formed at the negative electrode.
   C Bromine is formed at the negative electrode.
   D Bromine is formed at the positive electrode.

(Paper 1)

2 Which of the following statements about the transfer of charge during electrolysis is correct?
   A Electrons move through the electrolyte.
   B At the cathode, positive ions accept electrons.
   C Negative ions get reduced at the anode.
   D Ions move round the external circuit from the cathode to the anode.

(Paper 2)

3 Lead(II) bromide can be electrolysed using the apparatus shown below:

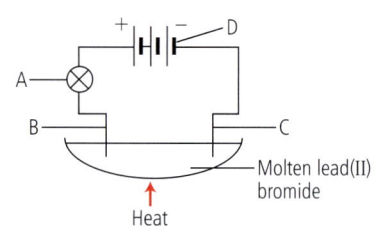

Tin can be _____ onto steel by electrolysis. The steel is made the _____ and the tin is made the _____. The _____ is an _____ solution of tin(II) chloride. During electrolysis, the tin anode gets _____ in size because tin loses _____ and tin(II) ions go into solution. These ions are _____ to the steel cathode. At the cathode, tin(II) ions _____ electrons and tin is _____.

7 Describe how sodium hydroxide and chlorine are manufactured from brine. Give details of:
   (a) the electrodes
   (b) the reactions at each electrode
   (c) why a solution of sodium hydroxide is formed.

(a) Which letter represents the anode? *[1]*
(b) State the products formed at
   (i) the anode and (ii) the cathode. *[1]*
(c) Complete the equation for this electrolysis:
   $PbBr_2(l) \longrightarrow$ _____ + _____ *[2]*
(d) The electrolyte is molten lead(II) bromide. What do you understand by the term *electrolyte*? *[1]*
(e) (i) Suggest a suitable substance which can be used to make the electrodes. *[1]*
   (ii) State two properties that a suitable electrode should have. *[2]*

(Paper 3)

4 The table below shows the electrical conductivity of substances A–F.

compound	A	B	C	D	E	F
conductivity/ S/m	0.7	0	0	1.1	0.8	0.7

(a) Which substance in the table is the best electrical conductor? *[1]*
(b) Steel-cored aluminium cables are used for conducting high-voltage electricity over long distances.
   (i) Give two reasons why aluminium is used for these cables. *[2]*
   (ii) What is the purpose of the steel core? *[1]*

(Paper 3)

5 An aqueous solution of lithium chloride is electrolysed using carbon electrodes.
(a) Explain why a solution of lithium chloride conducts electricity but solid lithium chloride does not conduct. *[2]*
(b) State the names of the products formed at
   (i) the anode and (ii) the cathode. *[2]*
(c) Write half-equations for the reactions at
   (i) the anode and (ii) the cathode. *[2]*
(d) Explain why lithium is not deposited at the cathode. *[2]*
(e) Suggest another compound that will give the same products at the anode and cathode when electrolysed. *[1]*

(Paper 4)

6 The electrolysis of a concentrated solution of sodium bromide produces bromine at the anode and hydrogen at the cathode.
(a) (i) Explain why sodium is not formed at the cathode. *[1]*
   (ii) Explain why bromine and not oxygen is formed at the anode. *[1]*
(b) Write half-equations for the reactions occurring at
   (i) the cathode and (ii) the anode. *[2]*
(c) At which electrode is oxidation taking place? Explain your answer. *[2]*
(d) Predict the products of electrolysis at the cathode and anode if a very dilute solution of sodium bromide is electrolysed. *[2]*

(Paper 4)

7 Aluminium is extracted by electrolysis. The electrolyte is a mixture of aluminium oxide and cryolite. The electrodes are made of graphite.
(a) Write an overall equation for this electrolysis. *[2]*
(b) Write half-equations for the reactions at
   (i) the anode and (ii) the cathode. *[2]*
(c) Explain the purpose of the cryolite in the electrolyte. *[2]*
(d) The graphite (carbon) anodes have to be renewed periodically. Explain why and write any relevant equations. *[2]*
(e) State the name of the ore containing aluminium oxide. *[1]*

(Paper 4)

8 When very dilute aqueous magnesium chloride is electrolysed, hydrogen and oxygen are formed as products.
 a Explain why hydrogen and oxygen are formed and not magnesium and chlorine. *[3]*
 b Describe the transfer of charge during this electrolysis in terms of:
   • the movement of electrons and ions.
   • oxidation and reduction. *[6]*

(Paper 4)

7 Chemical changes

7.1 Physical and chemical changes

LEARNING OUTCOMES

- Identify physical and chemical changes and understand the differences between them
- Describe the meaning of the terms *exothermic* and *endothermic*
- Interpret energy level diagrams showing exothermic and endothermic reactions

Physical changes

A physical change is one in which no new substance is formed. Melting, boiling, condensing and freezing are examples of physical change. Physical changes are easily reversible. For example, we can change liquid water to steam by heating the water to its boiling point. When the steam condenses, it returns to its liquid form. The temperature remains constant during changes of state (see Topic 1.2).

Dissolving is often thought of a physical change. We can dissolve salt (sodium chloride in water) and we can get the salt back again by evaporating the water. The salt is exactly the same substance and the same mass as we started with. Unlike melting or boiling, however, where the temperature remains constant, we do observe a temperature change when we dissolve salt in water.

Chemical changes

Chemical changes involve the formation of new substances during a chemical reaction.

- Heat is always taken in, or given out, during a chemical change.
- One or more new substances are formed.

Many chemical changes cannot easily be reversed. For example, when magnesium reacts with oxygen, the product is magnesium oxide. But we are unable to cool magnesium oxide to get magnesium and oxygen again – electrolysis would have to be used. Some chemical changes can be reversed by changing the conditions. These are called reversible reactions (see Topic 9.1).

Boiling and condensation are both physical changes. The water in the kettle boils to form steam. The steam condenses back to water droplets when it cools.

Exothermic or endothermic?

Processes that release heat energy to the surroundings are called **exothermic** processes. Dissolving magnesium chloride in water is exothermic. Many chemical reactions are **exothermic reactions**. When we add zinc to copper(II) sulfate solution in a test tube, the temperature of the reaction mixture increases. The energy released goes to warming up the surroundings. The surroundings include:

- The contents of the test tube (i.e. the reaction mixture).
- The air around the test tube.
- The test tube itself.
- Thermometers or stirring rods dipped into the test tube.

Chemical reactions which take in heat energy from the surroundings are called **endothermic reactions**. We say they absorb heat energy.

The heat absorbed is taken in by the reaction mixture and so lowers the temperature of the surroundings. The test tube gets cold. Examples of endothermic reactions are photosynthesis and reactions where continuous heating is needed.

Energy level diagrams

We can show exothermic and endothermic changes with the help of energy level diagrams. These diagrams show:

- The energy of the reactants and products on the vertical axis (y-axis). We do not usually write the exact values of the energy.
- The reactants and products, with the reactants on the left, the products on the right and the horizontal axis (x-axis) labelled as 'reaction pathway'.
- The enthalpy change is shown by an arrow.

The energy level diagram for an exothermic reaction shows:

- The energy of the reactants is higher than the energy of the products.
- The chemicals in the mixture release energy. This energy goes to heating up the surroundings, so the measured temperature increases.
- The arrow goes downwards to show that energy is released (given out).

The energy level diagram for an endothermic reaction shows:

- The energy of the reactants is lower than the energy of the products.
- The chemicals in the reaction mixture gain energy and the energy is absorbed from the surroundings. The measured temperature decreases.
- The arrow goes up to show that energy is absorbed (taken in).

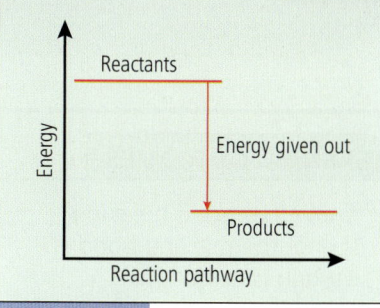

Figure 7.1.1 Exothermic reaction: the energy in the reactants is greater than the energy in the products

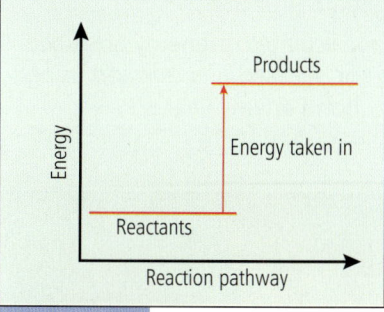

Figure 7.1.2 Endothermic reaction: the energy in the reactants is lower than the energy in the products

SUMMARY QUESTIONS

1 Copy and complete using the words below:

absorbs   exothermic   falls

reaction   surroundings

An endothermic reaction _____ heat from the _____. When this happens the temperature of the mixture _____. An _____ reaction gives out heat energy. When this happens the _____ warm up.

2 Draw an energy level diagram for this reaction:

Fe + S $\longrightarrow$ FeS   energy given out = 100 kJ/mol

3 a Give one difference between physical and chemical changes.
  b Give two examples of physical change and two examples of chemical change.

KEY POINTS

1 In physical change, no new substance is formed. Physical changes are reversible.

2 In chemical change, one or more new substances are formed. Some chemical changes cannot be reversed, though others can be reversed.

3 An exothermic change releases heat energy. The temperature of the surroundings increases.

4 An endothermic change absorbs heat energy. The temperature of the surroundings decreases.

5 In an energy level diagram for an exothermic reaction, the reactants have more energy than the products. In an energy level diagram for an endothermic reaction, the reactants have less energy than the products.

7.2 Energy transfer in chemical reactions

LEARNING OUTCOMES

- Know that bond breaking is endothermic and bond making is exothermic
- Draw and label energy level diagrams for exothermic and endothermic reactions using data provided
- Calculate the energy absorbed or released in a reaction using bond energy values

Keeping warm – we need exothermic reactions

STUDY TIP

Make sure you know that energy is given out (exothermic reaction) when new bonds are formed and taken in (absorbed) when bonds are broken.

Making and breaking bonds

If we break a thin stick, we put in energy to break it. Breaking a chemical bond is similar – we have to put energy in. So bond breaking is endothermic. When new bonds are formed, the opposite happens – energy is given out to the surroundings. So bond making is exothermic.

We can explain exothermic and endothermic reactions in terms of bond breaking and making:

Exothermic reaction: the energy taken in to break the bonds in the reactants is less than the energy given out when new bonds are made.

Endothermic reaction: the energy taken in to break the bonds in the reactants is more than the energy given out when new bonds are made.

The difference between the energy of the reactants and products is shown by the symbol ΔH (delta H). If heat energy is given out ΔH is given a negative sign. If heat energy is absorbed it is given a positive sign. For example in the zinc + copper(II) sulfate reaction:

$$Zn(s) + CuSO_4(aq) \longrightarrow Cu(s) + ZnSO_4(aq) \quad \Delta H = -212 \text{ kJ/mol}$$

Bond energies

Each type of bond has a particular amount of energy needed to break it. This is called the **bond energy**. Bond energy is the amount of energy needed to break one mole of a particular bond in one mole of gaseous atoms. The symbol for bond energy is E.

It needs 436 kJ to break bonds in one mole of hydrogen molecules. We can write this as:

$$E(H\text{---}H) = +436 \text{ kJ/mol}$$

It needs 498 kJ to break both the bonds in one mole of oxygen molecules. We can write this as:

$$E(O{=}O) = +498 \text{ kJ/mol}$$

Values of bond energies are always positive because they refer to bonds being broken.

Bond energy calculations

We can use bond energies to calculate how much energy is released or absorbed in a reaction. The 'balance sheet' method for doing this is shown in the worked example. Note that you have to take into account:

- the number of moles of each reactant and product in the stoichiometric equation
- the number of bonds of a particular type in each molecule, for example each molecule of water H_2O has two O—H bonds.

Worked example:

Calculate the energy change in the reaction:

$$2H_2(g) + O_2(g) \longrightarrow 2H_2O(g)$$

Bond energy values in kJ/mol: H—H 436; O=O 498; O—H 464

bonds broken (endothermic)		bonds formed (exothermic)	
2 H—H = 2 × 436 =	872 kJ	4 × O—H = 4 × 464	
1 O=O =	498 kJ	=	1856 kJ
total	1370 kJ	total	1856 kJ

The calculation shows that when all the bonds in hydrogen and water are broken, the energy change is +1370 kJ. The positive sign shows that the energy change is endothermic.

When new bonds are formed, the amount of energy released is the same as the amount of energy absorbed when the same type of bond is broken, but the sign is reversed. So, when two moles of water are formed from hydrogen and oxygen atoms, the energy change is −1856 kJ. The negative sign shows that the energy change is exothermic.

For the calculation above, the overall energy change is:

$$+1370 - 1856\,kJ = -486\,kJ$$

The negative sign shows that the energy change is exothermic.

SUMMARY QUESTIONS

1 When ethane burns in air to form carbon dioxide and water, heat energy is released. By referring to bond breaking and bond making, explain why energy is released.

2 When calcium carbonate is heated to a high temperature, the following reaction occurs:

$$CaCO_3(s) \longrightarrow CaO(s) + CO_2(g) \quad \Delta H = +572\,kJ/mol$$

Draw an energy level diagram for this reaction. Add as much relevant detail as possible.

3 Calculate the energy change in kJ when fluorine reacts with hydrogen to form hydrogen fluoride.

$$H_2 + F_2 \longrightarrow 2HF$$

Include the correct sign for the energy change in your answer.

Bond energy values in kJ/mol: H—H 426 ; F—F 158 ; H—F 568

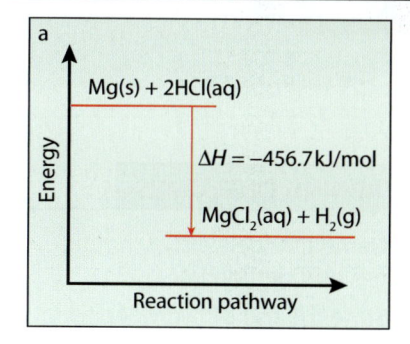

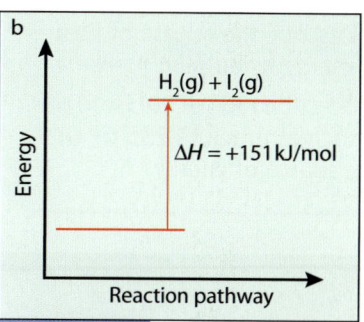

Figure 7.2.1 Energy level diagrams for (a) an exothermic reaction and (b) an endothermic reaction showing values for ΔH

Figure 7.2.2 Rearranging atoms

KEY POINTS

1 Bond breaking is endothermic and bond making is exothermic.

2 The energy change in a reaction is given by the symbol ΔH. For an exothermic reaction the value for ΔH is negative. For an endothermic reaction the value for ΔH is positive.

3 The energy absorbed or released in a reaction can be calculated using bond energy values.

7.3 Fuels and energy production

LEARNING OUTCOMES

- Name the fuels – coal, natural gas and petroleum
- State the use of hydrogen as a fuel
- Describe the release of heat energy by burning fuels
- Describe radioactive isotopes such as uranium-235 (^{235}U) as a source of energy

Wood is a useful fuel but pollutes the atmosphere when it is burned

STUDY TIP

Remember that burning is always exothermic.

DID YOU KNOW?

Much of the energy we use for transport comes from the bodies of tiny animals and plants which lived in the oceans millions of years ago.

The variety of fuels

A **fuel** is a substance that can be burned to release energy. Burning a fuel is an exothermic reaction. We use a variety of fuels for heating, lighting, cooking, for transport and for making electricity. Many of these are fossil fuels:

Coal was formed by the decay of plants in swampy areas in the absence of oxygen. Over millions of years the plant remains changed to form coal.

Petroleum or crude oil is a complex mixture of compounds containing carbon and hydrogen. It was formed from the bodies of tiny animals and plants that sank to the sea bed millions of years ago. Pressure from rocks formed above changed these tiny organisms into petroleum.

Natural gas is largely methane. This is often found underground trapped in layers of ice or near areas rich in petroleum.

PRACTICAL

Comparing the energy released

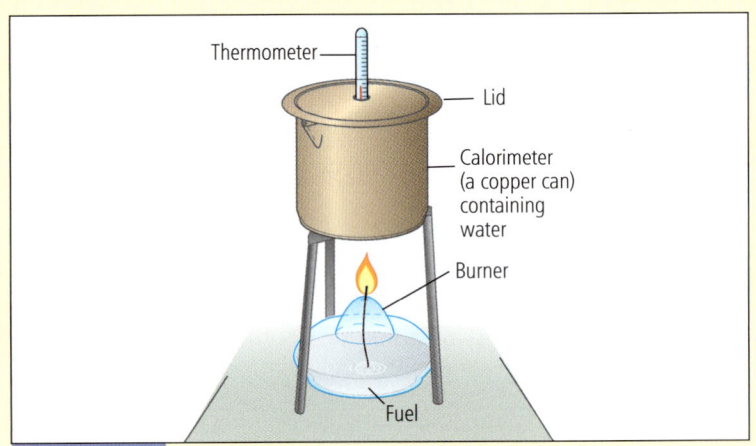

Figure 7.3.1 Comparing the energy released when liquid fuels are burned

1. We put a known amount of liquid fuel into the burner. When the fuel was set alight it heated up a measured volume of water in a calorimeter. The calorimeter was a copper can.

2. We can compare the energy given out by fuels by measuring the temperature rise when burning 1 g of each fuel, using the same volume of water.

Coal is very polluting and causes acid rain (see Topic 15.3) and increases global warming. Petroleum fractions and natural gas are less polluting but still contribute to global warming. Hydrogen is a fuel that is non-polluting. Although it can form explosive mixtures with air, it gives out a lot of energy when burned.

$C_8H_{18} + 12\frac{1}{2}O_2 \longrightarrow 8CO_2 + 9H_2O$ 48 kJ/g energy given out
'petrol'

$CH_4 + 2O_2 \longrightarrow CO_2 + 2H_2O$ 55 kJ/g energy given out
'natural gas'

$H_2 + \frac{1}{2}O_2 \longrightarrow H_2O$ 143 kJ/g energy given out
hydrogen

Energy from radioactivity

Many nuclear power stations use the radioisotope uranium-235 (^{235}U) as a nuclear fuel. This is a rather unusual use of the term *fuel* because no burning takes place.

The uranium is made into 'fuel rods'. These are lowered into a 'reactor' in a nuclear power station. The uranium-235 is bombarded with high-speed neutrons. The collisions cause the nucleus of the uranium-235 to split. When this happens a large amount of energy is released. More neutrons are produced, as well as atoms of lower mass such as thorium and radium. The 'new' neutrons then hit other uranium-235 atoms and split them.

If not stopped, this 'chain reaction' can cause an explosion. It is not allowed to reach this stage: 'control rods' are pushed into the nuclear reactor to absorb excess neutrons.

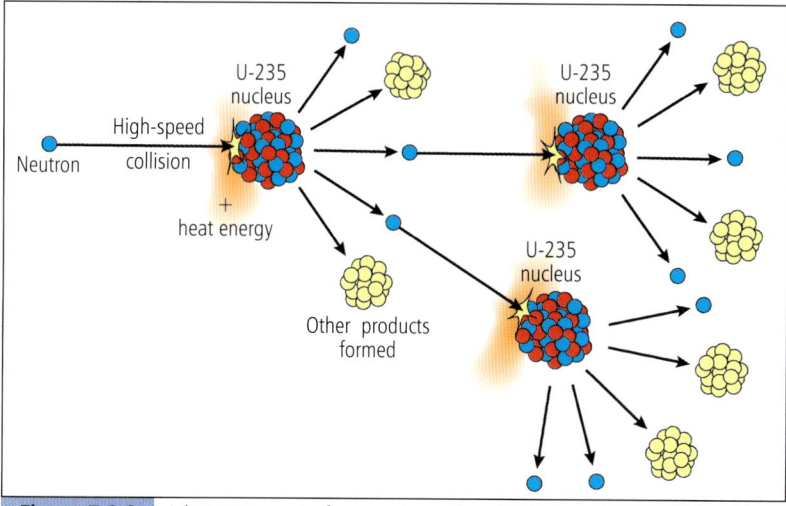

Figure 7.3.2 A large amount of energy is produced when neutrons collide with an atom of uranium-235

The energy given out as heat in the reaction is used to heat up a gas. This hot gas heats up water to form steam. The steam is used in a steam generator to produce electricity. Nuclear 'fuels' produce a lot more energy per gram than ordinary fuels and are not as polluting. However, their radioactive waste products are difficult to dispose of and many people can be exposed to radiation if there is a leak from the reactor.

KEY POINTS

1 Coal, natural gas and petroleum are fuels that are polluting when burned.

2 We can use a calorimeter and thermometer to compare the energy released when different fuels are burned.

3 The radioisotope uranium-235 is used in some nuclear power stations as a source of energy.

SUMMARY QUESTIONS

1 Copy and complete using the words below:

acid   coal   exothermic global   released transport

Fuels are used for heating, lighting and _____. The burning of fuels is an _____ reaction because energy is _____. Fossil fuels such as _____ and petroleum harm the environment by causing _____ rain and _____ warming.

2 Name a radioisotope used to produce electricity in a nuclear power station.

7.4 Energy from electrochemical cells

LEARNING OUTCOMES
- Describe the production of electrical energy from simple cells linked with the reactivity series

Small batteries are useful sources of energy

DID YOU KNOW?
A cell for medical uses can be as small as 0.65 cm long and 0.23 cm in diameter.

STUDY TIP
It is a common misunderstanding to confuse cells with electrolysis. In electrolysis an electric current is used to decompose the electrolyte. In a cell the different reactivity of the electrodes makes an electric current flow.

Simple electrochemical cells

In electrolysis we put in electrical energy to make a chemical change. This change is endothermic. Can we reverse this to make electrical energy from a chemical reaction? When we add zinc to copper(II) sulfate, an exothermic reaction takes place:

$$Zn(s) + Cu^{2+}(aq) \longrightarrow Cu(s) + Zn^{2+}(aq) \qquad \Delta H = -212 \text{ kJ/mol}$$

Instead of wasting all this heat energy, we can make the reaction into an electrochemical cell. A cell can do a useful job such as lighting a lamp or running an electric motor.

An electrochemical cell consists of two metals of different reactivity dipping into an electrolyte. The electrolyte can be an acid or alkali or a solution of a salt. In commercial cells the electrolyte is made into paste so that the electrolyte does not leak.

How does a simple electrochemical cell work?

The diagram shows an electrochemical cell made from a strip of zinc and a strip of copper dipping into an electrolyte of dilute sulfuric acid.

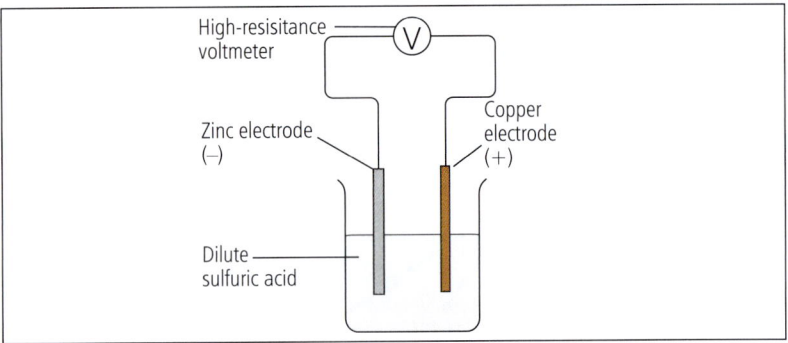

Figure 7.4.1 An electrochemical cell works because of the difference in reactivity of the metals

The more reactive metal in an electrochemical cell is better at releasing electrons. Zinc is more reactive than copper so the zinc atoms rather than the copper atoms lose their electrons. The zinc ions go into solution.

$$Zn(s) \longrightarrow Zn^{2+}(aq) + 2e^-$$

The zinc electrode becomes the negative electrode or negative pole of the cell. The electrons flow through the wire of the external circuit to the copper electrode.

At the positive electrode or positive pole, these electrons are released and combine with the hydrogen ions in the acid to form hydrogen gas.

$$2H^+(g) + 2e^- \longrightarrow H_2(g)$$

The difference in the ability of zinc and copper atoms to release electrons causes a voltage to be produced. This results in a flow of electrons from the negative electrode (pole) of the cell to the positive.

The electrical circuit is completed by the movement of ions in the electrolyte.

Getting the best voltage

Many electrochemical cells are small and portable but these do not deliver a high voltage. Batteries of several cells joined together give a higher voltage but take up more space. Cells that can be recharged many times, such as car batteries, are heavy and take up a lot a space. The voltage we get from an electrochemical cell depends on the metals chosen.

PRACTICAL

Which combination of metals gives the best voltage?

1 The voltmeter reading is taken with copper as the left-hand electrode. The voltmeter is put into the circuit so that it gives a positive reading with this combination of metals.

2 The right-hand electrode is then replaced by another metal and the voltmeter reading taken again.

3 This is repeated with different metals.

The table below shows the results.

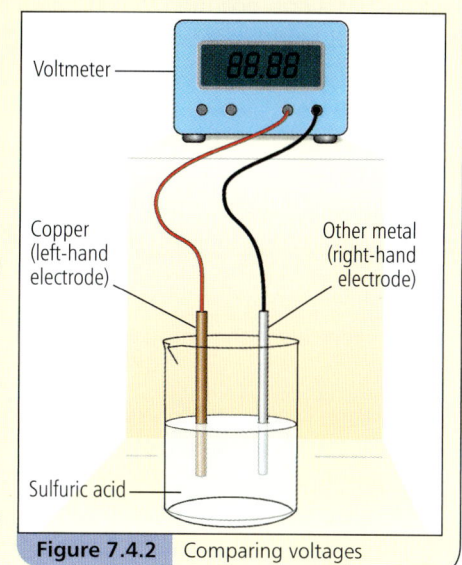

Figure 7.4.2 Comparing voltages

left-hand electrode	right-hand electrode	voltage/volts	order of reactivity of right-hand electrode
copper	magnesium	+2.7	most reactive
copper	zinc	+1.1	↑
copper	tin	+0.48	
copper	copper	0	
copper	silver	−0.46	least reactive

You can see from the table that the further apart the metals are in the reactivity series, the greater is the voltage. This is because the more reactive metal is better at releasing electrons to form ions. So a cell with electrodes of magnesium and copper gives a far greater voltage than one with electrodes of tin and copper.

The more reactive metal in the cell is always the negative electrode. When silver is used as the right-hand electrode, the voltmeter needle moves in the opposite direction. This is because copper is more reactive than silver so copper has become the negative electrode.

You can use a table of voltages like this to calculate the voltages of other combinations of metals. For example a cell made from magnesium and zinc has a voltage of 2.7 − 1.1 = 1.6 volts.

KEY POINTS

1 An electrolytic cell contains two electrodes of different reactivity dipping in an electrolyte.

2 The electrode higher in the reactivity series is the negative pole of the cell.

3 In an electrolytic cell the electrons move from the negative to the positive pole in the external circuit.

SUMMARY QUESTIONS

1 Copy and complete using the words below:

**connected   current
electrolyte   higher
negative   positive
reactivity**

An electrochemical cell contains two rods of metal with different ____ dipping into an ____. When ____ to each other by a wire, an electric ____ flows from the ____ electrode to the ____ electrode. The metal ____ in the reactivity series is the negative electrode of the cell.

2 Use the table (left) to suggest which of the following combinations of metals will give the highest cell voltage:

copper and tin; copper and zinc; magnesium and tin; zinc and tin

3 An electrochemical cell is made using rods of zinc and iron. The electrolyte is sulfuric acid. Write half equations for the reaction at:

a the positive pole
b the negative pole.

97

7.5 Fuel cells

LEARNING OUTCOMES
- Describe the use of hydrogen as a fuel in a fuel cell
- Describe the use of fuel cells in the production of electricity

Simple electrochemical cells lose their power after a time – they no longer produce a voltage. This happens because one of the reactants is used up. Electrical cells like car batteries are bulky – they take up a lot of room. Some electrochemical cells have to be recharged from time to time. In addition, many of the substances found in electrochemical cells are harmful and difficult to dispose of safely.

Hydrogen is a non-polluting fuel. When it burns in oxygen, water is the only product formed. We can use this reaction to supply electrical energy continuously. We do this by reacting hydrogen and oxygen in a **fuel cell**.

STUDY TIP
You do not need to remember details about the construction of a fuel cell, but you should be familiar with questions based on diagrams and relevant half-equations.

PRACTICAL

A model fuel cell

1. We electrolyse a solution of sodium hydroxide. Hydrogen forms at the cathode and oxygen forms at the anode.
2. We electrolyse the solution until both test tubes are filled with gas.
3. We then replace the power pack with a voltmeter. In the presence of the electrolyte, the hydrogen and oxygen react together. The voltmeter gives a reading. We have produced an electric current by combining hydrogen with oxygen.

How does a fuel cell work?

A fuel cell consists of two platinum electrodes and an electrolyte. The platinum is coated onto a porous material that allows gases to pass through it. Hydrogen gas and oxygen gas are bubbled through the porous electrodes where the reactions take place. Hydrogen gas is bubbled through the negative electrode and oxygen is bubbled through the positive electrode.

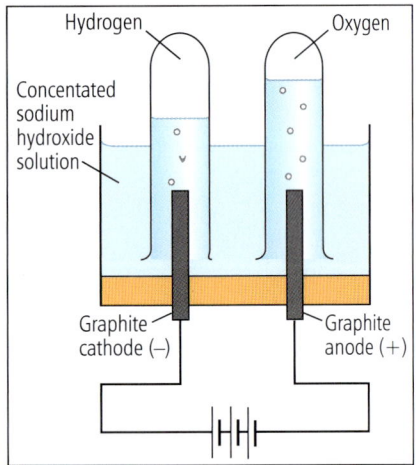

Figure 7.5.1 Electrolysing sodium hydroxide

Fuel cells are likely to eventually replace petrol and diesel engines in cars

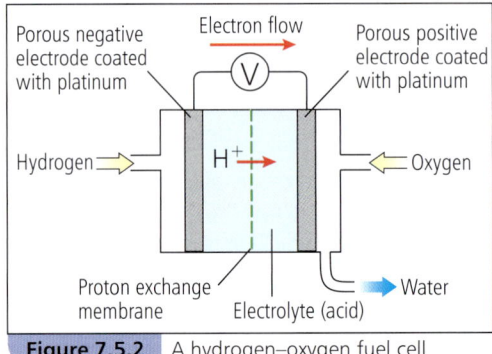

Figure 7.5.2 A hydrogen–oxygen fuel cell

There are two main types of fuel cell. One contains an acidic electrolyte, the other contains an alkaline electrolyte such as a concentrated solution of sodium hydroxide.

Acidic electrolyte:

At the negative electrode the hydrogen loses electrons and forms hydrogen ions in the electrolyte:

$$2H_2(g) \longrightarrow 4H^+(aq) + 4e^-$$

The released electrons move around the external circuit to the positive electrode. At the positive electrode oxygen gains electrons and reacts with hydrogen ions from the acid electrolyte.

$$O_2(g) + 4H^+(aq) + 4e^- \longrightarrow 2H_2O(l)$$

The hydrogen ions removed at the positive electrode are replaced by those produced at the negative electrode. So the concentration of the electrolyte remains constant. The overall reaction is:

$$2H_2(g) + O_2(g) \longrightarrow 2H_2O(l)$$

The water is removed.

Alkaline electrolyte:

At the negative electrode the hydrogen reacts with the hydroxide ions in the electrolyte and forms water:

$$2H_2(g) + 4OH^-(aq) \longrightarrow 4H_2O + 4e^-$$

At the positive electrode oxygen gains electrons and reacts with water to form hydroxide ions:

$$O_2(g) + 2H_2O(l) + 4e^- \longrightarrow 4OH^-(aq)$$

The hydroxide ions removed at the negative electrode are replaced by those produced at the positive electrode. So again the concentration of the electrolyte remains constant. The overall reaction is the same as for the acidic electrolyte.

What are the advantages of fuel cells?

Hydrogen fuel cells are used to provide electrical power in spacecraft. The water produced can be used for drinking. Fuel cells are increasingly used instead of petrol to power cars. Fuel cells have many advantages over batteries and petrol-driven engines:

- Water is the only product made – no pollutants are formed.
- They produce more energy per gram of fuel than other fuels.
- They are lightweight.
- They don't need recharging like batteries.
- Fuel cells operate with high efficiency.

Fuel cells seem to be the answer to many pollution problems. However, the hydrogen and oxygen needed for fuel cells to operate are usually produced using fossil fuels at present!

DID YOU KNOW?

Reduction always takes place at a cathode and oxidation at an anode. So, when using these terms with **cells** the anode is the − electrode and the cathode is the + electrode.

KEY POINTS

1 A hydrogen–oxygen fuel cell produces only water as a product.
2 A hydrogen–oxygen fuel cell is efficient and non-polluting.
3 When a fuel cell is working, hydrogen loses electrons at the negative electrode and oxygen gains electrons at the positive electrode.

SUMMARY QUESTIONS

1 Copy and complete using the words below:

**cathode   circuit   current
electrons   oxygen   two**

In a fuel cell, the hydrogen loses _____ at the negative electrode. The electrons move around the external _____ to the postive electrode creating an electric _____. At the positive electrode each _____ atom gains _____ electrons.

2 Write down two half equations for the reaction taking place at each electrode in a fuel cell containing an acidic electrolyte.

3 State three advantages of a fuel cell compared with a car battery.

SUMMARY QUESTIONS

1 State whether each of the following reactions is endothermic or exothermic:

(a) burning wood

(b) a reaction in which the temperature falls

(c) decomposing calcium carbonate by heating

(d) hydrogen exploding

2 Match each of the words on the left with one of the phrases on the right.

endothermic	burning in oxygen
exothermic	atoms of the same element having different numbers of neutrons
isotopes	an atom giving out energy as waves or particles
radioactive	a reaction which gives out heat energy
combustion	a reaction which absorbs heat energy

3 Copy and complete the paragraph using words from the list below:

carbon   decay   electricity   energy oxygen   pollutant   turbine   uranium water

Hydrogen and _____-235 are good sources of energy. Neither of these produces _____ dioxide as an atmospheric _____. When it burns, hydrogen reacts with _____ to produce _____ as the only product. Uranium-235 undergoes radioactive _____ to release _____ which is used to heat water. The steam formed drives a _____ to produce _____.

4 Match the phrases on the left with the words or phrases on the right.

Bond breaking is …	a fossil fuel
Bond making is …	a radioactive isotope
Coal is …	endothermic
An electrochemical cell is …	exothermic
Uranium-235 is …	a portable source of energy

PRACTICE QUESTIONS

1 Which one of these phrases best describes an endothermic reaction?

A The reaction releases heat energy.

B The reaction absorbs heat energy.

C The temperature of the reaction mixture increases.

D The surroundings get warmer.

(Paper 1)

2 Which one of these fuels is **not** a fossil fuel?

A Petroleum   **B** Natural gas

C Coal   **D** Hydrogen

(Paper 1)

3 When a certain volume and concentration of hydrochloric acid reacts with excess potassium hydroxide, 20 kJ of energy is released.

(a) State the name given to a chemical reaction that releases energy. [1]

(b) Describe an experiment to show that energy is released in this reaction. [3]

(c) Copy and complete the energy level diagram for this reaction. [3]

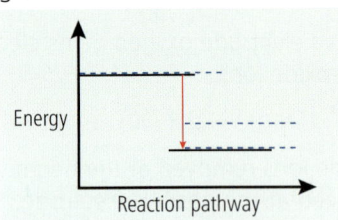

(d) Predict the amount of energy given out when the concentration of the hydrochloric acid is doubled. [1]

(Paper 3)

4 Equal amounts of three compounds are dissolved in the same amount of water in separate beakers and the temperature change observed. The table shows the results:

compound	initial temperature of water / °C	final temperature of solution / °C
potassium nitrate	25	18
sodium hydroxide	18	38
calcium chloride	22	40

(a) From the table, name one compound that:
   (i) releases energy when it dissolves
   (ii) absorbs energy when it dissolves. [2]
(b) State the name given to a chemical reaction that absorbs energy. [1]
(c) Which compound in the table shows the greatest temperature change when it dissolves in water? [1]
(d) When sodium hydroxide dissolves in water, estimate the temperature change when:
   (i) the amount of water is doubled
   (ii) the amount of sodium hydroxide is doubled. [2]

(Paper 3)

5 Methane burns to form carbon dioxide and water. The reaction is exothermic.
(a) What do you understand by the term *exothermic*? [1]
(b) Explain why this reaction is exothermic in terms of bond breaking and bond making. [3]
(c) The reaction can be represented by the equation:

$CH_4 + 2O_2 \rightarrow CO_2 + 2H_2O$

Draw an energy level diagram for this reaction. [3]
(d) Use the bond energies below to answer the following questions.
   O=O 498 kJ/mol   C=O 805 kJ/mol
   C—H 435 kJ/mol   O—H 464 kJ/mol
   (i) Calculate the energy needed to break the bonds in the reactants. [2]
   (ii) Calculate the energy released when the bonds in the products are formed. [2]
   (iii) Calculate the overall energy change for the reaction. [1]

(Paper 4)

6 Hydrogen can be used as a fuel.
(a) Explain why hydrogen is a better fuel than fossil fuels such as coal or petroleum. [3]
(b) Hydrogen burns in oxygen to form water:
   $2H_2 + O_2 \rightarrow 2H_2O$

Use the bond energies below to calculate the energy change for this reaction. [3]
   O=O 498 kJ/mol   H—H 436 kJ/mol
   O—H 464 kJ/mol
(c) Hydrogen and oxygen react in a fuel cell containing an electrolyte of potassium hydroxide. The reactions at the electrodes produce an electric current when the electrodes are connected to an external circuit.
   (i) Write a half equation for the reaction at the positive electrode. [1]
   (ii) Write a half equation for the reaction at the negative electrode. [1]
   (iii) Explain why an electric current is produced in the external circuit. [3]

(Paper 4)

7 A simple electrochemical cell is shown below.

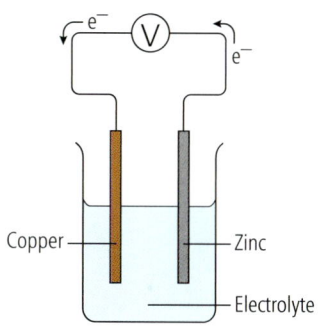

(a) Suggest a suitable electrolyte for this cell. [1]
(b) Explain why zinc is the negative pole of the cell and copper is the positive pole. [2]
(c) Explain why the electrons move in the direction shown in the diagram. [2]
(d) The voltage developed by this cell is 0.6 V. Use the electrochemical series below to answer the following questions.
   Mg   Zn   Fe   Pb   Cu   Ag
   most reactive ⟶ least reactive
   (i) What happens to the cell voltage when iron replaces zinc? [1]
   (ii) What happens to the voltage when silver replaces copper? [1]
   (iii) Explain why the cell does not produce a voltage when both electrodes are zinc. [1]

(Paper 4)

8 Rate of reaction

8.1 Investigating rate of reaction

LEARNING OUTCOMES

- Demonstrate knowledge and understanding of a practical method for investigating the rate of reaction involving gas evolution
- Devise and evaluate a suitable method for investigating rate of reaction

Supplement

Some reactions, such as an old car rusting, are very slow. Other reactions are very fast. When you mix solutions of silver nitrate and potassium iodide, you get a yellow precipitate straight away.

The speed of a reaction – usually called the **rate of reaction** – tells us how rapidly the products are formed from the reactants. Knowing about reaction rates is very important in the chemical industry. Chemical companies need to make their products as quickly and as cheaply as possible. Knowing how fast a reaction takes place helps them to do this.

Following the progress of a reaction

To find the rate of reaction we can either:

- measure how quickly the reactants are used up, or
- measure how quickly the products are formed.

Calculating the rate of reaction depends on measuring something that changes with time, for example, volume of gas, mass of the reaction mixture or amount of light transmitted through a solution.

We run at different rates. The chemical reactions in our bodies supply our muscles with food at the rate needed.

PRACTICAL

Following change in mass

You can use this method for reactions that give off a gas which is allowed to escape. As the reaction takes place the mass of the reaction mixture decreases. You record the mass at intervals of time. The loss of mass is equal to the mass of gas given off. When the total mass is plotted against time you get a graph like the one shown here.

Modern balances can be attached to a data logger and computer so that you can record the loss of mass continuously.

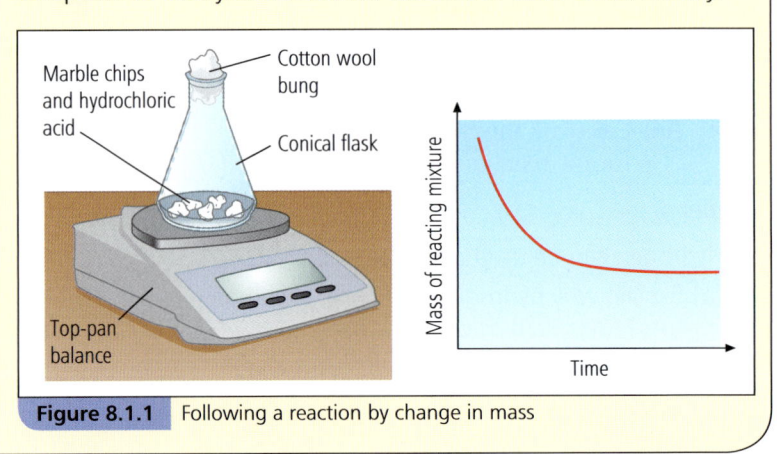

Figure 8.1.1 Following a reaction by change in mass

STUDY TIP

You should be able to evaluate methods for investigating rate of reactions. For example, change in mass is not suitable for investigating a reaction where hydrogen is given off. The change in balance reading will be too small.

PRACTICAL

Following change in the volume of gas given off

1 If a gas is given off in a reaction, you can collect it in a gas syringe. The volume of gas is recorded at intervals of time.

2 Then you can draw a graph of volume of gas against time.

If this method is used, it is important that there are no leaks in the apparatus so that no gas escapes.

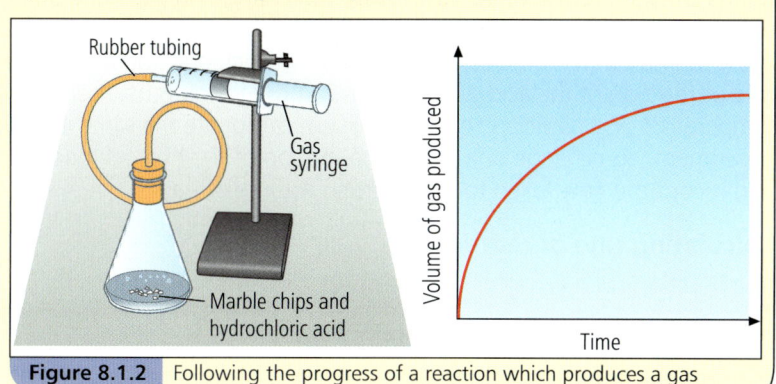

Figure 8.1.2 Following the progress of a reaction which produces a gas

A variety of methods

There are many methods for measuring rate of reaction. You can use any property that changes during a reaction. You could use a pH meter or electrical conductivity meter if the hydrogen ions are used up in the reaction. You could also record changes in pressure for reactions involving gases.

PRACTICAL

Following the progress of a precipitation reaction

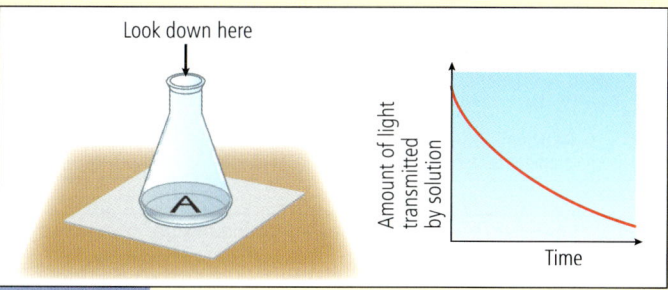

Figure 8.1.3 Following the progress of a precipitation reaction

If a precipitate is one of the products in a reaction, the solution goes cloudy. We can follow the reaction by placing the paper with a letter A on it underneath the flask containing one of the reactants. Start the reaction by adding the other reactant and recording the time taken for the letter A to 'disappear'.

We can also measure the amount of light passing through the solution using a colorimeter (or a light meter attached to a data logger and computer). This gives a graph as shown in Figure 8.1.3.

DID YOU KNOW?

Some reactions can be followed by recording tiny changes in the volume of a solution that take place during some chemical reactions.

KEY POINTS

1 We can follow the progress of a chemical reaction by measuring how fast the reactants are used up or how fast the products are formed.

2 We can use change in volume of gas, loss of mass of reactant or the time taken for a precipitate to make a letter 'disappear' to measure the rate of reaction.

SUMMARY QUESTIONS

1 Copy and complete using the words below:

**decrease   mass
mixture   rate**

The _____ of reaction can be followed by measuring the _____ in mass of the reaction _____. This is equal to the _____ of gas produced.

2 Sketch a graph to show how the volume of gas changes with time for this reaction:

Mg + 2HCl $\longrightarrow$ MgCl$_2$ + H$_2$

3 Put the following in order of increasing rate of reaction:

**cement setting
firework exploding
iron rusting**

103

8.2 Interpreting data

LEARNING OUTCOMES

- Interpret data obtained from experiments concerned with rate of reaction
- Know how to investigate and evaluate the effect of a given variable on reaction rate

Constants and variables

Many factors such as temperature, concentration of reagent and particle size can affect the rate of reaction. How can you compare the results of a series of experiments in a fair way? You need to keep some things constant (**controlled variables**) and change only the one that you are interested in (the **independent variable**). For example, if you want to find out how the concentration of hydrochloric acid affects the rate of reaction of the acid with calcium carbonate you must keep the temperature and particle size of the calcium carbonate the same in each experiment. You change only the concentration of the hydrochloric acid and record the time taken to get a certain volume of gas.

Calculating rate of reaction

In a reaction such as:

$$Mg + 2HCl \longrightarrow MgCl_2 + H_2$$

the time taken for the magnesium to disappear completely can be used as a measure of the rate of reaction. The average reaction rate can be worked out by using this equation:

$$\text{rate of reaction} = \frac{\text{amount of product formed or reactant used up}}{\text{time}}$$

A more accurate definition of rate of reaction is:

$$\text{rate of reaction} = \frac{\text{change in concentration of reactant or product}}{\text{time}}$$

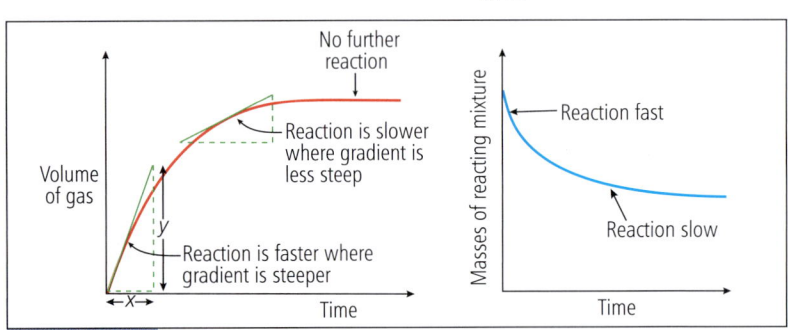

Some reactions such as rusting proceed very slowly

Figure 8.2.1 How volume of gas and mass of reaction mixture change with time

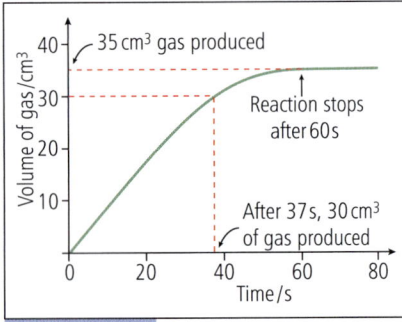

We often need to know how the reaction rate changes as the reaction proceeds. If we look at the graph of how the volume of gas given off in a reaction changes with time, the gradient (slope) of the graph gives us the reaction rate at any particular time. We see that the reaction is fastest near the start but then gets slower and slower until it finally stops.

If we look at a graph of loss of mass of the reaction mixture against time we can see a similar pattern.

From the left hand graph in Figure 8.2.1 we can also find out:

- how long it takes for a reaction to produce a given volume of gas
- the volume of gas produced in a given time.

Figure 8.2.2 Graph of volume of gas against time

Supplement

Limiting reactants (This is additional material which is useful but which you will not need for the exam)

When carrying out an experiment we sometimes use an excess of one of the reactants. The reactant that is not in excess is called the **limiting reactant**. The reaction stops when the limiting reactant is completely used up. You can work out which reactant is limiting by calculating which reactant has the least number of moles for reaction. You must also take into account the mole ratio of the reactants in the equation.

Example: In the reaction $Mg + 2HCl \longrightarrow MgCl_2 + H_2$

If 0.1 mol of magnesium reacts with of 0.1 mol of hydrochloric acid, the limiting reactant is hydrochloric acid because, if we look at the equation, for every mole of magnesium used we need two moles of hydrochloric acid. So there will still be $0.1/2 = 0.05$ mol of magnesium left when the hydrochloric acid has been used up.

Figure 8.2.3 shows how the volume of gas changes when hydrochloric acid, the limiting reactant, is present at several different concentrations.

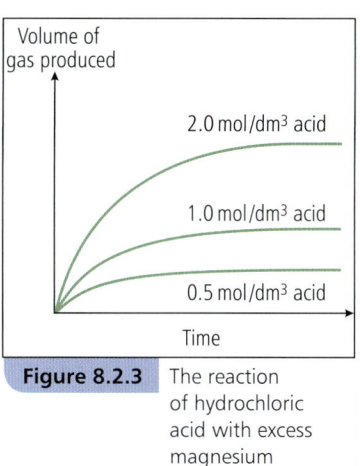

Figure 8.2.3 The reaction of hydrochloric acid with excess magnesium

> **STUDY TIP**
>
> Make sure that you know how to interpret the different parts of a graph of volume of gas released or loss in mass of the reactants against time.

SUMMARY QUESTIONS

1. Sketch a graph to show how the mass of the reaction mixture changes with time for the reaction:

   $MgCO_3 + 2HCl \longrightarrow MgCl_2 + CO_2 + H_2O$

   On your graph show:
   a. where the reaction had just stopped
   b. where the reaction is fastest
   c. where the reaction is very slow.

2. A student adds some pieces of calcium carbonate to some acid. At the start of the reaction many bubbles are seen. After 10 minutes some calcium carbonate is still present but the bubbles have stopped. Which reactant is in excess? Explain your answer.

3. Copy and complete using the words below:

   fast   slows   stops

   When excess calcium carbonate reacts with hydrochloric acid, the reaction is very _____ at first but then _____ down until it _____ completely.

KEY POINTS

1. Rate of reaction is calculated by dividing change in the amount of reactant or product by time.

2. As a reaction proceeds, the rate of reaction decreases as one or more of the reactants gets used up.

3. A reaction stops when one of the reactants is completely used up.

8.3 Surfaces and reaction rate

LEARNING OUTCOMES

- Describe the effect of particle size and catalysts on the rates of reactions
- Describe the explosive combustion of gases e.g. in mines, and fine powders e.g. in flour mills

Surface area and rate of reaction

If you want to make a fire quickly, you are more likely to succeed if you try to light small, thin pieces of wood rather than large pieces. We see a similar effect when solids react with solutions. A large lump of marble reacts slowly with hydrochloric acid but powdered marble reacts very quickly. Why is this?

The rate of reaction depends on how often the particles of acid collide with the particles on the surface of the marble. The greater the surface area of the marble, the more particles there are available to react. If we cut up the marble into smaller pieces, the surface area and the number of particles of marble which can react are both increased.

STUDY TIP

It is a common misconception to think that larger particles have a larger surface area than smaller ones. Think of a large cube cut up – by cutting, you are exposing more surfaces.

PRACTICAL

Investigating the effect of surface area

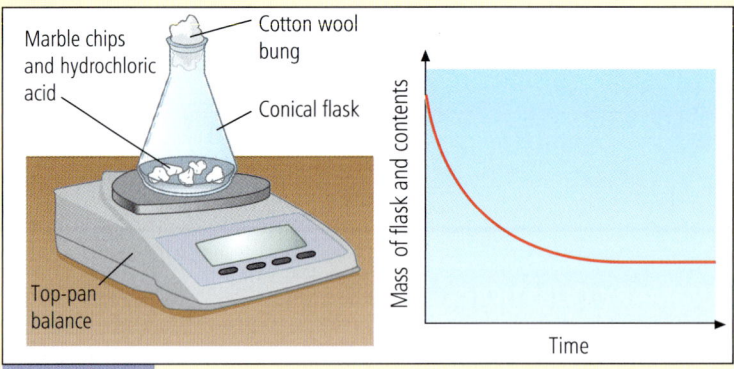

Figure 8.3.2 Investigating the effect of surface area

Marble is a form of calcium carbonate. You can investigate the effect of increasing surface area by reacting different sized marble chips with hydrochloric acid. You can carry out the reaction either by measuring the loss in mass as carbon dioxide is released or by recording the volume of carbon dioxide given off.

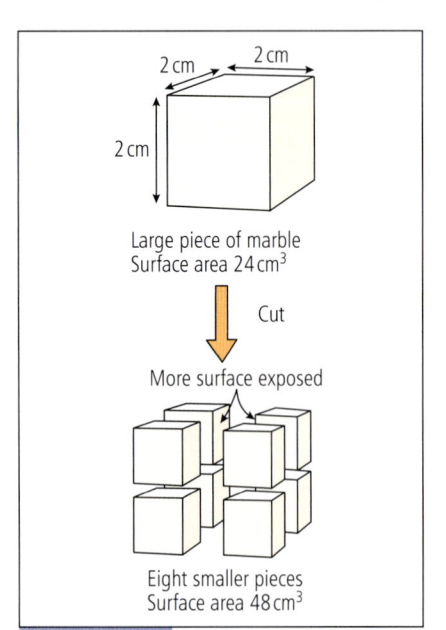

Figure 8.3.1 Cutting a cube into smaller ones increases the surface area

Large piece of marble Surface area 24 cm^3

Cut

More surface exposed

Eight smaller pieces Surface area 48 cm^3

Explosive reactions

Many industrial processes cause fine powders to get into the air. These powders are highly combustible. They burn very readily in air because of their very large surface area. A lit match or a spark from a machine can cause them to explode. Examples are flour in flour mills, wood dust in sawmills and coal dust in coal mines. In coal mines there is another hazard. The methane gas which is often present can form an explosive mixture with air.

PRACTICAL

Explosive milk!

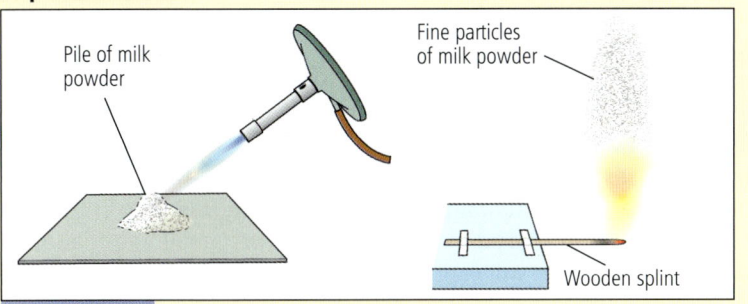

Figure 8.3.3 The combustion of milk powder

If you try to set fire to a pile of milk powder, it will just go black on the surface. If you sprinkle the milk powder onto a burning splint the flame will suddenly flare up as the milk powder 'explodes'.

Catalysts

A **catalyst** is a substance that speeds up a chemical reaction. The catalyst is not used up in the reaction – it remains chemically the same as it was at the start of the reaction. There are two types of catalyst: (i) solid catalysts and (ii) catalysts that work in solution.

A solid catalyst works by allowing the reactants to get close together on its surface so that less energy is needed to get the reaction to occur. Catalysts are generally used in the form of pellets or wire gauzes. This gives them a large surface area for reactions to occur on.

We need only tiny amounts of catalyst to speed up the reaction but they are often expensive. However, they can be used over and over again. Catalysts are important in speeding up the reactions in many important industrial chemical processes. In this way chemicals can be produced more quickly and at a lower temperature than by the uncatalysed reaction.

All living things contain particular types of catalysts called enzymes. These speed up all the chemical reactions in the body (see Topic 20.3).

SUMMARY QUESTIONS

1 Copy and complete using the words below:

    again   increases   rate   unchanged

A catalyst is a substance which _____ the _____ of a chemical reaction. The catalyst is _____ at the end of the reaction, so it can be used _____.

2 Factories that cut metals often have special fans to remove metal dust. Explain why metal dust in the atmosphere can be dangerous.

3 The inside of a catalytic converter in some cars has thousands of tiny beads coated with the catalyst. Explain why these beads are used rather than large lumps of catalyst.

DID YOU KNOW?

The earliest recorded explosion in a flour mill was in Italy in 1785.

Catalytic converters are important in reducing pollutant gases from cars

STUDY TIP

When defining a catalyst, the best definition is 'a substance that speeds up a reaction but remains chemically unchanged at the end of the reaction'. Phrases such as 'a substance which changes the rate of a reaction' are rather vague.

KEY POINTS

1 Increasing the surface area of a solid reactant increases the rate of reaction.

2 Smaller particles of solid have a larger surface area than larger ones with the same total volume.

3 A catalyst speeds up the rate of a chemical reaction but is not used up itself.

8.4 Concentration and rate of reaction

LEARNING OUTCOMES

- Describe and explain how concentration affects rate of reaction
- Describe and explain the effect of concentration in terms of collisions between reacting particles (Supplement)

How concentration affects the rate of reaction

The Taj Mahal, a beautiful building in India, is being eroded away! Over recent years the concentration of acids in rainwater all over the world has been steadily increasing. The acids in the air react with the marble and damage the surface of the building. The higher the concentration of acid in the air, the quicker a building made of marble or limestone will react.

PRACTICAL

How changing the concentration of acid affects reaction rate

1. The rate of reaction of marble chips (calcium carbonate) was carried out with hydrochloric acid of different concentrations.

$$CaCO_3 + 2HCl \longrightarrow CaCl_2 + CO_2 + H_2O$$

2. We can follow the reaction by measuring the increase in volume of carbon dioxide given off or by measuring the decrease in mass of the reaction mixture. The temperature, mass and size of marble chips are kept the same, as well as the volume of hydrochloric acid used. The concentration of hydrochloric acid is varied but it is always in excess.

3. For each concentration of hydrochloric acid used, record the volume of CO_2 gas produced at time intervals.

A graph of the results is shown.

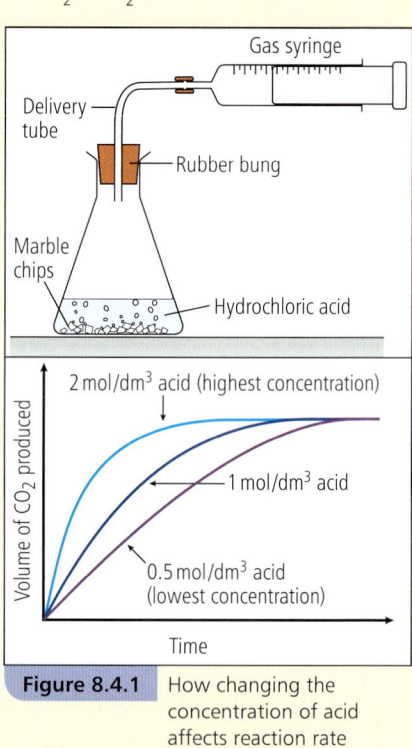

Figure 8.4.1 How changing the concentration of acid affects reaction rate

This limestone statue has been damaged by acid rain. The greater the concentration of acid in the rain, the greater the damage caused.

DID YOU KNOW?

Cars with steel bodies are likely to rust faster if you live near the sea because there is a higher concentration of salt in the atmosphere and in rainwater.

You can see that as the concentration of acid increases, the rate of reaction increases. The final volume of carbon dioxide released was the same in each experiment because hydrochloric acid was always in excess.

Using the collision theory (Supplement)

A concentrated solution has more particles of solute per unit volume than a dilute solution. In the reaction between calcium carbonate and hydrochloric acid, the important solute particles are the hydrogen ions in the hydrochloric acid (see Topic 10.4).

A reaction occurs when particles collide with enough energy. The more concentrated the hydrochloric acid, the more hydrogen ions there are in a given volume to collide and react with the carbonate particles in the calcium carbonate. The rate of reaction depends on the number of successful collisions per second. If there are more successful collisions per second, the rate of reaction is faster.

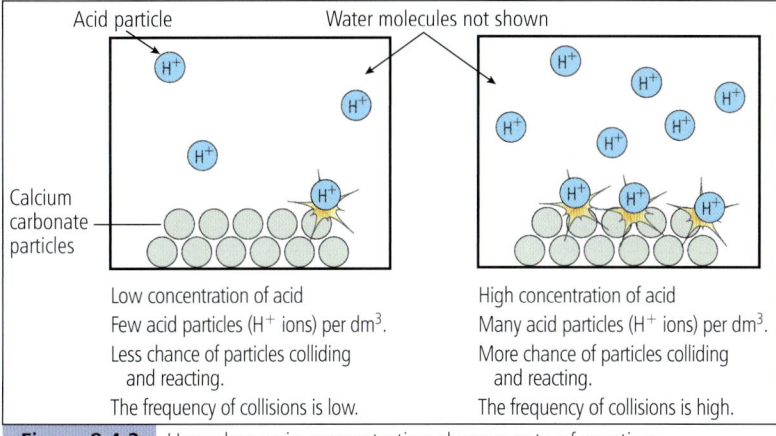

Figure 8.4.2 How change in concentration changes rate of reaction

> **STUDY TIP**
>
> When explaining the effect of concentration on reaction rate don't just refer to more collisions between the particles. It is the <u>more frequent</u> collision of the particles which is important.

In a reaction involving gases, increasing the pressure has a similar effect to increasing concentration in a liquid. Increasing the pressure pushes the particles closer together so that they collide with a greater frequency, so the rate of reaction increases.

SUMMARY QUESTIONS

1. Copy and complete using words from the list below:

   concentration   increases   rate

   The _____ of a chemical reaction _____ when the _____ of one or more of the reactants increases.

2. **a** When barium carbonate reacts with excess dilute hydrochloric acid, carbon dioxide gas is given off. Sketch a graph to show how the mass of the reaction mixture changes.

   b On the same set of axes, sketch the graph you would expect if you repeat the experiment using hydrochloric acid of half the concentration.

3. Copy and complete using words from the list below:

   collide   collision   concentrated   frequently   unit

   The effect of increasing reactant concentration on the rate of reaction can be explained by the _____ theory. When the reactants are more _____ there are more reactant particles present per _____ volume so they _____ more _____.

4. Use ideas from the collision theory to suggest why a reaction slows down as time goes on.

> **STUDY TIP**
>
> Remember that increasing the concentration of a reactant has no effect on the force with which the particles hit each other.

KEY POINTS

1. Increasing the concentration of reactants increases the rate of reaction.
2. Increasing the concentration of reactants increases the frequency of collision of the particles and so increases reaction rate.

8.5 Temperature and rate of reaction

LEARNING OUTCOMES

- Describe how temperature affects the rate of reaction
- Describe and explain how temperature affects the rate of reaction in terms of collision rate and number of molecules with sufficient activation energy

Supplement

STUDY TIP

When describing rates of reaction it is important to use words like 'faster' or 'slower' not just 'fast' or 'slow'.

DID YOU KNOW?

Food cooks more quickly at higher temperatures because the particles have more energy and there are more collisions per second between the molecules in the food.

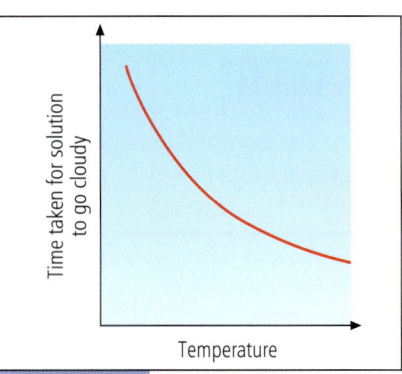

Figure 8.5.2 Results from the Practical in Figure 8.5.1

How temperature affects rate of reaction

We can change the rate of a chemical reaction by heating or cooling the reaction mixture. Food goes mouldy quicker when left out of the refrigerator. This is because the reactions which make food rot are faster at higher temperatures. Some surgical operations are carried out below normal room temperature to slow down the chemical reactions in the body.

PRACTICAL

Investigating the effect of temperature on reaction rate

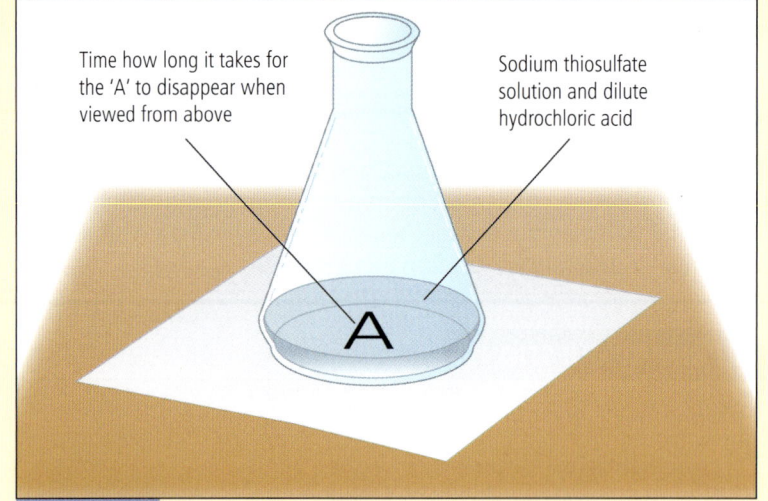

Figure 8.5.1 Reaction rate is affected by temperature

When we react sodium thiosulfate with hydrochloric acid a precipitate of sulfur is formed.

1. Place the flask of sodium thiosulfate on top of the letter A on the paper.
2. Add hydrochloric acid and record the time taken for the letter A to 'disappear'.
3. Repeat the experiment at different temperatures using the same volumes of acid and thiosulfate warmed up separately before mixing them together. Do not heat above 50 °C.

You can plot a graph of the time taken for the letter A to 'disappear' against temperature. The shorter the time taken for the letter A to 'disappear', the faster the reaction is. This is because rate of reaction is proportional to amount of product formed (or reactant used) divided by time. For every 10 °C rise in temperature, the rate of reaction approximately doubles.

Using the collision theory

There are two ways we can explain why increasing the temperature increases the rate of reaction. The second of these is the most important.

1. When we heat up a reaction mixture, the particles gain energy. When particles gain energy they move faster and collide more often. The frequency of collisions is increased. This results in an increased rate of reaction.
2. In order to react, particles must collide with a minimum amount of energy. This is called the activation energy. As the temperature gets higher, more and more particles have this minimum amount of energy to react when they collide. In other words, as the temperature is increased there is more chance of a collision between the reactant particles being successful. We say that the number of effective collisions in a given time increases as the temperature increases.

The faster you move, the more likely you are to bump into something – and the bump will be harder too!

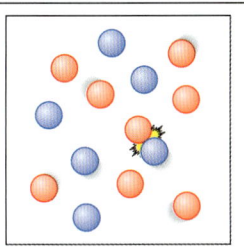

Lower temperature.
Particles have less energy.
They move more slowly and collide less frequently.
The collisions are not very effective.

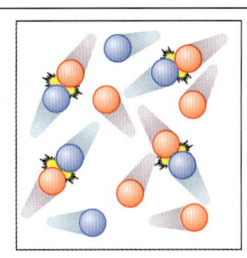

Higher temperature.
Particles have more energy.
They move faster and collide more frequently.
The collisions are very effective.

Figure 8.5.3 Increasing temperature increases reaction rate

STUDY TIP

Note that as the temperature increases, each particle collides with a greater force. It is also more accurate to say that 'there are more frequent collisions' than just 'more collisions'.

SUMMARY QUESTIONS

1 Copy and complete the sentence using this list:

   longer   rate   slower   time

   The _____ the _____ taken for a reaction to be complete, the _____ the _____ of reaction.

2 A student followed the rate of reaction of calcium carbonate with hydrochloric acid at different temperatures. At each temperature she recorded the volume of gas produced 30 seconds from the start of the reaction. Sketch a graph to show how the volume of gas changed with temperature.

3 Copy and complete the paragraph using this list:

   effective   energy   faster   increases   minimum   more

   When the temperature of a reaction mixture is increased the particles move _____ because the _____ of the particles increases. At a higher temperature there are also _____ particles with the _____ amount of energy to react when they collide. So the collisions are more _____ and the rate of reaction _____.

KEY POINTS

1 The higher the temperature the greater the rate of reaction.

2 The rate of reaction increases with an increase in temperature because:
   - more of the colliding molecules have energy equal to, or above, the activation energy.
   - there is an increase in collision rate.

8.6 Light-sensitive reactions

LEARNING OUTCOMES

- Describe and explain the role of light in photochemical reactions
- Describe the use of silver salts in photography as a process of the reduction of silver ions to silver atoms
- Describe the process of photosynthesis in simple terms

A few chemical reactions are started by ultraviolet or visible light. These are called **photochemical reactions**. We rely on the photochemical reactions in plants to provide the oxygen we breathe and the food we eat as well as for removing carbon dioxide from the atmosphere. The photochemical reaction in plants is called **photosynthesis**.

Not all photochemical reactions are useful. Compounds called chlorofluorocarbons (CFCs) are broken down by ultraviolet light and the products are responsible for the depletion of the ozone layer around the Earth.

Photosynthesis

Plants use the energy from sunlight to make glucose. This process is called photosynthesis.

$$6CO_2 + 6H_2O \rightarrow C_6H_{12}O_6 + 6O_2$$

carbon dioxide + water → glucose + oxygen

The glucose is turned into macromolecules called starch and cellulose (see Topic 20.2).

Photosynthesis is catalysed by the green pigments in plants called **chlorophylls**.

STUDY TIP

It is important to realise that light affects only a few reactions. The only ones you should know about are photosynthesis, the conversion of silver bromide to silver and the reaction of alkanes with chlorine (see Topic 18.1).

PRACTICAL

How light affects the rate of photosynthesis

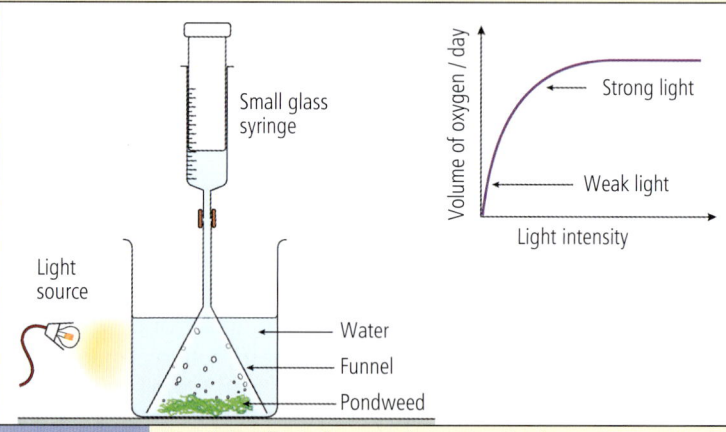

Figure 8.6.1 How light affects the rate of photosynthesis

When light shines on the pondweed, bubbles of oxygen slowly form on the leaves and then rise into the syringe. You can record the volume of oxygen collected over several days. Then you can repeat the experiment using stronger (more intense) light. The graph shows how the rate of photosynthesis changes with light intensity. The graph levels off because carbon dioxide becomes limiting.

The plant on the right looks less healthy than the plant on the left because it has not had enough light

Photography

The surface of black and white photographic film contains tiny crystals of silver bromide mixed with gelatine. When light shines on the film the silver bromide is 'activated'. Some of the silver bromide decomposes (breaks down) to form silver.

2AgBr	$\longrightarrow$	2Ag	+	Br_2
silver bromide (colourless in gelatine)		silver (black)		

The silver appears black in colour because the particles are very small.

In this reaction the silver ions in the silver bromide accept electrons from the bromide ions and become silver atoms. This is a **redox reaction** (see Topic 9.3). The silver ions are reduced because they accept electrons and the bromide ions are oxidised because they lose electrons (see Topic 9.3).

$2Ag^+ + 2e^- \longrightarrow 2Ag$ (a **reduction** reaction)

$2Br^- \longrightarrow Br_2 + 2e^-$ (an **oxidation** reaction)

The parts of the film exposed to stronger light appear black and the parts not exposed appear white. The greater the intensity (strength) of the light, the faster the reaction. A positive print is made by shining light through the negative onto a piece of photographic paper.

DID YOU KNOW?
Some spectacle lenses darken in bright light and become more transparent in dim light. This is because light reacts with the silver chloride in the lenses.

SUMMARY QUESTIONS

1 Match the words to the phrases:

photosynthesis…	…the catalyst in photosynthesis
chlorophyll…	…a reaction where electrons are lost
reduction…	…a reaction where electrons are gained
oxidation…	…a reaction that is started by light
photochemical reaction…	…a reaction where plants change carbon dioxide and water into glucose and oxygen

2 Copy and complete using the words below:

black   bromide   crystals   light   particles   silver

Photographic film contains tiny _____ of silver _____. When _____ shines on the film the _____ bromide breaks down to _____ of silver which are _____ in colour.

3 How does the rate of decomposition of silver bromide vary with the strength (intensity) of light?

PRACTICAL

A light-sensitive reaction

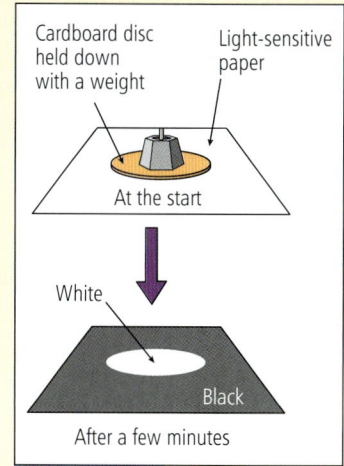

Figure 8.6.2 A light-sensitive reaction

Place a cardboard disc on a piece of light-sensitive photographic paper in the dark. (Or you can also use a piece of filter paper soaked first in silver nitrate then with potassium bromide.) The paper and disc are then left in the light. After a short time the paper will turn black but if you remove the disc you can still see a circle of white paper. This shows that light is needed for the reaction to happen.

KEY POINTS

1 The rate of some reactions is increased by increasing the intensity of light. These reactions are photochemical reactions.

2 Photosynthesis is a process by which plants change carbon dioxide and water into glucose and oxygen using chlorophyll as a catalyst.

3 In the photographic reaction silver ions are reduced to silver atoms.

SUMMARY QUESTIONS

1. Match the type of reaction on the left with the appropriate method for measuring reaction rate on the right:

a purple solution changes to a colourless solution	measure the volume of gas produced
hydrogen is released during a reaction	measure change in electrical conductivity
two solutions react slowly to form a precipitate	measure the light transmitted through a solution
there are more ions in solution in the reactants than in the products	see how long it takes for a letter 'A' under a flask to disappear

2. Write definitions of the following:
   (a) rate of reaction
   (b) catalyst
   (c) gradient

3. Sketch a graph to show how the total mass changes in the reaction:
   $CaCO_3(s) \rightarrow CaO(s) + CO_2(g)$

4. Complete the following phrases:
   (a) Increasing the surface area of a solid ____ the rate of reaction.
   (b) As a reaction proceeds the ____ of reaction ____.

5. What effect does each of the following have on the rate of reaction:
   (a) Diluting the reaction mixture
   (b) Using large lumps of solid rather than small ones.

6. Use your knowledge of the kinetic particle theory to describe and explain how increasing the pressure affects the rate of the following reaction:
   $2N_2O(g) \rightarrow 2N_2(g) + O_2(g)$

7. Write definitions of the following:
   (a) limiting reactant
   (b) photochemical reaction
   (c) photosynthesis.

8. Use the kinetic particle theory to suggest why food cooks more quickly when the temperature is higher.

PRACTICE QUESTIONS

1. Which one of the following statements about catalysts is true?
   A They are always nonmetals.
   B They do not take part in chemical reactions.
   C They have no effect on rate of reaction.
   D Their mass remains unchanged at the end of the reaction.

   (Paper 1)

2. The table shows the volume of oxygen given off when hydrogen peroxide decomposes at 40 °C in the presence of a catalyst.

Time/s	0	5	10	20	30	40	50	60
Volume of oxygen/cm³	0	22	34	48	56	59	60	60

   a) Plot a graph of the results with time on the x axis and volume of oxygen on the y axis. [3]
   b) On the same axes sketch a curve for the same reaction but carried out at a temperature of 50 °C. [2]
   c) What results would you expect if the catalyst was not present? Explain your answer. [2]

   (Paper 3)

3. A student compared how well different compounds catalysed the reaction between zinc and hydrochloric acid. The results are shown in the table.

Compound	Time taken for all the zinc to react/s
No catalyst	500
Copper(II)sulfate	150
Copper(II)chloride	175
Manganese(IV) oxide	390
Sodium chloride	500
Sodium sulfate	500

   a) What do you understand by the term *catalyst*? [1]
   b) Which is the best catalyst for this reaction? [1]
   c) What things must you keep constant in these experiments if it is to be a fair test? [3]
   d) Which compounds in the table are not catalysts? [1]

   (Paper 3)

4 The graph shows how the volume of carbon dioxide given off changes when hydrochloric acid of three different concentrations reacts with large pieces of calcium carbonate.

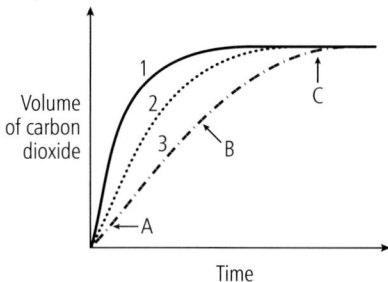

(a) Which line (1, 2 or 3) is the graph for the most concentrated acid? [1]

(b) In line 3, where is the rate of reaction fastest – at A, B or C? [1]

(c) Draw a diagram of the apparatus you can use to obtain these results. [3]

(d) Suggest two ways of increasing the rate of this reaction other than by increasing the concentration of acid. [2]

(Paper 3)

5 Magnesium ribbon reacts with hydrochloric acid. The equation is:

$$Mg + 2HCl \longrightarrow MgCl_2 + H_2$$

a) Suggest two methods that you could use to follow the progress of this reaction. [4]

b) For one of these methods describe how you can calculate the rate of reaction. [3]

c) Using the kinetic particle theory, explain how and why the rate of reaction changes when:

   (i) the concentration of hydrochloric acid is increased [3]

   (ii) the temperature of the reaction mixture is lowered [3]

   (iii) powdered magnesium is used rather than magnesium ribbon. [3]

(Paper 4)

6 A student investigated the reaction between 40 g calcium carbonate and 30 cm³ of 2.0 mol/dm³ hydrochloric acid.

$$CaCO_3 + 2HCl \longrightarrow CaCl_2 + CO_2 + H_2O$$

a) Draw a sketch graph to show how the loss in mass of the reaction mixture changes with time. [2]

b) By referring to the shape of your graph you have drawn in part a), explain how the rate of reaction decreases with time. [2]

c) Use the idea about colliding particles to explain why the rate of reaction decreases with time. [3]

d) **(i)** Calculate the number of moles of calcium carbonate and hydrochloric acid at the start of the reaction. [2]

   (ii) Was the hydrochloric acid or the calcium carbonate in excess? Explain your answer. [3]

e) In another experiment the student investigated how the rate of this reaction varies with temperature. The results are shown in the table.

Temperature / °C	20	30	40	50
Time taken for a piece of calcium carbonate to dissolve / s	64	32	16	8

   (i) Draw a suitable graph to display these results. [3]

   (ii) From your graph, predict how many seconds it would take the piece of calcium carbonate to dissolve at 60 °C. [1]

(Paper 4)

7 Crystals of silver bromide were brushed onto a sheet of damp filter paper. A cardboard letter 'H' was placed on top of the paper.

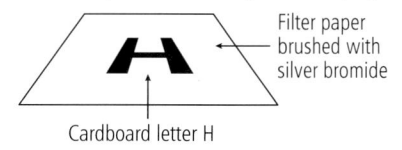

The filter paper and cardboard 'H' were left in the light for some time. After a few hours the filter paper around the 'H' had turned black but underneath the 'H' it was still white.

a) Explain these results. [4]

b) How can you make the filter paper go black more quickly? [1]

c) Write an ionic equation to show what is happening to the silver ions when the light strikes the silver bromide on the paper. [1]

(Paper 4)

9 Chemical reactions

9.1 Reversible reactions

LEARNING OUTCOMES

- Understand that some chemical reactions can be reversed by changing the conditions
- **Supplement** Demonstrate knowledge and understanding of the concept of equilibrium

Why do the bubbles in a bottle of fizzy drink suddenly appear when you take the top off the bottle? Carbon dioxide gas is dissolved in the drink under pressure. When you take the top off the bottle the pressure is released so the carbon dioxide bubbles out of solution. When you put the top back on, the pressure builds up again and the bubbling stops. Reactions like this, which can go in either direction, are called **reversible reactions**.

$$CO_2(g) \underset{\text{decrease in pressure}}{\overset{\text{increase in pressure}}{\rightleftharpoons}} CO_2(aq)$$

The sign $\rightleftharpoons$ is the symbol for a reversible change. It is used for reactions which we call **equilibrium** reactions. Industrially important equilibrium reactions include the Haber process (see Topic 16.2) and the Contact process (see Topic 16.4).

Heating hydrated salts

When you heat crystals of blue copper(II) sulfate they break down or decompose to form a white powder.

$$CuSO_4 \cdot 5H_2O(s) \xrightarrow{\text{heat}} CuSO_4(s) + 5H_2O(l)$$

hydrated copper(II) sulfate (blue) → anhydrous copper(II) sulfate (white)

Blue copper(II) sulfate crystals turn white when heated

The blue copper(II) sulfate crystals have water as part of their structure. We say that the copper(II) sulfate is **hydrated**. We call the water in the salt, **water of crystallisation**. When the blue crystals are heated the water of crystallisation is lost. We are left with white anhydrous copper(II) sulfate. **Anhydrous** means 'without water'. We can reverse this reaction by adding water back to the white copper(II) sulfate.

$$CuSO_4(s) + 5H_2O(l) \longrightarrow CuSO_4 \cdot 5H_2O(s)$$

Because this reaction is reversible, we can write both reactions in the same equation.

$$CuSO_4 \cdot 5H_2O(s) \underset{\text{backward or reverse reaction}}{\overset{\text{forward reaction}}{\rightleftharpoons}} CuSO_4(s) + 5H_2O(l)$$

STUDY TIP

Make sure that you understand the terms <u>hydrated</u>, <u>anhydrous</u> and <u>water of crystallisation</u>.

DID YOU KNOW?

Cobalt chloride can be used as a humidity indicator because its colour changes according to the amount of water linked to the cobalt ion.

Equilibrium

Supplement

When we change blue copper(II) sulfate to white copper(II) sulfate the two reactions are separate. We don't heat and add water at the same time. In some reactions, however, both the forward and reverse reactions are going on at the same time. We call these equilibrium reactions. Equilibrium reactions have particular features:

- The reactants or products must not escape from the reaction mixture. We call this a closed system.

116

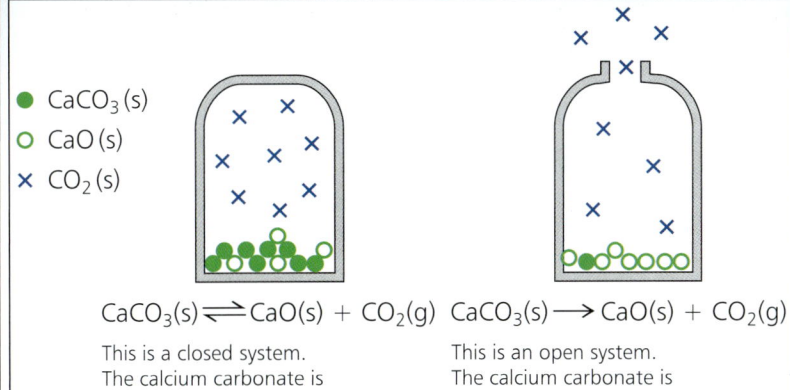

- CaCO₃ (s)
- CaO (s)
- × CO₂ (s)

$CaCO_3(s) \rightleftharpoons CaO(s) + CO_2(g)$
This is a closed system. The calcium carbonate is decomposing to calcium oxide but the carbon dioxide is not lost.

$CaCO_3(s) \rightarrow CaO(s) + CO_2(g)$
This is an open system. The calcium carbonate is decomposing to calcium oxide and the carbon dioxide is lost.

Figure 9.1.1 Comparison of a closed system and an open system

- At equilibrium the reactants are continually being changed to products and the products are being changed back to reactants. We say that this is a dynamic equilibrium.
- At equilibrium the concentration of the reactants and products does not change. This is because the rate of the forward reaction is the same as the rate of the reverse reaction.
- The equilibrium can be approached from either direction. We can start with only the reactants or only the products. Whichever we start with, we end up with fixed concentrations of reactants and products in the equilibrium mixture. For example, when we heat hydrogen and iodine in a sealed tube:

$$H_2(g) + I_2(g) \rightleftharpoons 2HI(g)$$
hydrogen iodine   hydrogen iodide

we get the same fixed concentration of hydrogen, iodine and hydrogen iodide at equilibrium whether we start from hydrogen and iodine or from hydrogen iodide.

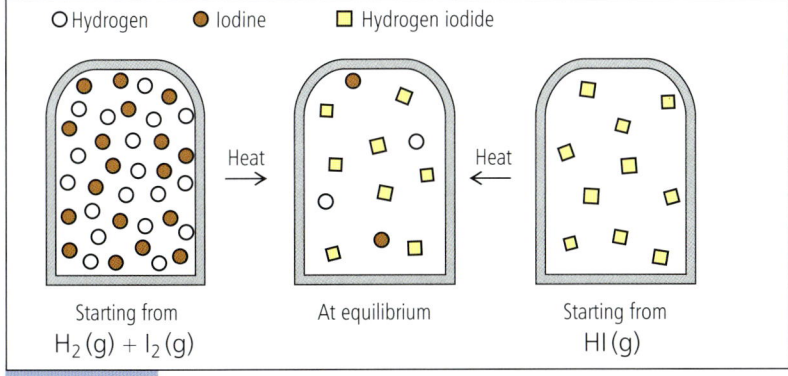

○ Hydrogen   ● Iodine   □ Hydrogen iodide

Starting from $H_2(g) + I_2(g)$ — Heat → At equilibrium ← Heat — Starting from HI(g)

Figure 9.1.2 The concentrations of hydrogen, iodine and hydrogen iodide are constant whether we start from reactants or products

The position of equilibrium tells us how far the reaction goes in favour of reactants or products. If the concentration of products is greater than the concentration of the reactants, we say that the position of equilibrium is to the right – it favours the products. If the concentration of reactants is greater than the concentration of the products, we say that the position of equilibrium is to the left – it favours the reactants.

STUDY TIP

It is important to realise that at equilibrium the rate of the forward reaction is equal to the rate of the reverse reaction.

KEY POINTS

1. In a reversible reaction, the products can react to form the original reactants again.
2. An equilibrium reaction can take place only in a closed system.
3. At equilibrium the concentrations of reactants and products remain fixed.
4. At equilibrium, the forward reaction happens at the same rate as the reverse reaction.

SUMMARY QUESTIONS

1. Copy and complete using the words below:

   anhydrous   blue   crystallisation   hydrated   reversible   white

   When we heat _____ copper(II) sulfate it loses its water of _____. Its colour changes from _____ to _____. When we add water to _____ copper(II) sulfate it turns blue. This is a _____ reaction.

2. State three features of an equilibrium reaction.

3. What do you understand by the terms:
   a dynamic equilibrium
   b closed system?

9.2 Shifting the equilibrium

LEARNING OUTCOMES

- Predict how changing the concentration of reactants or products alters the equilibrium reaction
- Predict how changing the pressure alters the position of equilibrium
- Predict how temperature alters the position of equilibrium

Changing the concentration of reactants or products or changing the temperature or pressure has an effect on the equilibrium reaction. The reaction tries to oppose the changes that you make. Catalysts do not affect the position of equilibrium – they just speed up the forward and reverse reactions equally.

Changing the concentration

When the concentration of a reactant is increased, the equilibrium moves to the right. Adding more reactants unbalances the equilibrium. So the equilibrium moves to the right to form more products until the equilibrium is restored. In this way it keeps the relative concentrations of products and reactants the same as before.

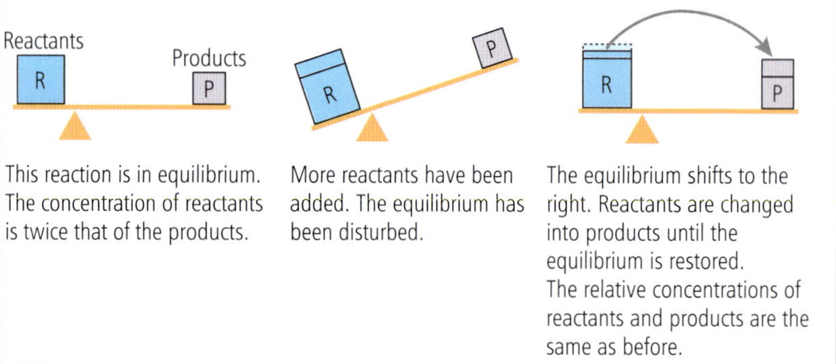

Equilibrium is all about balance

This reaction is in equilibrium. The concentration of reactants is twice that of the products.

More reactants have been added. The equilibrium has been disturbed.

The equilibrium shifts to the right. Reactants are changed into products until the equilibrium is restored. The relative concentrations of reactants and products are the same as before.

Figure 9.2.1 Equilibrium is a balancing act!

When the concentration of a product is increased, the equilibrium moves to the left. Products are changed to reactants until equilibrium is restored.

DEMONSTRATION

Changing the direction of a reaction

You pass chlorine gas through a tube containing the brown liquid iodine monochloride. As more chlorine is passed through, the brown liquid turns into yellow crystals. When you tip out the chlorine the yellow crystals turn back into the brown liquid.

By changing the concentration of chlorine we have made the reaction go backwards or forwards.

$$Cl_2(g) + ICl(l) \rightleftharpoons ICl_3(s)$$
$$\text{iodine monochloride (brown liquid)} \quad \text{iodine trichloride (yellow crystals)}$$

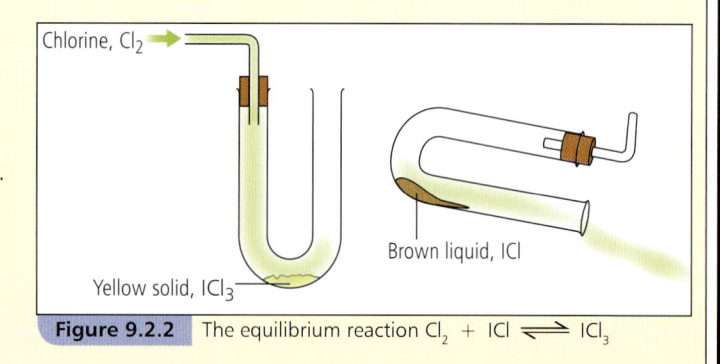

Figure 9.2.2 The equilibrium reaction $Cl_2 + ICl \rightleftharpoons ICl_3$

A high concentration of chlorine pushes the reaction from left to right. The forward reaction is favoured. Removing the chlorine favours the reverse reaction.

Changing the pressure

Pressure can only affect reactions where there is a gas in the equation. Increasing the pressure moves the equilibrium to the side with the smaller volume of gas. So in the reaction:

$$2SO_2(g) + O_2(g) \rightleftharpoons 2SO_3(g)$$

there are three volumes or three moles of gas on the left and only two on the right. So increase in pressure moves the reaction to the right. More SO_2 and O_2 combine to form SO_3. This happens because increasing the pressure squashes the molecules closer together – you are increasing their concentration. The reaction mixture tries to overcome this by moving the equilibrium to the right so that the overall number of molecules is reduced. Decreasing the pressure has the opposite effect. It pushes the reaction to the left. If there are equal volumes or moles of gas on both sides of the equation, increasing the pressure has no effect.

Changing the temperature

If a reaction is exothermic (giving out heat) in the forward reaction, it will be endothermic (taking in heat) in the reverse reaction. When we produce ammonia from nitrogen and hydrogen the forward reaction is exothermic, so the reverse reaction is endothermic.

$$N_2(g) + 3H_2(g) \underset{\text{endothermic}}{\overset{\text{exothermic}}{\rightleftharpoons}} 2NH_3(g)$$

The table shows the effect of temperature on the percentage yield of ammonia at 300 atmospheres pressure.

temperature / °C	350	400	450	500
percentage yield of ammonia	65	49	37	20

You can see that the yield of ammonia gets lower as the temperature increases. So for an exothermic reaction when temperature increases the equilibrium shifts in favour of the reverse reaction. It favours the endothermic change where heat is taken in.

In an endothermic reaction the opposite happens. Increasing the temperature moves the reaction to the right to favour the products. It favours the exothermic change where heat is given out.

KEY POINTS

In an equilibrium reaction:

1 Increasing the concentration of a reactant moves the reaction in the direction of the products until the equilibrium balance is restored.
2 Increasing the pressure moves the reaction in the direction of the lower number of gas molecules.
3 For an endothermic reaction, increasing the temperature moves the reaction to the right in the direction of the products. For an exothermic reaction increasing the temperature moves the reaction to the left.

STUDY TIP

Remember that if the equilibrium conditions are changed the reaction always tries to act in the opposite direction.

DID YOU KNOW?

The bonding of oxygen to the pigment in red blood cells is an equilibrium reaction. The position of the equilibrium depends on the concentration of oxygen in the blood.

SUMMARY QUESTIONS

1 Copy and complete using the words below:

balanced concentration
disturbed equilibrium
products right

In an _____ reaction, when you increase the _____ of a reactant, the reaction moves to the _____. The equilibrium is _____ so more of the reactants are changed to _____. This continues until the concentrations of reactants and products are _____ correctly.

2 Describe how each of the following changes affects this equilibrium reaction:

$$CO(g) + 2H_2(g) \rightleftharpoons CH_3OH(g)$$

a increasing the pressure
b increasing the concentration of $CO(g)$
c decreasing the concentration of $CH_3OH(g)$.

9.3 Redox reactions

LEARNING OUTCOMES

- Define oxidation and reduction in terms of oxygen loss or gain
- Understand the significance of oxidation states
- Define oxidising agent and reducing agent

When we burn magnesium in oxygen, magnesium oxide is formed:

$$2Mg(s) + O_2(g) \longrightarrow 2MgO(s)$$

We say that magnesium has been oxidised – it has gained oxygen. **Oxidation** is the gain of oxygen. When we remove oxygen from a compound we say that the compound has been reduced. **Reduction** is the loss of oxygen.

DEMONSTRATION

The reduction of copper(II) oxide

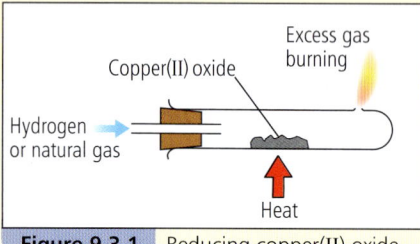

Figure 9.3.1 Reducing copper(II) oxide

We pass hydrogen or natural gas over heated copper(II) oxide. The black copper(II) oxide changes to pink copper. We have removed the oxygen from the copper(II) oxide. So the copper(II) oxide has been reduced by the hydrogen.

When you look at the reaction between copper(II) oxide and hydrogen, you can see that both oxidation and reduction have taken place at the same time.

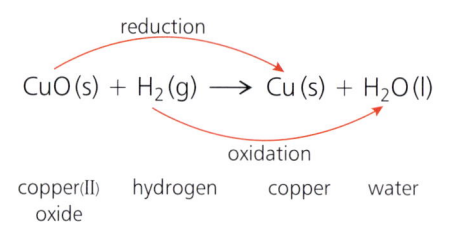

$$\text{CuO(s)} + \text{H}_2\text{(g)} \longrightarrow \text{Cu(s)} + \text{H}_2\text{O(l)}$$

(reduction / oxidation)

copper(II) oxide · hydrogen · copper · water

The copper(II) oxide has been reduced because it has lost oxygen. The hydrogen has been oxidised because it has gained oxygen. Because the reduction and oxidation have taken place together, we call this a **redox reaction**.

Redox reactions are involved when the fuel in these rockets burn

DID YOU KNOW?

The energy used to power some of the rocket boosters in the space shuttle is provided by the redox reaction between ammonium chlorate(VII) and aluminium powder.

STUDY TIP

Remember OIL RIG – Oxidation Is Loss (of electrons), Reduction Is Gain (of electrons)

The hydrogen is the **reducing agent** because it has removed the oxygen from the copper(II) oxide. It has become oxidised. The copper(II) oxide is the **oxidising agent** because it gives its oxygen to hydrogen to form water. It has been reduced.

An oxidising agent is a substance which oxidises another substance in a redox reaction. A reducing agent is a substance which reduces another substance in a redox reaction.

Where hydrogen takes part in a reaction we can also define reduction as addition of hydrogen to a compound and oxidation as removal of hydrogen from a compound. This is often used in organic reactions, for example:

$$C_2H_4 + H_2 \longrightarrow C_2H_6$$
$$\text{ethene} \quad \text{hydrogen} \quad \text{ethane}$$

In this reaction we say that the hydrogen has reduced ethene to ethane.

Oxidation states

We often put a Roman numeral after the name of an element in a compound. This number is called the **oxidation state**. It refers to a particular element in the compound. For example:

iron(II) chloride   iron(III) chloride   copper(I) oxide   copper(II) oxide

You can see that there are two types of iron chloride. So we have to find a way of telling these apart. Oxidation states help us to do this. Oxidation states tell us about:

- the type of ion present in the compound. For example, copper(I) oxide has Cu^+ ions and copper(II) oxide has Cu^{2+} ions
- how oxidised an element in a compound is. For example, the formula for manganese(II) oxide is MnO but the formula for manganese(IV) oxide is MnO_2. This shows us that manganese in manganese(IV) oxide is more oxidised than in manganese(II) oxide – it has more oxygen.

Oxidation states can also be used for non-metallic elements. For example: $KClO_3$ is potassium chlorate(V) and $KClO_4$ is potassium chlorate(VII). The chlorine in potassium chlorate(VII) is more oxidised because it is associated with more oxygen.

Other useful examples are copper(II) sulfate, potassium manganate(VII) and potassium dichromate(VI).

Electron transfer in redox reactions

We saw in Topic 6.3 that we can write half-equations for electrode reactions. We do this by balancing the equations by adding one or more electrons to either side of the equation. We can now give a wider definition of oxidation and reduction that does not involve oxygen or hydrogen:

Oxidation is loss of electrons. Reduction is gain of electrons. So when we electrolyse molten sodium chloride we can split up the overall equation into two parts to show what happens to each ion:

overall equation: $2NaCl(l) \longrightarrow 2Na(l) + Cl_2(g)$

sodium: $2Na^+ + 2e^- \longrightarrow 2Na$

The sodium ions have been reduced because they have gained electrons.

chloride: $2Cl^- \longrightarrow Cl_2 + 2e^-$

The chloride ions have been oxidised because they have lost electrons.

KEY POINTS

1. Oxidation is the gain of oxygen by a substance. Reduction is the loss of oxygen from a substance.
2. Redox reactions involve both oxidation and reduction at the same time.
3. Oxidising agents oxidise another substance. Reducing agents reduce another substance.
4. Oxidation is loss of electrons. Reduction is gain of electrons.

SUMMARY QUESTIONS

1. Copy and complete using the words below:

   agent   oxidised   reduced sulfur

   When sulfur burns in oxygen _____ dioxide is formed. The sulfur is _____ because it gains oxygen. When copper(II) oxide reacts with hydrogen, the copper(II) oxide is _____ to copper. Hydrogen is the reducing _____ in this reaction.

2. Identify the reducing agent in each of the following equations:

   a $2Na + O_2 \longrightarrow Na_2O_2$
   b $PbO + C \longrightarrow Pb + CO$
   c $CuO + CO \longrightarrow Cu + CO_2$

3. The equation for the electrolysis of molten calcium chloride is:

   $CaCl_2(l) \longrightarrow Ca(l) + Cl_2(g)$

   Write two half-equations for this reaction. State which shows an oxidation and which shows a reduction.

9.4 More about redox reactions

LEARNING OUTCOMES

- Understand the use of oxidation number changes in redox reactions
- Define redox in terms of electron transfer
- Identify redox reactions by changes in oxidation state and by the colour changes using potassium manganate(VII) and potassium iodide

These sunglasses get darker and lighter because of redox reactions

STUDY TIP

Make sure that you don't confuse oxidising agents and reducing agents with the compounds that are being oxidised or reduced. An oxidising agent oxidises another substance. When it does this the oxidising agent itself is reduced.

Analysing redox reactions

We can identify many redox reactions by the fact that we can write half-equations for them.

For example, when zinc reacts with copper(II) sulfate the equation is:

$$Zn(s) + CuSO_4(aq) \longrightarrow ZnSO_4(aq) + Cu(s)$$

We can divide this into two half-equations to show what happens to each element:

$$Zn(s) \longrightarrow Zn^{2+}(aq) + 2e^-$$

The zinc atoms have lost electrons. They have been oxidised to zinc ions:

$$Cu^{2+}(aq) + 2e^- \longrightarrow Cu$$

The copper ions have gained electrons. They have been reduced to copper atoms. The zinc has been oxidised by copper ions. So a copper ion acts as an oxidising agent. The copper ions have been reduced by zinc. So zinc is acting as a reducing agent.

More about oxidation states

We can use oxidation states to follow what happens in a redox reaction. Oxidation states are sometimes called **oxidation numbers**. We use a set of rules to find the oxidation state of an element in a compound:

- An element that is uncombined has an oxidation state of zero. For example, $Zn = 0$, $S = 0$.
- For ionic compounds, the oxidation state is the same as the charge on the ion. So sodium ion $= +1$, aluminium ion $= +3$, oxide $= -2$, chloride $= -1$.
- The total oxidation state of all the elements in a compound is zero. For example, in $MgCl_2$:

  oxidation number of Mg $= +2$

  oxidation number of $2Cl^- = 2 \times -1 = -2$.

- For elements that form more than one ion, the oxidation state can be worked out from the formula of the oxide.

Elements in covalent compounds can also be given oxidation states but the rules are more complicated.

We can use changes in oxidation state to define oxidation and reduction:

An increase in oxidation state of an element is oxidation. A decrease in oxidation state of an element is reduction.

Consider this reaction:

$$Zn + Cu^{2+} \longrightarrow Zn^{2+} + Cu$$

oxidation states: $0 \quad +2 \quad\quad +2 \quad 0$

reduction: $Cu^{2+} \rightarrow Cu$

oxidation: $Zn \rightarrow Zn^{2+}$

122

Zinc has increased its oxidation state from 0 to +2 – it is oxidised to zinc ions. Copper has decreased its oxidation state from +2 to 0 – copper(II) ions are reduced to copper. You can also see that a reducing agent decreases the oxidation state of another compound – zinc reduces copper(II) ions to copper. An oxidising agent increases the oxidation state of another compound – copper(II) ions oxidise zinc to zinc ions.

Colour changes in redox reactions

We can use particular reagents as oxidising or reducing agents. Potassium manganate(VII) in acidic solution is a good oxidising agent. When it oxidises a substance its colour changes from purple to colourless. This can be used as a test for reductants (reducing agents).

> **DID YOU KNOW?**
>
> Potassium manganate(VII) and a chemical called a polyol are included in survival kits. When these are mixed there is a redox reaction which can be used to start a fire.

PRACTICAL

Reacting iron(II) ions with potassium manganate(VII)

You add a solution of potassium manganate(VII) drop by drop to the solution containing iron(II) ions. The solution changes colour from light green to yellow as the iron(II) ions are oxidised by the potassium manganate(VII). A yellow solution containing iron(III) ions has been formed. The purple potassium manganate(VII) becomes colourless when it reacts. So you only see the colour of the iron(III) ions. The reaction is complete when you see a purple colour in the beaker.

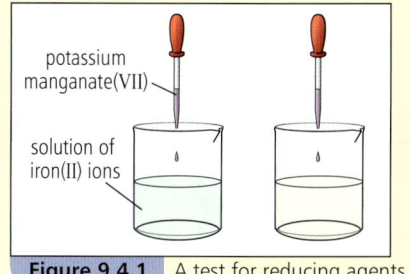

Figure 9.4.1 A test for reducing agents using potassium manganate(VII)

Potassium iodide in acidic solution of potassium manganate (VII) is a good reducing agent. When it reduces a substance its colour changes from colourless to brown. This can be used as a test for oxidising agents. For example, it reduces hydrogen peroxide to water:

$$2I^- + 2H^+ + H_2O_2 \longrightarrow 2H_2O + I_2$$

(potassium iodide) (acid) (hydrogen peroxide)   (iodine)
(colourless) (colourless) (colourless)   (brown)

SUMMARY QUESTIONS

1. Copy and complete using the words below:

   decrease   increases   oxidation   redox   state

   Changes in oxidation _____ can be used to follow what happens in a _____ reaction. Oxidation occurs when an element _____ its _____ state. Reduction can be defined as a _____ in oxidation state.

2. What is the oxidation state of:
   a. aluminium in $AlCl_3$
   b. K in KCl
   c. Br in NaBr
   d. O in MgO?

3. Excess potassium iodide solution is added to a solution of potassium manganate(VII). State the colour change you would observe in this reaction.

KEY POINTS

1. In a redox reaction involving ions, two half-equations can be written, one showing oxidation and the other showing reduction.

2. An increase in oxidation state of an element is oxidation. A decrease in oxidation state is reduction.

3. The colour changes of potassium manganate(VII) and potassium iodide can be used to test for oxidising and reducing agents.

SUMMARY QUESTIONS

1 Match the words on the left with the phrases on the right.

hydrated	a reaction that can go in the forward or backward direction
reversible reaction	addition of oxygen or removal of hydrogen
oxidation	without water
anhydrous	removal of oxygen or addition of hydrogen
reduction	with water added

2 State whether the underlined elements or compounds are oxidised or reduced when they react.
   (a) $\underline{PbO} + C \longrightarrow Pb + CO$
   (b) $\underline{S} + O_2 \longrightarrow SO_2$
   (c) $CuO + \underline{H_2} \longrightarrow Cu + H_2O$
   (d) $\underline{CO_2} + C \longrightarrow 2CO$

Supplement

3 (a) State the name of one oxidant (oxidising agent) and state its colour change when it reacts.
   (b) State the name of a reductant (reducing agent) and state its colour change when it reacts.

4 State three characteristics of an equilibrium reaction.

5 How do each of the following affect this endothermic reaction?

$$CaCO_3 \rightleftharpoons CaO + CO_2$$

   (a) Increasing the concentration of carbon dioxide
   (b) Decreasing the temperature
   (c) Allowing the carbon dioxide to escape
   (d) Increasing the pressure

6 Give three different definitions of oxidation.

7 For each of the following reactions, state whether oxidation or reduction is taking place.
   (a) $Cu^{2+} + 2e^- \longrightarrow Cu$
   (b) $Fe^{2+} \longrightarrow Fe^{3+} + e^-$
   (c) $2H^+ + 2e^- \longrightarrow H_2$

PRACTICE QUESTIONS

1 Blue copper(II) sulfate has the formula $CuSO_4.5H_2O$. Which phrase best describes blue copper(II) sulfate?
   A Hydrated salt
   B Anhydrous salt
   C Crystalline element
   D Dehydrated salt

(Paper 1)

2 Which word best describes the reaction:

$$N_2(g) + 3H_2(g) \rightleftharpoons 2NH_3(g)$$

   A Hydrate
   B Reversible
   C Irreversible
   D Complete

(Paper 1)

3 Which one of the following statements about this reaction is correct?

$$N_2(g) + 3H_2(g) \rightleftharpoons 2NH_3(g) \quad \Delta H -92.4 \text{ kj/mol}$$

   A Increasing the temperature shifts the equilibrium to the left.
   B Removing the ammonia shifts the equilibrium to the left.
   C Decreasing the pressure shifts the equilibrium to the right.
   D Decreasing the concentration of hydrogen shifts the equilibrium to the right.

(Paper 2)

4 When blue copper(II) sulfate is heated it turns white.

$$CuSO_4.5H_2O \longrightarrow CuSO_4 + 5H_2O$$

8 Which is the oxidant (oxidising agent) in each of the following equations?
   (a) $Cl_2 + 2KBr \longrightarrow 2KCl + Br_2$
   (b) $Zn + CuSO_4 \longrightarrow ZnSO_4 + Cu$
   (c) $5Fe^{2+} + MnO_4^- + 8H^+ \longrightarrow 5Fe^{3+} + Mn^{2+} + 4H_2O$

(a) (i) How can you reverse this reaction? *[1]*
   (ii) Write an equation for the reverse reaction. *[1]*
(b) Complete the following sentences using words from the list.

   alcohol   anhydrous   blue   crystallisation   hydrated   water   white

   When _____ copper(II) sulfate is heated it loses its water of _____ and turns _____. White copper(II) sulfate is called _____ copper(II) sulfate. *[4]*
(c) Copper(II) sulfate is reduced by zinc.
   (i) Copy and complete the equation for this reaction.
   $$CuSO_4 + Zn \longrightarrow \_\_\_\_\_ + ZnSO_4$$ *[1]*
   (ii) What do you understand by the term *reduction*? *[1]*
   (iii) What does the symbol (II) in copper(II) sulfate show? *[1]*

(Paper 3)

5 A mixture of hydrogen and iodine is put into a closed tube and heated at 400 °C.
$$H_2(g) + I_2(g) \rightleftharpoons 2HI(g)$$
(a) At first the rate of the forward reaction decreases with time. Suggest a reason for this. *[2]*
(b) Why does the rate of the backward reaction increase with time? *[1]*
(c) After a time equilibrium is reached. How do the rates of the forward and the backward reaction compare at equilibrium? *[1]*
(d) What effect would increasing the concentration of iodine have on this equilibrium? Explain your answer. *[2]*
(e) Increasing the pressure does not have any affect on this equilibrium. Explain why not. *[1]*
(f) The reaction is exothermic. Predict the effect of increasing the temperature on this equilibrium. Explain your answer. *[2]*

(Paper 4)

6 Aqueous chlorine reacts with an aqueous solution of potassium iodide:
$$Cl_2(aq) + 2KI(aq) \longrightarrow 2KCl(aq) + I_2(aq)$$
(a) Which is the oxidising agent in this reaction? Explain your answer. *[2]*
(b) Write an ionic equation for this reaction. *[2]*
(c) Use your ionic equation to help you write two half-equations for this reaction. *[2]*
(d) Chlorine reacts with iodine to form a brown liquid called iodine monochloride, ICl. In the presence of excess chlorine, iodine monochloride reacts further to form yellow crystals of iodine trichloride, ICl_3.
$$Cl_2(g) + ICl(l) \rightleftharpoons ICl_3(s)$$
   (i) What is the meaning of the symbol $\rightleftharpoons$? *[1]*
   (ii) Describe and explain what you would observe when the chlorine gas supply is turned off. *[3]*
   (iii) What effect does a decrease in pressure have on this equilibrium reaction? Give an explanation for your answer. *[2]*

(Paper 4)

7 Oxygen combines with both nitrogen and nitrogen(II) oxide, NO(g).
$$N_2(g) + O_2(g) \rightleftharpoons 2NO(g)$$
$$2NO(g) + O_2(g) \rightleftharpoons 2NO_2(g)$$
(a) Explain why both these reactions are redox reactions. In each case identify the reductant. *[3]*
(b) Describe and explain how an increase in pressure affects:
   (i) the reaction of oxygen with nitrogen *[2]*
   (ii) the reaction of oxygen with nitrogen(II) oxide. *[2]*
(c) The reaction of nitrogen with oxygen is endothermic. Describe how an increase in temperature affects this equilibrium. Explain your answer. *[2]*

(Paper 4)

10 Acids and bases

10.1 How acidic?

LEARNING OUTCOMES

- Describe neutrality and relative acidity and alkalinity in terms of pH measured using universal indicator paper

Many **acids** exist naturally in plants and foods. Citric acid is found in oranges and lemons, ethanoic acid is found in vinegar and methanoic acid is present in nettles. The common laboratory acids are:

hydrochloric acid HCl   sulfuric acid H_2SO_4 nitric acid HNO_3

All acids form hydrogen ions, H^+, when dissolved in water. It is the hydrogen ions that make a solution acidic.

Alkalis are the opposite of acids in the way they react. Common laboratory alkalis are:

sodium hydroxide NaOH   calcium hydroxide $Ca(OH)_2$ ammonia NH_3

Alkalis form hydroxide ions, OH^-, when they dissolve in water.

Substances that are neither acidic nor alkaline are **neutral**. Pure water is neutral.

The pH scale

We use the **pH scale** to show us if a solution is acidic, alkaline or neutral. We can also use this scale to find out exactly how acidic or alkaline a substance is.

STUDY TIP

Remember that a lower acidity gives a higher pH and a higher acidity gives a low pH. Halow sounds rather like hello!

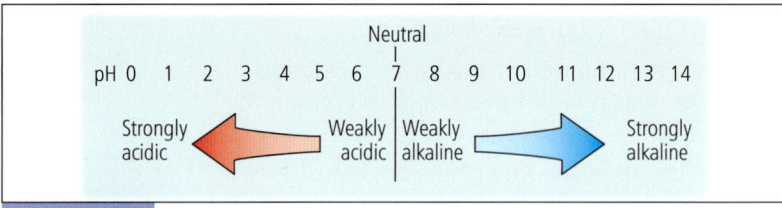

Figure 10.1.1 The pH scale

The pH scale runs from 0 to 14. Acids have a pH below 7. Alkalis have a pH above 7. The lower the pH, the more acidic the solution is. A solution with a lower pH has a higher concentration of hydrogen ions. The higher the pH, the more alkaline the solution is. A higher pH means a higher concentration of hydroxide ions. We can measure pH using **universal indicator** paper or a pH electrode connected to a pH meter.

DID YOU KNOW?

The pH scale was thought up by the Danish chemist Sørensen in 1909. pH stands for 'potenz H(ydrogen)'. This means power of the hydrogen ion concentration.

PRACTICAL

Comparing the pH of household products

1. Put the pH electrode into a solution of a substance such as vinegar, lemon juice, washing powder or soap.
2. Record the reading on the pH meter.

You can also use a pH sensor connected to a data logger and computer to see how the pH changes as you add an acid to an alkali in the beaker.

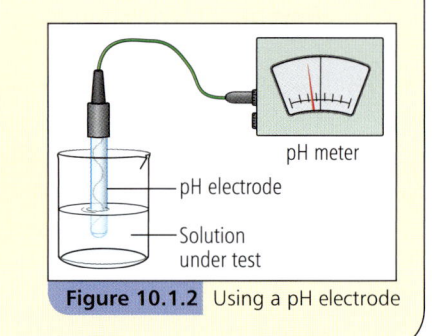

Figure 10.1.2 Using a pH electrode

Using universal indicator

An acid–alkali **indicator** is a chemical which changes colour when we add an acid or an alkali. Universal indicator is a mixture of indicators which shows a range of colours depending on the pH.

You can use universal indicator solution or universal indicator paper to measure the pH. When you take a drop of the test solution and place it on universal indicator paper, the paper turns a particular colour. The colour of the indicator paper is then matched against a colour chart showing the pH for different colours.

The colour of universal indicator changes with the pH

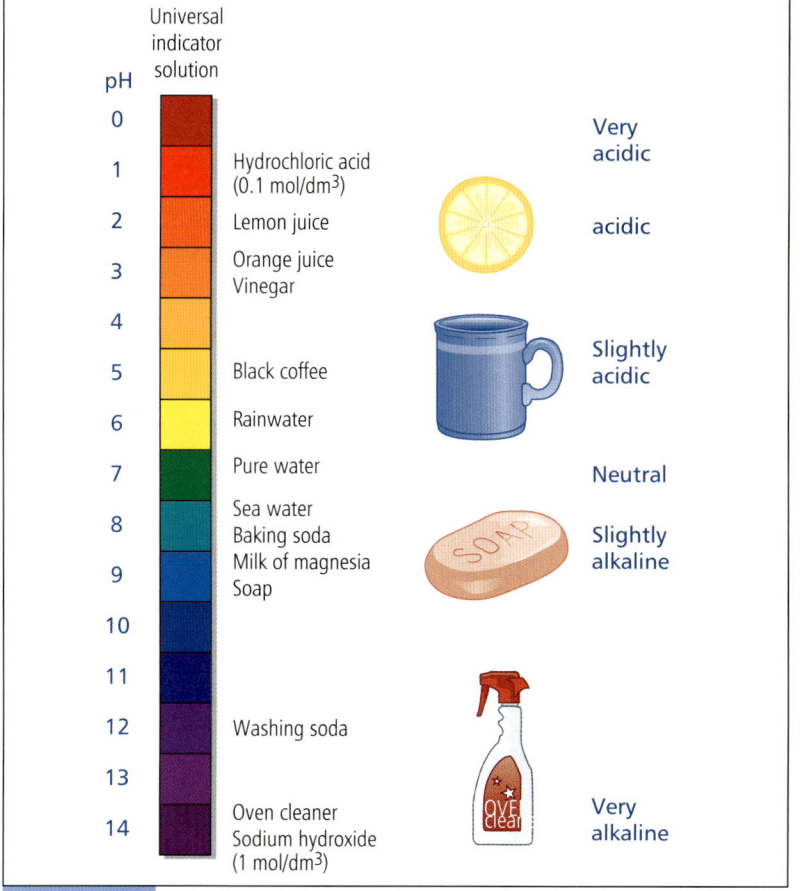

Figure 10.1.3 The colour of universal indicator and pH of some common substances

KEY POINTS

1. The pH scale is used to show the acidity or alkalinity of a solution.
2. Solutions with a pH below 7 are acidic. Solutions with a pH above 7 are alkaline.
3. Universal indicator can be used to find the pH of a solution.

SUMMARY QUESTIONS

1. Copy and complete using the words below:

   **acidic   alkaline   high
   neither   neutral   scale
   seven   universal**

   The pH _____ shows how acidic or _____ a solution is. Strongly _____ solutions have a low pH. Strongly alkaline solutions have a _____ pH. A solution that is _____ acidic nor alkaline is called a _____ solution. It has a pH of _____. The pH of a solution can be found using _____ indicator or a pH meter.

2. Which of these solutions are (i) acidic (ii) alkaline (iii) neutral?
   a   pH 6
   b   pH 13
   c   pH 7
   d   pH 8

3. Describe how you can use universal indicator paper to distinguish between a solution of sodium hydroxide and a solution of ethanoic acid.

10.2 Properties of acids

LEARNING OUTCOMES

- Describe the characteristic properties of acids as reactions with metals, bases and carbonates and effect on litmus and methyl orange

DID YOU KNOW?

Bromothymol blue can be used to distinguish between most acids and bases. It is blue above pH 7.6 and yellow below pH 6.0. Most other indicators can only distinguish between strong acids and strong bases. This is because they do not change colour close to pH 7. For example, methyl orange is red below pH 3.2, but changes to yellow at pH 4.4. So, an acid with a pH of 5 will show the same colour as a base.

Acids are found in many foods we eat. They also help us to start our cars! That's because car batteries contain sulfuric acid.

STUDY TIP

When acids react with carbonates, water is produced – as well as a salt and carbon dioxide. For supplement, you must be able to write the symbol equations.

The litmus test

Litmus is an **indicator** that has two colours, red and blue. If a solution is acidic it turns blue litmus red. You can use litmus as a solution or as litmus test paper. If you use it as a test paper, it should be damp.

The simplest definition of an acid states that an acid is a substance that dissolves in water to form hydrogen ions.

$$HCl(g) + aq \longrightarrow H^+(aq) + Cl^-(aq)$$

hydrogen chloride   water   hydrochloric acid

We can show that water plays an important part by first dissolving hydrogen chloride gas in a solvent called methylbenzene. The solution does not turn blue litmus red. This shows that the solution is not acidic. When we shake this solution with water the litmus does turn red. This shows that an acid has been formed once water has been added.

Chemical properties of acids

Reaction of acids with metals

Many metals react with dilute acids to form a salt and hydrogen. A **salt** is a compound formed when a metal or an ammonium group (NH_4) replaces hydrogen in an acid. Some examples are:

- hydrochloric acid, HCl, forms chlorides, for example magnesium chloride, $MgCl_2$
- sulfuric acid, H_2SO_4, forms sulfates, for example zinc sulfate, $ZnSO_4$
- nitric acid, HNO_3, forms nitrates, for example lithium nitrate, $LiNO_3$.

However nitric acid does not always give off hydrogen with metals unless it is very dilute. Metals such as silver do not react with acids. Others such as copper react in a different way but only with concentrated acids.

PRACTICAL

The reaction of magnesium with hydrochloric acid

You add magnesium to hydrochloric acid. You can collect the hydrogen gas over water. When the tube is full of gas, you can test the gas to see if it is hydrogen. (Hydrogen gives a 'pop' when a lighted splint is put in the mouth of the test tube.)

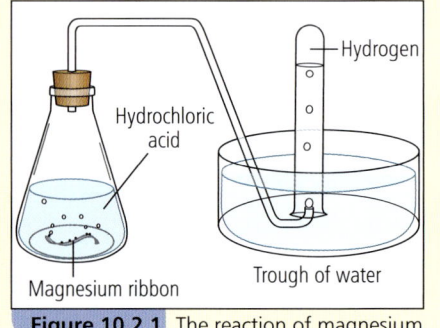

Figure 10.2.1 The reaction of magnesium with hydrochloric acid

The equation for the reaction of magnesium with hydrochloric acid is:

$$Mg(s) + 2HCl(aq) \longrightarrow MgCl_2(aq) + H_2(g)$$

magnesium + hydrochloric acid ⟶ magnesium chloride + hydrogen

Supplement

> The reaction of acid with metal is an example of a redox reaction. The metal loses electrons to form metal ions. The metal is oxidised. The hydrogen ions gain electrons to form hydrogen gas. The hydrogen ions are reduced.

Reaction of acids with metal oxides

Acids react with metal oxides to form a salt and water: For example:

$$CuO(s) + H_2SO_4(aq) \longrightarrow CuSO_4(aq) + H_2O(l)$$

copper(II) oxide + sulfuric acid ⟶ copper(II) sulfate + water

metal oxide + acid ⟶ salt + water

Reaction of acids with metal hydroxides or aqueous ammonia

Acids react with many metal hydroxides to form a salt and water. For example:

$$NaOH(aq) + HNO_3(aq) \longrightarrow NaNO_3(aq) + H_2O(l)$$

sodium hydroxide + nitric acid ⟶ sodium nitrate + water

Aqueous ammonia reacts in a similar way to metal hydroxides. But water does not appear in the equation because we do not usually write aqueous ammonia to show the hydroxide ions in solution. For example:

$$NH_3(aq) + HCl(aq) \longrightarrow NH_4Cl(aq)$$

aqueous ammonia + hydrochloric acid ⟶ ammonium chloride

These are examples of **neutralisation** reactions. When the acid reacts completely with the alkali we say it has neutralised the alkali.

Reaction of acids with carbonates

Carbonates react with acids to form a salt, water and carbon dioxide. The carbon dioxide bubbles off as a gas. For example:

$$CuCO_3(s) + H_2SO_4(aq) \longrightarrow CuSO_4(aq) + H_2O(l) + CO_2(g)$$

copper(II) carbonate + sulfuric acid ⟶ copper(II) sulfate + water + carbon dioxide

metal carbonate + acid ⟶ salt + water + carbon dioxide

Hydrogencarbonates react in a similar way. For example:

$$NaHCO_3(s) + HCl(aq) \longrightarrow NaCl(aq) + H_2O(l) + CO_2(g)$$

sodium hydrogencarbonate + hydrochloric acid ⟶ sodium chloride + water + carbon dioxide

Sodium hydrogencarbonate is used in baking powder together with powdered tartaric acid. In the presence of moisture, carbon dioxide is given off which makes a cake rise.

KEY POINTS

1 Acids are substances that form hydrogen ions when dissolved in water.

2 Many metals react with acids to form a salt and hydrogen.

3 Acids react with metal oxides and hydroxides to form a salt and water.

4 Acids react with carbonates to form a salt, water and carbon dioxide.

SUMMARY QUESTIONS

1 Copy and complete using the words below:

**dissolves   hydrogen
hydroxides   oxides
salt   water**

An acid is a substance that _____ in water to form _____ ions. Acids react with metals to form a metal _____ and hydrogen. When acids react with metal _____ or _____ a salt and _____ are formed.

2 Name the salts formed when:

  a sulfuric acid reacts with zinc

  b calcium oxide reacts with hydrochloric acid

  c magnesium carbonate reacts with nitric acid.

3 Write word equations for the reaction of:

  a magnesium with (very dilute) nitric acid

  b copper(II) carbonate with hydrochloric acid

  c sodium hydroxide with sulfuric acid.

10.3 Bases

LEARNING OUTCOMES

- Describe the characteristic properties of bases as reactions with acids and with ammonium salts and effect on litmus and methyl orange
- Describe and explain the importance of controlling acidity in soil

Many substances we use in the home are **bases**. Dishwasher tablets, washing powder and cleaning liquids all contain bases. Antacid tablets used to treat indigestion often contain the bases magnesium oxide or magnesium hydroxide. But what exactly are bases? A base is a substance that can react with an acid. Bases can be oxides, hydroxides or carbonates of metals. Ammonia is a base even though it does not contain a metal.

If a base is soluble in water, we call it an **alkali**. Alkalis dissolve in water to form hydroxide ions. An alkali turns damp red litmus paper blue. So aqueous ammonia and aqueous sodium hydroxide turn red litmus blue. Alkalis turn methyl orange yellow.

Chemical reactions of bases

Reaction with acids

Metal oxides and hydroxides react with acids to form a salt and water.

$$CaO(s) + 2HCl(aq) \longrightarrow CaCl_2(aq) + H_2O(l)$$
calcium oxide + hydrochloric acid → calcium chloride + water

$$NaOH(aq) + HNO_3(aq) \longrightarrow NaNO_3(aq) + H_2O(l)$$
sodium hydroxide + nitric acid → sodium nitrate + water

Metal carbonates react with acids to form a salt, water and carbon dioxide.

$$MgCO_3(s) + H_2SO_4(aq) \longrightarrow MgSO_4(aq) + H_2O(l) + CO_2(g)$$
magnesium carbonate + sulfuric acid → magnesium sulfate + water + carbon dioxide

When a base reacts with an acid to form a salt we call the reaction a **neutralisation reaction**.

Reaction of alkalis with ammonium salts

Ammonium salts are much more **volatile** than metal salts. When warmed with an alkali, ammonium salts decompose to form a metal salt, ammonia and water.

$$KOH(aq) + NH_4Cl(aq) \xrightarrow{heat} KCl(aq) + NH_3(g) + H_2O(l)$$
potassium hydroxide + ammonium chloride → potassium chloride + ammonia + water

Controlling soil acidity

Soil may become acidic after a number of years. This can be due to:

- acid rain (see Topic 15.3)
- bacteria and fungi rotting the vegetation so that it releases acids
- use of fertilisers containing ammonium salts.

Alkalis are found in many cleaning products

DID YOU KNOW?

When wood ashes are burned, the product is alkaline. The word alkali comes from the Arabic 'al-khali' which means burned ashes.

PRACTICAL

Reacting ammonium sulfate with an alkali

1. When you warm a solution of an ammonium salt with sodium hydroxide solution a gas is given off.
2. You test this gas with damp red litmus paper. The litmus paper turns blue. This shows that an alkaline gas is present. If you are heating an ammonium salt the gas will be ammonia.

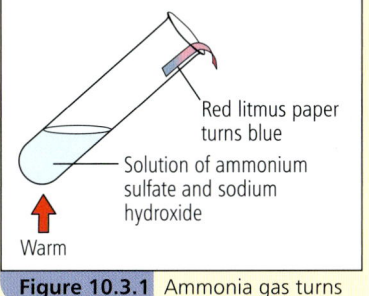

Figure 10.3.1 Ammonia gas turns red litmus blue

STUDY TIP

You must be able to tell the difference between the molecule ammonia, NH_3, and ammonium salts which contain the NH_4 (ammonium) group. It is a common error to write ammonia chloride.

Many crop plants such as onions, cabbages and beans grow better if the soil is neutral. If soil acidity drops below pH 5.5 many plants will not grow well.

We can remove excess acidity from the soil by adding crushed limestone – calcium carbonate. This neutralises the acid. The calcium carbonate and the products are neutral.

$$CaCO_3(s) + 2H^+(aq) \longrightarrow Ca^{2+}(aq) + CO_2(g) + H_2O(l)$$
calcium carbonate — acid — calcium ions — carbon dioxide — water

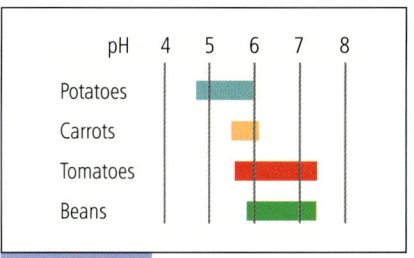

Figure 10.3.2 Crop plants grow best in these pH ranges

Farmers often add lime (calcium oxide) to the soil. This also neutralises excess acid.

$$CaO(s) + 2H^+(aq) \longrightarrow Ca^{2+}(aq) + H_2O(l)$$
calcium oxide — acid — calcium ions — water

We must be careful not to add too much lime to the soil. Lime is strongly alkaline when it dissolves in water. Most plants do not survive alkaline conditions. If the soil is too alkaline, farmers spray the soil with manure or even with very dilute sulfuric acid.

SUMMARY QUESTIONS

1. Copy and complete using the words below:

    acids   ammonia   hydroxide   salt   soluble

    Alkalis are _____ bases. Examples of alkalis are _____ and sodium _____. Alkalis react with _____ to form a _____ and water.

2. Write word equations for the reactions of:
    a. ammonium sulfate with potassium hydroxide
    b. calcium oxide with hydrochloric acid
    c. magnesium hydroxide with nitric acid.

3. Explain why farmers sometimes need to put crushed limestone on their soil.

KEY POINTS

1. A base is a substance that can neutralise an acid.
2. An alkali is a soluble base.
3. Bases react with acids to form a salt and water.
4. Ammonia gas is released when ammonium salts are heated with an alkali.
5. Crushed limestone or lime are added to neutralise excess acidity in the soil.

10.4 More about acids and bases

LEARNING OUTCOMES

- Define acids and bases in terms of proton transfer
- Understand the difference between strong and weak acids and bases
- Write ionic equations for acid–base reactions

More about neutralisation

In Topic 4.4 we learned how to write ionic equations by cancelling the spectator ions. So what is the ionic equation for the neutralisation of an alkali by an acid? Let us take for example the reaction:

$$NaOH(aq) + HCl(aq) \longrightarrow NaCl(aq) + H_2O(l)$$
sodium hydroxide — hydrochloric acid — sodium chloride — water

Separating into ions then cancelling the spectator ions, we have:

$$\cancel{Na^+(aq)} + OH^-(aq) + H^+(aq) + \cancel{Cl^-(aq)} \longrightarrow \cancel{Na^+(aq)} + \cancel{Cl^-(aq)} + H_2O(l)$$

This leaves us with a very simple ionic equation:

$$OH^-(aq) + H^+(aq) \longrightarrow H_2O(l)$$

If you carry out the same process with any combination of acid and alkali, you get the same ionic equation. This makes sense if you think about it: every acid has hydrogen ions dissolved in water and every alkali has hydroxide ions dissolved in water. So these two ions combine to make water.

Acids, bases and protons

A hydrogen ion is formed by the removal of the single electron from a hydrogen atom. Nearly all hydrogen atoms have a single proton and no neutrons in their nuclei. So a hydrogen ion is nothing more than a proton. We can define acids and bases more generally by seeing what happens to the hydrogen ions (protons) when acids and bases react.

An acid is a **proton donor** – it gives a proton to a base.

$$H^+(aq) + OH^-(aq) \longrightarrow H_2O(l)$$
acid — base — water

A base is a **proton acceptor** – it removes protons from an acid. You can see this in the equation above – the hydroxide ion has accepted a proton from the acid to form water.

We can see how this works in a more general case, the reaction of an ammonium salt with sodium hydroxide.

$$NH_4Cl(aq) + NaOH(aq) \longrightarrow NH_3(g) + NaCl(aq) + H_2O(l)$$
ammonium chloride — sodium hydroxide — ammonia — sodium chloride — water

The ionic equation is:

$$NH_4^+(aq) + OH^-(aq) \longrightarrow NH_3(g) + H_2O(l)$$

In this equation there are no hydrogen ions but there is a transfer of protons. The base, OH^-, has accepted a proton from the ammonium ion, NH_4^+. So the ammonium ion must be acting as an acid.

proton transfer

$$NH_4^+(aq) + OH^-(aq) \longrightarrow NH_3(g) + H_2O(l)$$
acid — base

DID YOU KNOW?

Traces of vinegar, a weak acid, have been found in 5000-year-old Egyptian tombs. Over the centuries, people have drunk apple cider vinegar, thinking that it will benefit their health. Some have suggested that cider vinegar helps you to lose weight. Recent studies, however, have shown that drinking large amounts of apple cider vinegar may be harmful because it reduces bone density and lowers the amount of potassium absorbed by the body.

Strong and weak acids

The laboratory acids hydrochloric, sulfuric and nitric acids are all strong acids. A strong acid is completely ionised when it dissolves in water. There are no molecules of the acid present.

$$HCl(g) + aq \longrightarrow H^+(aq) + Cl^-(aq)$$

Weak acids are generally organic acids, such as citric acid and ethanoic acid. A weak acid is only partly ionised when it dissolves in water. There are lots of acid molecules present but very few ions.

$$\underset{\substack{\text{ethanoic acid} \\ \text{lots of molecules}}}{CH_3COOH(l)} + aq \rightleftharpoons \underset{\substack{\text{ethanoate ion} \\ \text{very few ions}}}{CH_3COO^-(aq)} + H^+(aq)$$

We can tell if an acid is strong or weak by measuring its electrical conductivity, pH and rate of reaction with metals or metal carbonates.

- The electrical conductivity of a solution depends on the concentration of the ions present – the more ions, the greater the conductivity. A strong acid conducts much better than a weak acid of the same concentration. This is because there is a greater concentration of ions in the strong acid compared with the weak acid.
- A strong acid has a lower pH than a weak acid of the same concentration. This is because there is a greater concentration of hydrogen ions in the strong acid compared with the weak acid.
- A strong acid reacts faster than a weak acid. This is also because there is a greater concentration of hydrogen ions in the strong acid compared with the weak acid.

Strong and weak bases

All hydroxides of alkali metals are strong bases. Strong bases are completely ionised in water.

For example, sodium hydroxide: $NaOH(s) + aq \longrightarrow Na^+(aq) + OH^-(aq)$

Ammonia is a weak base because it is only partly ionised in water. There are very few ammonium ions.

$$NH_3(g) + H_2O(l) \rightleftharpoons NH_4^+(aq) + OH^-(aq)$$

Compared with strong bases, weak bases have:

- a lower electrical conductivity
- a less alkaline (lower) pH
- a slower rate of reaction.

Measuring the pH of a solution

> **STUDY TIP**
>
> It is incorrect to use the words 'strong' and 'weak' when referring to the concentration of acids or alkalis. Use 'concentrated' or 'dilute'. Strong and weak refer to the degree of ionisation of the acid or base, not the concentration.

> **DID YOU KNOW?**
>
> When an acid neutralises an alkali in an aqueous solution, the ionic equation is
>
> $H^+(aq) + OH^-(aq) \longrightarrow H_2O(l)$

SUMMARY QUESTIONS

1 Copy and complete using the words below:

   base   dissolves   hydrogen   ions   proton

   When an acid _____ in water, hydrogen _____ are formed. A _____ ion is a proton. An acid is a _____ donor. It gives its protons to a _____.

2 Describe one way to tell if an acid is strong or weak. What do you understand by the terms *strong acid* and *weak acid*?

KEY POINTS

1 An acid is a proton donor. A base is a proton acceptor.
2 Strong acids and bases are completely ionised in water.
3 Weak acids and bases are partly ionised in water.

10.5 Oxides

LEARNING OUTCOMES

- Classify oxides as either acidic or basic according to their metallic or non-metallic character
- Classify other oxides as neutral or amphoteric

Oxides are compounds of metals or non-metals with oxygen. There are four types of oxide. We tell the difference between these four types by their typical chemical reactions.

Basic oxides

Most metal oxides are basic oxides. Many basic oxides are formed by the direct combination of a metal with oxygen. Basic oxides react with acids to form a salt and water. For example:

$$CaO(s) + 2HCl(aq) \longrightarrow CaCl_2(aq) + H_2O(l)$$

calcium oxide + hydrochloric acid ⟶ calcium chloride + water
(lime)

Basic oxides do not react with alkalis.

Many basic oxides do not react with water. But those from Group I and many from Group II in the Periodic Table react to form a metal hydroxide. An alkaline solution is formed which turns red litmus blue.

$$BaO(s) + H_2O(l) \longrightarrow Ba(OH)_2(aq)$$

barium oxide + water ⟶ barium hydroxide

Acidic oxides

Most non-metal oxides are acidic oxides. Many are formed by direct reaction with oxygen.

$$S(s) + O_2(g) \longrightarrow SO_2(g)$$

sulfur   sulfur dioxide

Acidic oxides react with alkalis to form a salt and water. For example:

$$CO_2(g) + 2NaOH(aq) \longrightarrow Na_2CO_3(aq) + H_2O(l)$$

carbon dioxide + sodium hydroxide ⟶ sodium carbonate + water

Some acidic oxides react with bases such as metal oxides when heated. For example:

$$SiO_2 + CaO \xrightarrow{heat} CaSiO_3$$

silicon(IV) oxide + calcium oxide ⟶ calcium silicate

Many acidic oxides react with water to form acidic solutions. For example:

$$SO_3(g) + H_2O(l) \longrightarrow H_2SO_4(aq)$$

sulfur trioxide + water ⟶ sulfuric acid

Spreading lime, a basic oxide, on the soil decreases its acidity

DID YOU KNOW?

Various oxides of iron are used as the colour in cosmetics such as eye shadow.

Neutral oxides

Neutral oxides do not react with acids or bases. Examples of neutral oxides are nitrogen(I) oxide, N_2O, nitrogen(II) oxide, NO, and carbon monoxide, CO. Most are the lower oxides of non-metals. For example:

- carbon monoxide is a neutral oxide but carbon dioxide, CO_2, is an acidic oxide
- nitrogen(II) oxide is a neutral oxide but nitrogen dioxide (or nitrogen(IV) oxide), NO_2, is an acidic oxide.

Amphoteric oxides

The word amphoteric means 'both of them'. **Amphoteric oxides** have both acidic and basic properties. The oxides of aluminium and zinc are examples. They form salts when they react with acids. They also react with alkalis to form complex salts.

Examples:

$$ZnO(s) + 2HNO_3(aq) \rightarrow Zn(NO_3)_2(aq) + H_2O(l)$$
zinc oxide + nitric acid → zinc nitrate + water

$$ZnO(s) + 2NaOH(aq) \rightarrow Na_2ZnO_2(aq) + H_2O(l)$$
zinc oxide + sodium hydroxide → sodium zincate + water

$$Al_2O_3(s) + 6HCl(aq) \rightarrow 2AlCl_3(aq) + 3H_2O(l)$$
aluminium oxide + hydrochloric acid → aluminium chloride + water

$$Al_2O_3(s) + 2NaOH(aq) \rightarrow 2NaAlO_2(aq) + H_2O(l)$$
aluminium oxide + sodium hydroxide → sodium aluminate + water

The zincates and aluminates have the ending -ate to show that their ions are compound ions containing oxygen – rather like sulfates, carbonates and nitrates. Zincate ions are ZnO_2^{2-} and aluminate ions are AlO_2^-. In some books you may see these ions written $Zn(OH)_4^{2-}$ and $Al(OH)_4^-$. Notice that sodium zincate and aluminate are soluble in water.

STUDY TIP

For the core, you will be expected to know the word equations and simple symbol equations for the reactions of acidic and basic oxides. For the supplement you may be asked to write symbol equations.

SUMMARY QUESTIONS

1 Copy and complete using the words below:

   acidic   alkalis   litmus   react   salt

   Most oxides of non-metals are _____ oxides. Acidic oxides react with _____ to form a _____ and water. Some acidic oxides _____ with water to form solutions of acids. These solutions turn blue _____ red.

2 How can you show that calcium oxide is a basic oxide?

3 Write symbol equations for the reaction of zinc oxide with:
   a hydrochloric acid
   b aqueous sodium hydroxide.

KEY POINTS

1 The oxides of most metals are basic oxides. The oxides of most non-metals are acidic oxides.

2 Nitrogen(I) oxide, nitrogen(II) oxide and carbon monoxide are neutral oxides.

3 The oxides of aluminium and zinc are amphoteric – they react with both acids and alkalis.

SUMMARY QUESTIONS

1. Match the reactants on the left with the products on the right.

acid + carbonate	salt + water
acid + hydroxide	salt + ammonia + water
acid + metal	salt + hydrogen
ammonium salt + alkali	salt + water + carbon dioxide

2. Copy and complete these word equations:
   (a) zinc + hydrochloric acid ⟶ ____ + ____
   (b) sulfuric acid + magnesium oxide ⟶ ____ + ____
   (c) ____ + ____ ⟶ calcium nitrate + carbon dioxide + water
   (d) ammonium sulfate + ____ ⟶ potassium sulfate + ammonia + water

3. Explain why it is important to farmers to be able to control the acidity of their soil. In your answer refer to:
   - the pH of the soil
   - the use of named bases
   - neutralisation.

4. Match the substances on the left with both the pH in the middle and the acidity on the right.

vinegar	pH 0	strongly acidic
dishwasher powder	pH 12	weakly acidic
soap	pH 7	weakly alkaline
distilled water	pH 4.5	neutral
concentrated hydrochloric acid	pH 7.5	strongly alkaline

Supplement

5. Classify the following oxides as acidic, basic, neutral or amphoteric.
   (a) zinc oxide
   (b) magnesium oxide
   (c) sulfur dioxide
   (d) phosphorus(V) oxide
   (e) nitrogen(I) oxide

6. Write balanced equations for the reaction of:
   (a) sodium carbonate with nitric acid
   (b) magnesium with hydrochloric acid.

PRACTICE QUESTIONS

1. A solution has a pH of 9. The solution is best described as:
   A strongly alkaline
   B weakly acidic
   C weakly alkaline
   D strongly acidic.

   (Paper 1)

2. The equation between ammonium ions and hydroxide ions is shown by the equation:

   $NH_4^+(aq) + OH^-(aq) \longrightarrow NH_3(g) + H_2O(l)$

   Which one of these statements about this reaction is true?
   A Ammonium ions are acting as a base.
   B A proton is transferred from the hydroxide ions to the ammonium ions.
   C Ammonium atoms are acting as proton acceptors.
   D Hydroxide ions are acting as proton acceptors.

   (Paper 2)

3. Magnesium sulfate can be made by neutralising magnesium oxide with an acid.
   (a) What do you understand by the term *acid*? [1]
   (b) Name the acid used to make magnesium sulfate. [1]
   (c) Copy and complete the equation for this reaction:

   MgO + ____ ⟶ $MgSO_4$ + ____ [2]

   (d) What type of oxide is magnesium oxide? Give a reason for your answer. [2]
   (e) (i) Suggest one other method of making magnesium sulfate. [1]
   (ii) Write a word equation for this reaction. [1]

   (Paper 3)

4. Ammonia is a gas that is very soluble in water.
   (a) From the following list choose the most likely pH of an aqueous solution of ammonia.
   pH 2   pH 5   pH 7   pH 9   pH 13 [1]
   (b) What effect does aqueous ammonia have on red litmus paper? [1]

(c) Ammonia is released when ammonium sulfate is warmed with sodium hydroxide solution. Write a word equation for this reaction. [2]

(d) Ammonia reacts with hydrochloric acid to form a salt.
  (i) Copy and complete the equation for this reaction.
  ____ + ____ $\longrightarrow$ NH$_4$Cl [2]
  (ii) Name the salt formed in this reaction. [1]

(e) Ammonia can be used to control the acidity of the soil. Suggest one problem in using ammonia to control soil acidity. [1]

(Paper 3)

5 Hydrochloric acid is a strong acid.
  (a) What effect does hydrochloric have on:
    (i) blue litmus paper
    (ii) universal indicator paper? [2]
  (b) Hydrochloric acid reacts with magnesium.
    (i) Write a word equation for this reaction. [1]
    (ii) Name the salt formed in this reaction. [1]
  (c) Hydrochloric acid also reacts with magnesium carbonate. Copy and complete the symbol equation for this reaction.
    MgCO$_3$ + __HCl $\longrightarrow$ MgCl$_2$ + ____ + ____ [3]
  (d) When hydrochloric acid reacts with sodium sulfite, sulfur dioxide is produced. What type of oxide is sulfur dioxide? [1]
  (e) Describe how the pH of the mixture changes as hydrochloric acid is added to a solution of potassium hydroxide until hydrochloric acid is in excess. [2]

(Paper 3)

6 Ethanoic acid is a weak acid.
  (a) Suggest how you can use universal indicator paper to show that ethanoic acid is a weak acid. [2]
  (b) Describe two other methods you could use to show that ethanoic acid is a weak acid. For each method, describe what you would observe. [4]
  (c) Write a balanced equation to show the reaction of ethanoic acid with magnesium. [2]

(d) Ethanoic acid reacts with potassium hydroxide:
  CH$_3$COOH + KOH $\longrightarrow$ CH$_3$COOK + H$_2$O
  Refer to ideas about proton transfer to explain how this equation shows that ethanoic acid is an acid and potassium hydroxide is a base. [2]

(e) Potassium hydroxide is a strong base. What do you understand by the term *strong base*? [1]

(Paper 4)

7 Calcium oxide reacts with hydrochloric acid to form calcium chloride and water.
  (a) Write a balanced equation for this reaction. [2]
  (b) Calcium oxide is a base. What do you understand by the term *base*? [1]
  (c) Farmers often add calcium oxide to the soil. Explain why they do this. [2]
  (d) Write an equation to show how the oxide ion reacts with one type of ion in hydrochloric acid to form water. [2]
  (e) Zinc oxide reacts with both acids and alkalis.
    (i) What is the name given to an oxide that reacts with both acids and alkalis? [1]
    (ii) Write a balanced equation for the reaction of zinc oxide with sodium hydroxide. [2]

(Paper 4)

8 Ammonia is a weak base.
  (a) What do you understand by the term *weak base*? [2]
  (b) Write a balanced equation to show the reaction of ammonia with sulfuric acid. [2]
  (c) When ammonium salts are warmed with hydroxide ions, ammonia is released.
    NH$_4^+$ + OH$^-$ $\longrightarrow$ NH$_3$ + H$_2$O
    Explain how ammonium ions are acting as an acid in this reaction. [2]
  (d) When ammonia reacts with water an equilibrium mixture of reactants and products is formed.
    NH$_3$ + H$_2$O $\rightleftharpoons$ NH$_4^+$ + OH$^-$
    Describe each of the reactants and products as either acids or bases. Give reasons for your answers. [4]

(Paper 4)

11 Making and identifying salts

11.1 Making salts (1)

LEARNING OUTCOMES

- Demonstrate knowledge and understanding of preparation, separation and purification of salts from a metal or insoluble base
- Suggest a method of making a given salt from a suitable starting material given appropriate information

How do we make salts?

We often use **salts** in our homes. We put sodium chloride on our food to bring out its taste, we may use bath salts to help us relax in the bath and some of the medicines we take are salts. Aspirin is made into a salt to make it more soluble.

In Topic 10.2 we learned that a salt is a compound formed when a metal or an ammonium group replaces hydrogen in an acid. Many salts are soluble in water although a few, such as lead chloride, are insoluble. There are four ways of making salts:

- reacting a metal with an acid
- reacting an insoluble base with an acid
- neutralising an alkali with an acid by the titration method
- by precipitation.

Not all these methods are suitable for making a particular salt. So we have to choose the method that best fits the type of salt we want to make. We also have to choose the best method of purifying the salt from those we learnt about in Topic 1.9.

Salts from metals

We can make salts by reacting acids with metals. But this method is suitable only for metals above hydrogen in the **reactivity series**. So we can make salts of magnesium, zinc, aluminium and iron in this way. For example, to make zinc sulfate we carry out the following reaction:

$$Zn(s) + H_2SO_4(aq) \longrightarrow ZnSO_4(aq) + H_2(g)$$
$$\text{zinc} + \text{sulfuric acid} \longrightarrow \text{zinc sulfate} + \text{hydrogen}$$

We cannot use this type of reaction for making salts of copper, lead and silver which are too close to or below hydrogen in the reactivity series. Also it is not a good idea to prepare salts of very reactive metals such as sodium and potassium using this method. The reaction of these metals with the acid is too violent – a titration method is more suitable in these cases.

This is the procedure:

- Add the metal to the acid in a flask so that the metal is in excess. The acid is the limiting reactant.
- Warm the flask gently to complete the reaction.
- Filter off the excess metal. The filtrate is a solution of the metal salt.
- Put the filtrate into an evaporating basin and evaporate the water until the crystallisation point is reached. Then you allow the salt to crystallise at room temperature.
- Filter off the crystals and wash them with a tiny amount of solvent so they don't dissolve.
- Dry the crystals between sheets of filter paper.

Salts from insoluble bases

We can make salts of many metals by reacting an insoluble base with an acid. We use this method for making salts of metals that are low in the reactivity series.

For example, to make copper(II) sulfate we carry out the following reaction:

$$CuO(s) + H_2SO_4(aq) \rightarrow CuSO_4(aq) + H_2O(l)$$

copper(II) oxide + sulfuric acid → copper(II) sulfate + water

The method is exactly the same as for making salts by reacting a metal with an acid, for example making copper(II) sulfate from copper(II) oxide.

PRACTICAL

Making copper(II) sulfate from copper(II) oxide

1. Add insoluble copper(II) oxide to sulfuric acid and stir. Warm gently.
2. The solution turns blue as the reaction occurs, showing that copper(II) sulfate is being formed.
3. When the reaction is complete, filter the solution to remove excess copper(II) oxide.
4. Evaporate some of the water from the filtrate and leave to crystallise.

Figure 11.1.1 Copper(II) sulfate from copper(II) oxide

DID YOU KNOW?

The 'War of the Pacific' from 1879 to 1883 involved the countries of Bolivia, Chile and Peru. Each country wanted control of the part of the Atacama Desert that had large nitrate deposits.

STUDY TIP

When you make a salt using excess metal or metal oxide, you first must filter off the excess solid reactant. You should be able to describe how to make a salt.

Salts come in many different colours and crystal forms

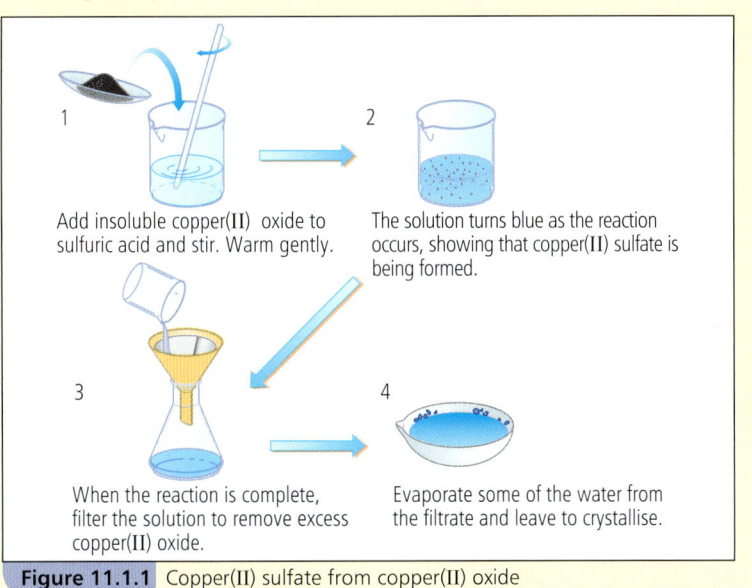

SUMMARY QUESTIONS

1 Copy and complete using the words below:

filter   filtrate   insoluble   limiting   neutralised   oxide

You can make a metal salt by reaction of an _____ metal oxide with an acid. During this reaction the acid is _____ by the metal _____. The acid is the _____ reactant. You _____ off the excess metal oxide. The _____ is a solution of the metal salt.

2 Write word equations for these reactions:
   a $Fe_2O_3(s) + 6HCl(aq) \rightarrow 2FeCl_3(aq) + 3H_2O(l)$
   b $Mg(s) + H_2SO_4(aq) \rightarrow MgSO_4(aq) + H_2(g)$

3 Write instructions for making the salt calcium chloride from calcium oxide.

KEY POINTS

1 The salt of a metal above hydrogen in the reactivity series is made by reaction of the metal with an acid.

2 Salts can be made by the reaction of an insoluble base with an acid.

3 When making a salt from a metal or metal oxide the acid is the limiting reactant.

11.2 Making salts (2): Titration method

LEARNING OUTCOMES

- Demonstrate the preparation, separation and purification of a salt from a soluble base
- Suggest a method of making a given salt from a suitable starting material given appropriate information

We can use **titration** to make a soluble salt from a soluble base and an acid. This method is used to make salts of the Group I metals and ammonium salts.

We use an acid–alkali titration to find out how much acid is needed to react exactly with a solution of an alkali. We use an **indicator** to find when the acid has just reacted with all the alkali. We call this the **end point** of the titration. At the end point the indicator changes colour. The indicator we choose depends on whether we use a strong or weak acid or alkali:

- For a strong acid and alkali we can use any indicator.
- If we are making a salt from a weak alkali, for example ammonia, we titrate with a strong acid. We use methyl orange indicator. This goes from orange to red when there is excess acid.
- If we are making a salt from a weak acid, for example ethanoic acid, we use a strong base. We use phenolphalein indicator. This goes from pink to colourless when there is excess acid.

How to carry out a titration

You use the following sequence to carry out a titration:

1. Measure a known volume of alkali into the titration flask using a **volumetric pipette**. First wash the pipette with a little of the alkali you are using.
2. Add a few drops of indicator solution to the alkali in the flask.
3. Fill a clean **burette** with acid. First wash the burette with a little of the acid you are using.
4. Record the burette reading.
5. Open the burette tap and let the acid flow into the flask. Keep swirling the flask gently to make sure that the acid and alkali mix and react.
6. Keep adding the acid slowly until the indicator changes colour. This is the end point.
7. Record the reading on the burette. The final reading minus the initial reading is called the *titre*. The first time you do this gives you the rough titre or 'range-finder' titre.
8. Repeat this process at least three times. You can add the acid rapidly until you are a few cm³ from the end point. Ideally, these will be within 0.1 cm³ of each other. Then add the acid drop by drop so that you can get an accurate titre.
9. If you are doing calculations to find the concentration of the alkali in the flask, you take the average of the accurate titres. You can ignore any titres that appear to be inconsistent.

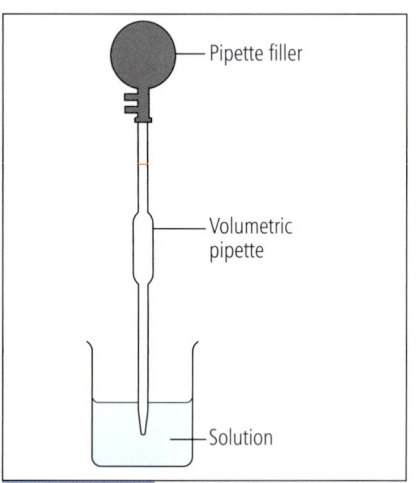

Figure 11.2.1 Filling a volumetric pipette

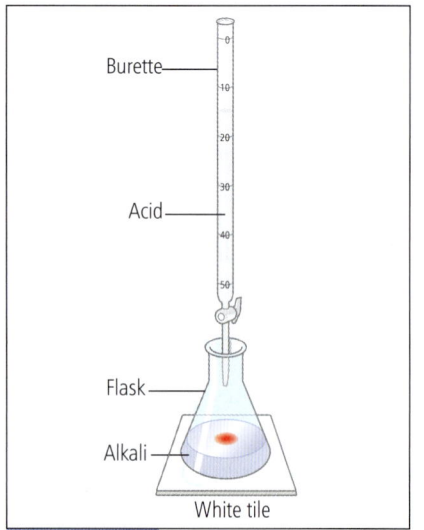

Figure 11.2.2 Titration apparatus

A soluble salt from an acid and alkali

We can make a soluble salt by titrating an alkali with an acid. For example, if we want to make potassium sulfate we carry out the reaction:

$2KOH(aq) + H_2SO_4(aq) \rightarrow K_2SO_4(aq) + H_2O(l)$

potassium hydroxide + sulfuric acid → potassium sulfate + water

We first carry out a titration using an indicator to find the correct volumes of solution to mix. Then we carry it out without an indicator to prepare a sample of the salt uncontaminated by indicator.

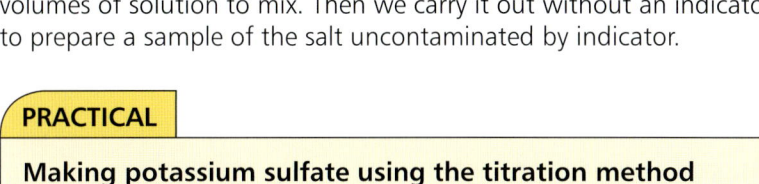

PRACTICAL

Making potassium sulfate using the titration method

1. Put the alkali in the flask with a few drops of indicator.
2. Add acid from the burette until the indicator changes colour. You record the volume of acid added.
3. Repeat the experiment without indicator in the flask. Add the volume of acid you recorded in the last step.
4. Put the solution from the flask into an evaporating basin and evaporate the water until the crystallisation point is reached.
5. Allow crystals to form. You then filter these. The crystals can then be carefully washed and dried with filter paper.

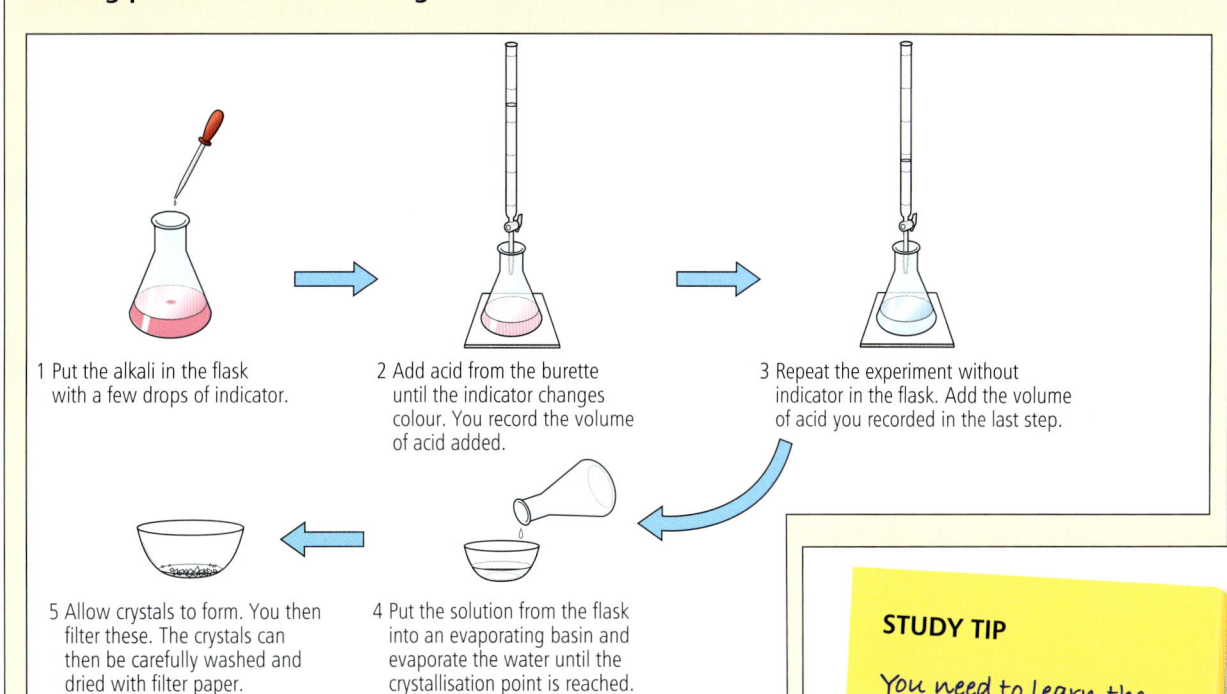

Figure 11.2.3 Making potassium sulfate using sulfuric acid

STUDY TIP

You need to learn the procedure for a titration. Remember that the acid usually goes in the burette and the alkali in the flask.

SUMMARY QUESTIONS

1 Copy and complete using the words below:

alkali   burette   indicator   repeated   soluble

A titration method is used to make a _____ salt from an acid and an _____. The acid is added to the alkali using a _____ until the _____ in the flask changes colour. The process is then _____ without using the indicator.

2 Name an acid and alkali you can use to make these salts:
   a ammonium chloride
   b sodium chloride
   c ammonium nitrate

3 Name the salts formed from these acids and alkalis:
   a sodium hydroxide + nitric acid
   b ammonia + sulfuric acid
   c lithium hydroxide + nitric acid.

KEY POINTS

1 A titration is used to make a soluble salt from an acid and an alkali.

2 Salts made by the titration method include salts of Group I elements and ammonium salts.

3 When making a salt using the titration method, the titration is first carried out using an indicator and then repeated without an indicator.

11.3 Making salts (3): precipitation

LEARNING OUTCOMES

- Demonstrate knowledge and understanding of the preparation of insoluble salts by precipitation

Soluble or insoluble?

This page gives additional information to help you recognise which compounds are soluble and which are insoluble.

The salts that we have discussed in Topics 11.1 and 11.2 are soluble in water. Some salts are insoluble. We make these salts by mixing two soluble compounds. The solid obtained when solutions of two soluble compounds are mixed is called a **precipitate**.

If we are going to make salts by precipitation, we must know which compounds are soluble in water and which are insoluble. Fortunately there are some rules which help us to do this. We call these rules the solubility rules. These rules are shown in the table:

soluble compounds	insoluble compounds
all salts of Group I elements	
all nitrates	
all ammonium salts	
most chlorides, bromides and iodides	chlorides, bromides and iodides of silver and lead
most sulfates	sulfates of calcium, barium and lead
Group I hydroxides and carbonates are soluble (calcium hydroxide is slightly soluble)	most hydroxides and carbonates
Group I and II oxides react with water	most metal oxides

If we want to make an insoluble salt, for example lead chloride, we:
- identify the ions present in the insoluble salt – lead and chloride
- use our solubility rules to choose soluble compounds including these ions – for example, lead nitrate for the lead and sodium chloride for the chloride
- add one solution to the other
- filter off the precipitate then wash and dry the solid.

STUDY TIP

Make sure that you know what types of compound are soluble or insoluble. Without this knowledge you will not be able to select precipitation as the correct method to make a particular salt.

PRACTICAL

Making an insoluble salt

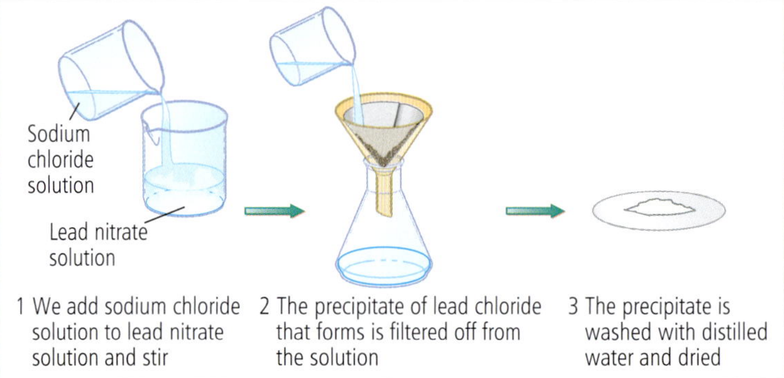

1 We add sodium chloride solution to lead nitrate solution and stir

2 The precipitate of lead chloride that forms is filtered off from the solution

3 The precipitate is washed with distilled water and dried

Figure 11.3.1 Making lead chloride

What happens in a precipitation reaction?

We can explain why a solid precipitates by looking at the reaction between lead nitrate and sodium chloride as an example.

$$Pb(NO_3)_2(aq) + 2NaCl(aq) \longrightarrow PbCl_2(s) + 2NaNO_3(aq)$$

We know that in a solution of an ionic compound the ions are free to move. So our solution of lead nitrate contains lead ions and nitrate ions that are separate from one another. They are able to move freely and randomly throughout the water. The water molecules help to keep them in solution. A similar thing happens with our solution of sodium chloride.

Lead chloride, the insoluble salt, is precipitated when we mix the solutions. The lead ions in solution have a greater attraction for the chloride ions than the water molecules that keep them in solution. So the lead ions and chloride ions come together in large numbers and form a three-dimensional ionic lattice. The sodium ions and nitrate ions remain in solution. They are the **spectator ions**.

Water treatment plants use iron sulfate or aluminium sulfate to precipitate some unwanted compounds in the impure water

Key: Lead ions ● Chloride ions ▬ Nitrate ions ▪ Sodium ions ○

Lead ions and nitrate ions are free to move in solution.

Sodium ions and chloride ions are free to move in solution.

When mixed, lead chloride forms an ionic lattice and sodium and nitrate ions remain in solution.

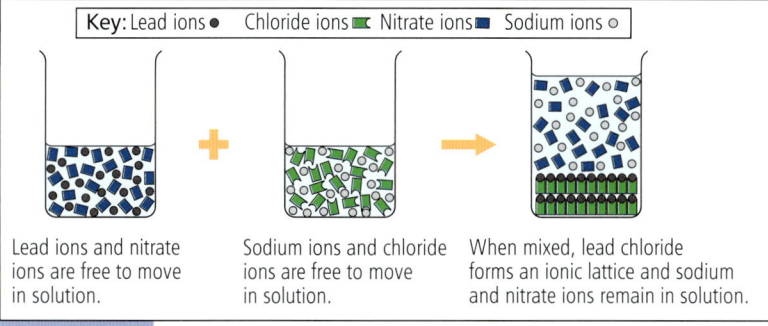

Figure 11.3.2 How a precipitate forms. The water molecules are not shown. They are represented by the blue background.

SUMMARY QUESTIONS

1 Copy and complete using the words below:

   **attracted   insoluble   ion
   lattice   precipitation   spectator**

   We can make an _____ salt by mixing two solutions of soluble salts. This type of reaction is called _____. When the solutions react the two types of _____ that form the precipitate are _____ to each other and form a giant _____. The ions that do not take part in the reaction are called _____ ions.

2 Which of these compounds are insoluble in water?
   a Potassium bromide
   b Silver bromide
   c Sodium hydroxide
   d Calcium sulfate
   e Lead iodide

3 Write symbol equations for these reactions. Include state symbols.
   a Lead nitrate reacting with potassium iodide.
   b Iron(II) chloride, $FeCl_2$(aq), reacting with sodium hydroxide.

DID YOU KNOW?

The tiny creatures that make up coral reefs concentrate calcium and carbonate ions around them. When the concentration of these ions is high enough, calcium carbonate precipitates and forms a protective layer around them.

KEY POINTS

1 Insoluble salts are made by mixing solutions of two soluble salts.

2 All salts of Group I elements, ammonium salts and nitrates are soluble in water.

3 Most chlorides, bromides and iodides are soluble. Those of lead and silver are insoluble.

11.4 What's that gas?

LEARNING OUTCOMES

- Describe the tests for hydrogen and oxygen
- Describe tests for ammonia and chlorine
- Describe a test for carbon dioxide using limewater
- Describe a test for sulfur dioxide using potassium manganate(VII)

STUDY TIP

A common mistake is to confuse the tests for hydrogen and oxygen. It may help you to remember that 'lighted' (splint) has an 'h' in it for hydrogen and 'glowing' (splint) has an 'o' in it for oxygen.

DID YOU KNOW?

109 million tonnes of ammonia are produced in the world each year. China produces about a third of this.

The litmus turns blue because an alkaline gas is given off. This is probably ammonia.

Many chemical reactions produce a gas as one of the products. Before we can identify a particular gas we have to collect it. The way we do this depends on:

- the density of the gas – is it heavier or lighter than air?
- the solubility of the gas in water – is it soluble or insoluble?

We can use a gas syringe to collect any gas. But it is easier to identify a gas if we collect it in a test tube. There are three ways we can collect a gas in a test tube:

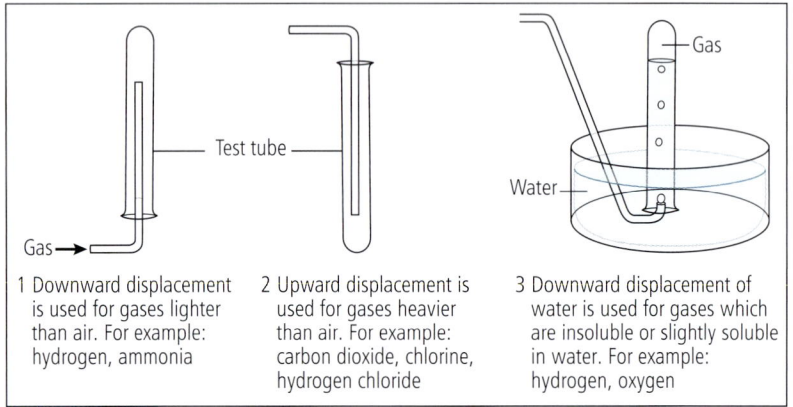

1 Downward displacement is used for gases lighter than air. For example: hydrogen, ammonia

2 Upward displacement is used for gases heavier than air. For example: carbon dioxide, chlorine, hydrogen chloride

3 Downward displacement of water is used for gases which are insoluble or slightly soluble in water. For example: hydrogen, oxygen

Figure 11.4.1 Methods for collecting gases

After collecting the gas in the test tube you put a bung on the tube so that the gas does not escape before you identify it.

Identifying hydrogen

You put a lighted splint at the mouth of the test tube. If the gas is hydrogen it burns with a squeaky 'pop' sound. The hydrogen is reacting with oxygen in the air to cause a small explosion when a flame or spark is present.

Identifying oxygen

You put a glowing splint into the test tube. If the gas is oxygen the splint will relight. The splint is made of wood and wood is a fuel. Fuels burn better in oxygen than in air – there is no nitrogen to dilute the oxygen. So the splint will burn much better in pure oxygen – so much so that the glowing splint will relight.

Using the litmus test for ammonia

We can tell if a gas is acidic or alkaline by holding a piece of damp litmus paper at the mouth of the test tube. If the gas is alkaline it will turn red litmus paper blue. The gas is almost certainly ammonia if there is a strong sharp smell as well. If the gas given off in a reaction is acidic it will turn damp blue litmus paper red.

Identifying carbon dioxide

If we think that a gas given off in a reaction is carbon dioxide, we can bubble it through limewater. If carbon dioxide is present, the limewater turns milky or cloudy. A simpler way to test for carbon dioxide is to simply put a drop of limewater on the end of a flattened glass rod and hold it above the reaction mixture. But take care that the drop does not fall off!

Limewater is a solution of calcium hydroxide. This solution is colourless. But when you bubble carbon dioxide through it, a fine white precipitate of calcium carbonate is formed:

$$Ca(OH)_2(aq) + CO_2(g) \longrightarrow CaCO_3(s) + H_2O(l)$$

calcium hydroxide + carbon dioxide $\longrightarrow$ calcium carbonate + water

Carbon dioxide is an acidic oxide. So it reacts with a base to form a salt and water. If you bubble the carbon dioxide through the limewater for too long the limewater goes colourless again. This is because the calcium carbonate dissolves to form soluble calcium hydrogencarbonate.

Figure 11.4.2 Testing for carbon dioxide

Identifying chlorine

Chlorine is a poisonous green gas. So if you think chlorine is going to be released you should carry out the test in a fume cupboard. You put damp litmus paper or universal indicator paper at the mouth of the test tube. The indicator paper turns white – it is bleached.

Identifying sulfur dioxide

Sulfur dioxide is a colourless gas with a strong acidic smell. It is poisonous, so tests for this gas should be carried out in a fume cupboard. We use an acidified solution of potassium manganate(VII), which is purple in colour, to test for sulfur dioxide. When sulfur dioxide is bubbled through acidified aqueous potassium manganate(VII), the solution turns from purple to colourless.

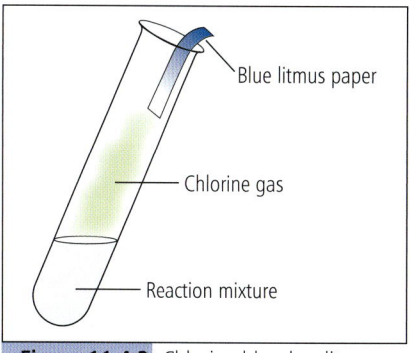

Figure 11.4.3 Chlorine bleaches litmus paper

SUMMARY QUESTIONS

1 Copy and complete using the words below:

   collecting   displacement   heavier   insoluble   water

   There are several methods of _____ a gas. If a gas is _____ than air you collect it by upward _____ of air. If a gas is _____ in water you can collect it over _____.

2 Describe and explain the test for ammonia.

3 Describe the differences between the test for oxygen and the test for hydrogen.

KEY POINTS

1 Oxygen relights a glowing splint.

2 Hydrogen gives a squeaky pop with a lighted splint.

3 Chlorine bleaches damp litmus paper.

4 Carbon dioxide turns limewater milky.

5 Ammonia turns red litmus paper blue

6 Sulfur dioxide turns acidified potassium manganate(VII) colourless.

11.5 Testing for cations

LEARNING OUTCOMES

- Describe tests for these ions in aqueous solution: Al^{3+}, Ca^{2+}, Cr^{3+}, Cu^{2+}, Fe^{2+}, Fe^{3+}, NH_4^+ and Zn^{2+}
- Describe flame tests for Li^+, Na^+, K^+ and Cu^{2+}

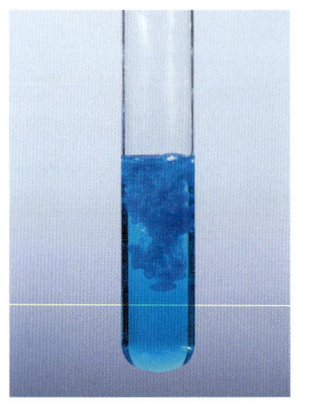

A light blue precipitate forms when we add sodium hydroxide solution to a solution containing Cu^{2+} ions

DID YOU KNOW?

Copper(II) ions are used to clean ponds and keep them free of algae (green slime).

Many compounds look similar in the laboratory. If you have a white powder it could be sodium chloride, magnesium sulfate or aluminium oxide. Even coloured compounds may appear colourless when they are dissolved at low concentration in aqueous solution. For example: iron(II) sulfate has light green crystals but when dissolved in water it appears colourless unless you make a very concentrated solution.

So how can we identify a substance in solution? One way to do this is to use aqueous sodium hydroxide or aqueous ammonia. These alkalis can be used to identify positive ions in compounds. Positive ions are often called **cations** because they move to the **cathode** when an ionic solution is electrolysed.

If you have a solid that you want to identify it is best to dissolve it in a little water first and use this aqueous solution for the test. The procedure for identifying an unknown cation is:

- Put a small amount of the solution you want to identify into a test tube.
- Add a few drops of aqueous sodium hydroxide.
- Observe the colour of any precipitate formed.
- Add excess aqueous sodium hydroxide and shake the test tube.
- Record whether or not the precipitate dissolves, and any colour changes.

This procedure can be repeated using aqueous ammonia instead of sodium hydroxide.

The table shows the results if particular cations are present.

You can see that sodium hydroxide and ammonia react in a similar way with some of the ions. However we can use these two alkalis to distinguish the colourless solutions containing aluminium and zinc ions. If the alkalis are not in excess the precipitates formed are metal hydroxides. The equations for all these reactions are similar.

metal cation	result with aqueous sodium hydroxide	result with aqueous ammonia
aluminium, Al^{3+}	white precipitate soluble in excess (colourless solution)	white precipitate insoluble in excess
calcium, Ca^{2+}	white precipitate insoluble in excess	no precipitate or very slight white precipitate
copper(II), Cu^{2+}	light blue precipitate insoluble in excess	light blue precipitate soluble in excess (dark blue solution)
chromium(III), Cr^{3+}	grey–green precipitate soluble in excess green solution	grey-green precipitate insoluble in excess partly dissolves on standing to form violet solution
iron(II), Fe^{2+}	grey-green precipitate insoluble in excess	grey-green precipitate insoluble in excess
iron(III), Fe^{3+}	reddish-brown precipitate insoluble in excess	reddish-brown precipitate insoluble in excess
zinc, Zn^{2+}	white precipitate soluble in excess (colourless solution)	white precipitate soluble in excess (colourless solution)

For copper(II) ions the equation is:

$$Cu^{2+}(aq) + 2OH^-(aq) \rightarrow Cu(OH)_2(s)$$

copper(II) ions + hydroxide ions ⟶ copper hydroxide

The aluminium and zinc ions dissolve in excess sodium hydroxide because they form soluble aluminates and zincates (see Topic 10.5).

We can also test for ammonium ions using sodium hydroxide solution. When a solution containing ammonium ions is heated with sodium hydroxide solution, ammonia gas is given off. This turns red litmus blue.

$$NH_4^+(aq) + OH^-(aq) \rightarrow NH_3(g) + H_2O(l)$$

ammonium ions + hydroxide ions ⟶ ammonia + water

Flame tests

A flame test can be used to identify some cations, especially those in compounds containing elements from Groups I and II. The procedure is:

- Clean a platinum or nichrome wire by dipping it in concentrated hydrochloric acid.
- Place a sample of the compound on the end of the wire.
- Hold the wire on the edge of a non-luminous (blue) Bunsen flame.
- Note any change in the colour of the flame.

The typical flame test colours for some metal ions are shown in the table.

Metal ion	Flame colour
Lithium (Li^+)	red
Sodium (Na^+)	yellow
Potassium (K^+)	lilac
Copper(II) (Cu^{2+})	blue-green

STUDY TIP

When testing for metal ions using sodium hydroxide, make sure that you identify three things: (i) if there is a precipitate (ii) the colour of the precipitate (iii) what happens when you add excess sodium hydroxide.

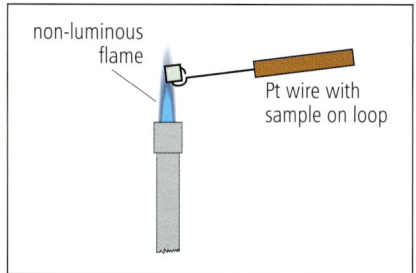

Figure 11.5.1 Carrying out a flame test

SUMMARY QUESTIONS

1 Copy and complete using the words below:

**ammonia   cations   distinctive   hydroxide
precipitate   white   zinc**

A solution containing metal _____ can be identified using aqueous sodium _____ or aqueous _____. A _____ is formed. This often has a _____ colour. Zinc and aluminium ions both form _____ precipitates with aqueous ammonia but only the precpitate from _____ dissolves in excess ammonia.

2 State the colour of the precipitate when solutions containing these ions react with a few drops of sodium hydroxide:

　a iron(III)　　b zinc　　c copper

3 How you can distinguish between a solution containing zinc ions and a solution containing calcium ions?

KEY POINTS

1 We can use aqueous sodium hydroxide and aqueous ammonia to identify unknown cations.

2 When sodium hydroxide is heated with a solution containing ammonium ions, ammonia gas is produced.

3 Aluminium hydroxide and zinc hydroxide precipitates dissolve in excess sodium hydroxide but only zinc ions dissolve in excess aqueous ammonia.

11.6 Testing for anions

LEARNING OUTCOMES

- Describe the tests for halide ions
- Describe the tests for carbonates, nitrates and sulfates and sulfites

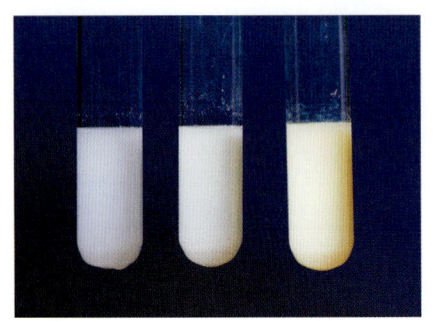

In the last topic we saw that we can test an unknown substance to identify the positive ion present. We used an alkali to do this. We can also carry out particular tests to identify negative ions. Negative ions are called **anions**. If we want to find the type of anion present in an unknown compound we have to use a variety of tests.

When we have completed all our tests for both cations and anions we can identify the unknown compound. We can also carry out further tests to confirm our conclusions. The whole process of finding out what elements are present in a compound is called **qualitative analysis**.

Identifying chlorides, bromides and iodides

We call the simple ions of the Group VII elements **halides**. Chlorides, bromides and iodides are all halides. Halides can be identified using aqueous silver nitrate. The procedure is as follows.

To a small volume of the halide solution in a test tube:

- add an equal volume of dilute nitric acid
- add a few drops of aqueous silver nitrate
- observe the colour of the precipitate.

Chlorides give a white precipitate.

Bromides give a cream precipitate.

Iodides give a pale yellow precipitate.

The precipitates are the silver halides, for example:

$$Ag^+(aq) + Br^-(aq) \rightarrow AgBr(s)$$

silver ions + bromide ions → silver bromide

In the presence of sunlight the silver chloride precipitate goes greyish-purple very quickly. The silver bromide goes greyish-purple slowly. This **photochemical reaction** was discussed in Topic 8.6.

Chloride, bromide and iodide ions give different colour precipitates with silver nitrate

STUDY TIP

Remember that you add nitric acid and silver nitrate in the test for halide ions. If you add hydrochloric acid you will be adding chloride ions!

Identifying carbonate ions

We add dilute acid to the unknown compound. The unknown compound can be either a solid or a solution. If a carbonate is present we will see effervescence (bubbles of gas). We test to see if the gas given off is carbon dioxide using limewater.

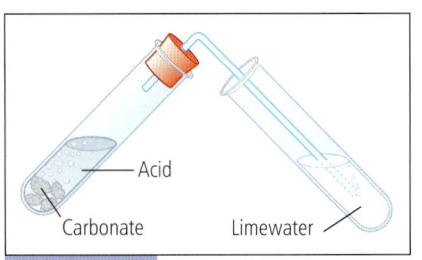

Identifying nitrates

The identification of nitrates makes use of the test for ammonia. The procedure is:

- Put an aqueous solution of the unknown compound into a test tube. Add aqueous sodium hydroxide, then aluminium foil and warm gently.

Figure 11.6.1 The test for carbonates

- Test the gas given off with a piece of damp red litmus paper placed at the mouth of the test tube. If ammonia is given off, the litmus paper will turn blue. So the compound is likely to be a nitrate.

Identifying sulfates

Barium chloride or barium nitrate solution is used to test for sulfates.

The procedure is:

- Put an aqueous solution of the unknown compound into a test tube.
- Add an equal volume of dilute hydrochloric acid and then add an aqueous solution of a soluble barium salt. This can be barium chloride or barium nitrate.
- If a white precipitate is formed the compound is a sulfate.

The equation for this reaction is:

$$Ba^{2+}(aq) + SO_4^{2-}(aq) \longrightarrow BaSO_4(s)$$

barium ions   sulfate ions   barium sulfate (white precipitate)

Indentifying sulfites

Most sulfites are insoluble. The exceptions are sulfites of the Group II elements and ammonium sulfite. Sulfites contain the ion SO_3^{2-}. So the formula for sodium sulfite is Na_2SO_3 and the formula for calcium sulfite is $CaSO_3$.

To test for the presence of a sulfite ion, we add dilute hydrochloric acid and warm gently. If sulfur dioxide is given off, we know that a sulfite is present. Sulfur dioxide turns acidified potassium manganate(VII) from purple to colourless. The test is most easily done by placing a piece of filter paper soaked in acidified potassium manganate(VII) above the test tube containing the acid and sulfite.

SUMMARY QUESTIONS

1 Name the anions present in the following precipitates:
  a Nitric acid was added to a solution followed by aqueous silver nitrate. A pale yellow precipitate was observed.
  b Hydrochloric acid was added to a solution followed by aqueous barium chloride. A white precipitate was observed.

2 State the tests to identify these anions and give the result if the test is positive.
  a nitrate
  b carbonate
  c chloride

3 A student wants to identify the compound in solution Y. To one sample of Y she adds nitric acid followed by aqueous silver nitrate. A pale yellow precipitate is formed. To another sample of Y she adds a solution of sodium hydroxide. A blue precipitate is formed. Identify the compound in solution Y.

KEY POINTS

1 Carbonates can be identified by adding a dilute acid to an unknown compound then testing the gas produced with limewater.

2 Halides are identified by the colour of the precipitate obtained when silver nitrate is added.

3 Nitrates are identified by adding sodium hydroxide and aluminium foil, warming gently, then testing for the release of ammonia gas.

4 Sulfates are identified by adding acidified barium chloride to the unknown compound. A white precipitate indicates the presence of a sulfate.

5 Sulfites are identified by adding dilute hydrochloric acid, warming and testing for sulfur dioxide.

SUMMARY QUESTIONS

1 Match the ions or molecules on the left with the correct test reagents on the right.

iron(II) ions	silver nitrate
iodide ions	acidified barium nitrate
sulfate ions	heat with aluminium powder and sodium hydroxide
carbon dioxide	sodium hydroxide
nitrate ions	limewater

2 Copy and complete using words from the list below:

crystallise evaporating excess filtered filtrate salt sulfuric water

Zinc sulfate is a _____ that can be made by reacting _____ zinc with _____ acid. The excess zinc is _____ off. The _____ of zinc sulfate solution is then put into an _____ basin. Some of the _____ is boiled off and the solution is left to _____.

3 Both hydrogen and oxygen can be identified by a test involving a wooden splint.
   a State the differences between the tests for these two gases.
   b State the differences in the results obtained from these tests.

4 Draw a flow chart to show how to make crystals of sodium chloride from aqueous solutions of hydrochloric acid and sodium hydroxide.

5 Describe three different methods of making the soluble salt calcium chloride. For each method describe:
   • suitable reagents that can be used
   • which reagent, if any, is in excess.

6 Which of the following compounds are soluble in water and which are insoluble?
   a silver chloride   b sodium bromide
   c calcium carbonate d sodium carbonate
   e lead sulfate   f magnesium chloride
   g lead nitrate   h iron(II) hydroxide

PRACTICE QUESTIONS

1 Sodium hydroxide is added to solution M. A reddish-brown precipitate is formed. Solution M contains:
   A iron(II) ions
   B iron(III) ions
   C copper(II) ions
   D silver(I) ions.

   (Paper 1)

2 Which one of these methods is used to prepare the salt ammonium chloride from aqueous ammonia?
   A Precipitation reaction
   B Adding ammonia to an insoluble oxide
   C Adding aqueous ammonia to a metal
   D Titration

   (Paper 1)

3 Which one of the following ions gives a red colour in the flame test?
   A copper
   B lithium
   C potassium
   D sodium

   (Paper 1)

7 Suggest the best method for making each of the following salts. Choose from:
   i   titration
   ii precipitation
   iii adding an insoluble metal or metal compound to an acid.
   a Lead bromide from lead nitrate and potassium bromide
   b Potassium nitrate from nitric acid and potassium hydroxide
   c Iron(III) hydroxide from iron(III) chloride and sodium hydroxide
   d Copper(II) sulfate from copper(II) carbonate and sulfuric acid.

4 Some of the instructions for making crystals of magnesium sulfate are given below.
- Heat excess magnesium oxide with sulfuric acid.
- Filter off the excess magnesium oxide.
- Put the filtrate into an evaporating basin.

(a) Complete the instructions to produce pure dry crystals of magnesium sulfate. [3]

(b) Why was excess magnesium oxide used? [1]

(c) (i) Describe a test for sulfate ions. [1]

(ii) Describe what you observe in this test if sulfate ions are present. [1]

(Paper 3)

5 M is a white powder that dissolves in water to form a solution with a pH of 3.5.

(a) Describe the effect of a solution of M on blue litmus paper. [1]

(b) When a solution of M is added to solid L, a colourless gas is given off. How can you show that this gas is carbon dioxide? [2]

(c) The solution formed contains calcium ions. How can you show that the solution contains calcium ions? [3]

(d) Identify solid L. Give reasons for your answer. [3]

(Paper 3)

6 A student wants to make the soluble salt potassium chloride from potassium hydroxide using a titration method.

(a) Suggest a suitable compound that the student could add to potassium hydroxide to make potassium chloride by this method. [1]

(b) Draw a labelled diagram of the apparatus required to carry out a titration. [3]

(c) Describe how to carry out a titration. [3]

(d) Describe how to make colourless crystals of potassium chloride using this method. [3]

(e) How can you show that the crystals contain chloride ions? [3]

(Paper 3)

7 Lead iodide, PbI_2, and barium sulfate, $BaSO_4$, are insoluble salts.

(a) Suggest two compounds that you can use to make lead iodide. [2]

(b) Write an ionic equation for this reaction. Include state symbols. [3]

(c) Describe the method used to make dry lead iodide crystals. [3]

(d) How can you show that the crystals contain iodide ions? [3]

(e) Barium sulfate can be prepared by the following reaction:

$$Ba(NO_3)_2(aq) + Na_2SO_4(aq) \longrightarrow BaSO_4(s) + 2NaNO_3(aq)$$

(i) Write an ionic equation for this reaction. [2]

(ii) How can you show that a solution of barium nitrate contains nitrate ions? [3]

(Paper 4)

8 Most iron(II) and iron(III) salts are soluble in water.

(a) (i) How can you distinguish between solutions of iron(II) and iron(III) salts using sodium hydroxide? [2]

(ii) Explain why addition of ammonia to iron(II) and iron(III) salts gives the same results as in part (i). [2]

(b) Crystals of iron(II) chloride gradually change from light green to yellowish brown when left in the air.

(i) Suggest why this colour change happens. [2]

(ii) Write an ionic equation for this reaction. [1]

(c) Very dilute solutions of iron(II) chloride and iron(II) sulfate have similar colours. Explain how you can distinguish solutions of these compounds. Give full details of the tests you can carry out as well as the expected results. [3]

(d) (i) Describe how you can make the soluble salt iron(II) sulfate from iron. [4]

(ii) Write an equation for the reaction you have chosen in part (i). Include state symbols. [3]

(Paper 4)

12 The Periodic Table

12.1 The Periodic Table

LEARNING OUTCOMES

- Describe the arrangement of the elements in the Periodic Table
- Describe the change from metallic to non-metallic character across a period
- Describe the relationship between the group number, number of outer electrons and metallic/non-metallic character
- Identify trends in the Periodic Table

We have already seen that the elements in the Periodic Table are arranged in order of their **proton number** (**atomic number**). In the Periodic Table the elements are arranged so that elements with similar properties fall under each other in vertical columns. We call these vertical columns **groups**.

We call the horizontal rows **periods**. You will notice that hydrogen is all on its own and not given a group. We do not put it in Group I. Although it has one electron in its single shell it has totally different properties from the Group I metals.

Metals are on the left-hand side of the Periodic Table and non-metals are on the right. But the dividing line between metals and non-metals is not very clear. If you look at the dividing line between metals and non-metals you will see that it looks like a staircase. Just either side of this dividing line some elements show a combination of metallic and non-metallic properties. We call these elements metalloids.

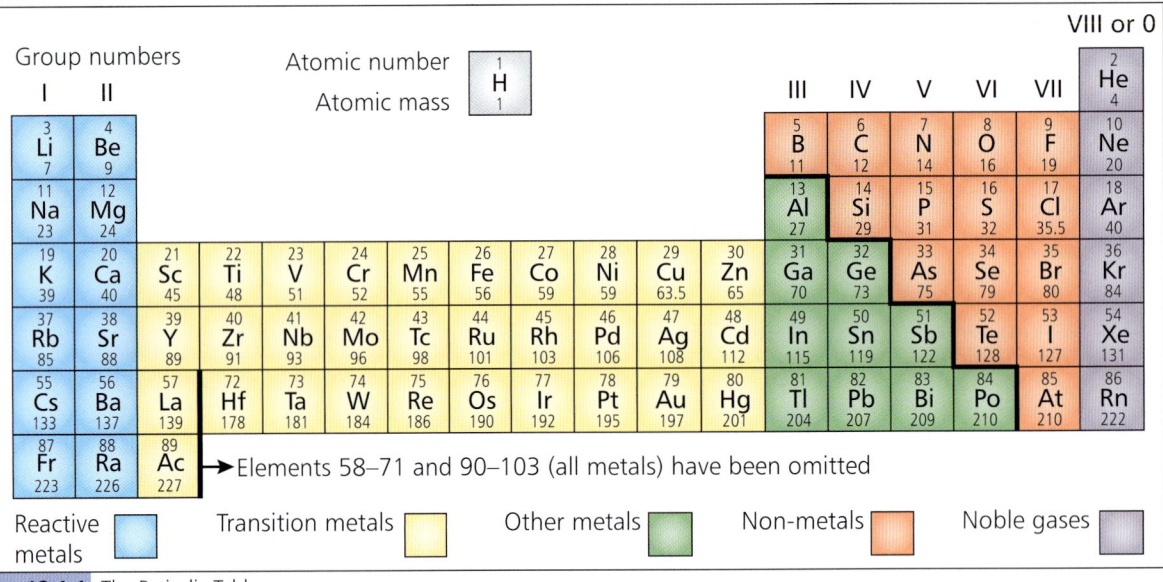

Figure 12.1.1 The Periodic Table

Trends down the groups

Elements in the same group have similar chemical properties. This is due to the fact that each element in the group has the same number of electrons in its outer shell. These are the **valency electrons**. Group I elements have one valency electron, Group II elements have two and so on.

Within each group we can identify trends in physical and chemical properties down the group. In Group I the elements get more reactive down the group but in Group VII they get less reactive down the group. In many groups there is also a trend from being less metallic at the top of the group to more metallic at the bottom. This is very obvious in Group IV where carbon at the top is a non-metal but tin and lead at the bottom of the group are metals.

DID YOU KNOW?

Mendeleev was addicted to the game of 'patience', which involves playing cards placed vertically and horizontally. Perhaps he used this as an idea when arranging the elements in the Periodic Table.

Outer electrons and the Periodic Table

The number of outer electrons in the Periodic Table varies with the Group number.

Group	I	II	III	IV	V	VI	VII	VIII
number of outer shell elecrons	1	2	3	4	5	6	7	8

Groups I–III are metals (except boron). Their atoms form ions by losing electrons.

Groups IV and V have non metals at the top and metals at the bottom (see Fig 12.1.1).

Groups VI and VII are mainly non-metals. The atoms of groups VI and VII form ions by gaining electrons.

Trends across a period

As you read down the Periodic Table each period has one more electron shell than the one before it. Hydrogen and helium are the only two elements in Period 1 because the first shell can hold a maximum of only two electrons.

The second period starts with lithium. As you read across Period 2 each successive element has one more electron in its outer shell until you reach neon with the maximum number of eight electrons. The third period is similar. The fourth period is complicated by the transition elements which form a block of metals in the middle of the Periodic Table.

There are also trends across a period. For Periods 2 and 3 the melting and boiling points tend to increase up to Group IV and then decrease again towards Group VIII. This reflects the different structures of the elements in each group. So for Period 2: on the left we have metallic structures; in the middle we have a giant covalent structure (carbon) in Group IV; on the right we have non-metals with molecular structures.

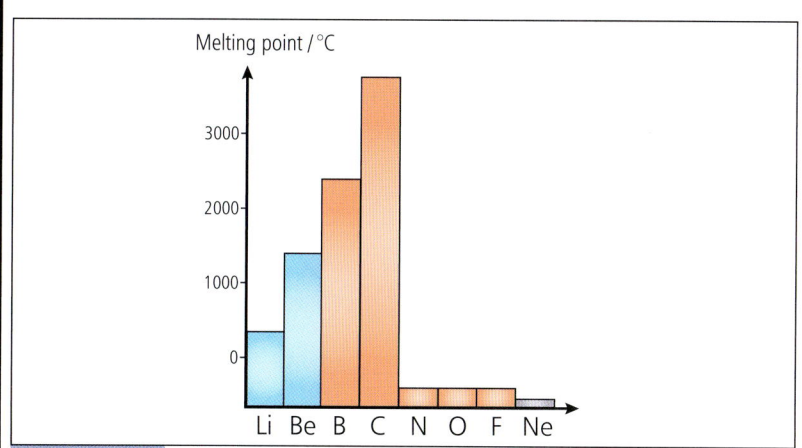

Figure 12.1.2 How the melting points vary across Period 2

As you read across a period there are also differences in chemical properties. We have the basic oxides on the left and the acidic oxides on the right. Most metal oxides are basic oxides and most non-metal oxides are acidic oxides.

STUDY TIP

You need to know where the metals and non-metals appear in the Periodic Table. You do not have to remember exactly the dividing line between metals and non-metals in Groups III to VI.

KEY POINTS

1 Elements in the same group of the Periodic Table often have similar properties.

2 Metallic character decreases across a period but increases down a group.

3 There are trends in properties down a group as well as across a period.

SUMMARY QUESTIONS

1 Copy and complete using the words below:

**chemical eight electrons
groups proton outer**

In the Periodic Table the elements are arranged in order of increasing _____ number. Across Periods 2 to 6 the number of outer _____ increases to a maximum of _____. Elements in many _____ have similar _____ properties because they have the same number of _____ electrons.

2 Describe how:

a metallic character

b melting point

change across a period.

12.2 The Group I metals

LEARNING OUTCOMES

- Describe the trends in physical properties of the Group I metals
- Describe how lithium, sodium and potassium react with water
- Predict the properties of other Group I metals

The Group I metals are called the **alkali metals**. They are a family of metals with similar chemical properties. They are rather unusual metals because they are soft and have fairly low melting points.

Lithium, sodium and potassium are the only Group I metals you will see in the school laboratory, so we will concentrate on these three. These metals have to be stored under oil to stop them reacting with oxygen in the air. When cut, they show a silvery surface that oxidises very quickly.

Physical properties of the alkali metals

The table shows some physical properties of lithium, sodium and potassium.

metal	electronic structure	density / g per cm³	melting point / °C	boiling point / °C	hardness
lithium	2,1	0.53	181	1342	fairly soft
sodium	2,8,1	0.97	98	883	soft
potassium	2,8,8,1	0.86	63	760	very soft

There are several trends down the group:

- The melting points and boiling points decrease down the group.
- The metals get softer down the group.
- There is a general increase in density down the group. At first sight it seems that the densities show no trend. But if we include the other Group I elements we can see that there is one:

density in g/cm³: Li 0.53; Na 0.97; K 0.86; Rb 1.53; Cs 1.88

It is sodium that upsets the pattern.

We can use these trends to predict the physical properties of other alkali metals. For example, we can predict that the melting point of rubidium will be lower than the melting point of potassium by about 20–30 °C. This gives a melting point for rubidium of about 33–43 °C. Its actual melting point is 39 °C.

Figure 12.2.1 The alkali metals

Alkali metals are stored in oil

Potassium (on the left) reacts more rapidly with water than lithium (on the right)

DEMONSTRATION

Reacting alkali metals with water

Small pieces of lithium, sodium and potassium are dropped into a large trough of water one by one. You observe what happens. When each reaction has finished a few drops of universal indicator are added to the trough.

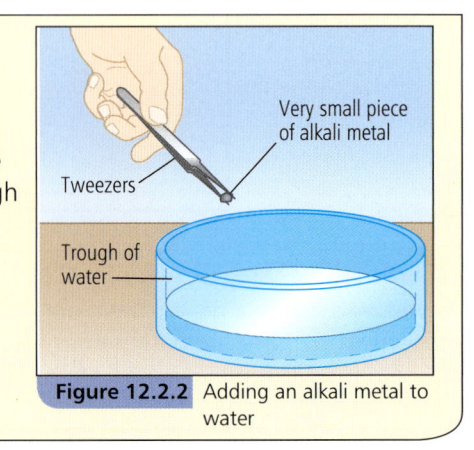

Figure 12.2.2 Adding an alkali metal to water

The reaction of the alkali metals with water

When we add the alkali metals to water, bubbles are formed and we hear a fizzing sound. We can make other observations too.

The results are shown in the table:

lithium	fizzes slowly, a few bubbles	disappears slowly	moves slowly on the surface	remains solid, no flames
sodium	fizzes quickly, many bubbles	disappears quickly	moves quickly on the surface	melts into a liquid ball, no flames
potassium	fizzes violently, even more bubbles	disappears very quickly	moves very quickly on the surface	melts into a liquid ball, violet flame

You can see that the reactions get more vigorous down the group. So you can predict that the reaction of rubidium with water will be very violent, bursting into flame very quickly, with lots of bubbles. It may even explode!

The reactions of the alkali metals with water are very similar – it's just the reactivity that is very different. The bubbles and fizzing are caused by hydrogen gas which is released in the reaction. We observe a flame with potassium because the reaction is violent enough to make the hydrogen catch light. Sparks are often seen as well.

When we add universal indicator to the trough, the solution turns purple. This shows that an alkali has been formed. Alkalis have OH^- ions which have come from the metal hydroxides formed. That's why the Group I metals are called alkali metals. We can write similar equations for each metal reacting with water. For example:

$$2Li(s) + 2H_2O(l) \rightarrow 2LiOH(aq) + H_2(g)$$
lithium + water → lithium hydroxide + hydrogen

$$2K(s) + 2H_2O(l) \rightarrow 2KOH(aq) + H_2(g)$$
potassium + water → potassium hydroxide + hydrogen

SUMMARY QUESTIONS

1 Copy and complete using the words below:

alkali   fire   fizzes   hydrogen   hydroxide   increases   rapidly   surface

The Group I metals are also called _____ metals. They react readily with water to produce _____ gas and an alkaline solution of a metal _____. Their reactivity _____ down the group. Lithium _____ slowly on the _____ of the water but potassium fizzes _____ and the hydrogen produced catches _____.

2 The boiling point of sodium is 883°C. The boiling point of potassium is 760°C. Predict the boiling point of rubidium.

3 Write a balanced equation for the reaction of sodium with water.

STUDY TIP

When you describe observations, concentrate on what you see, hear, smell or feel by touch.

STUDY TIP

When you compare Group I metals, remember that they have 'similar properties' NOT 'the same properties'. The properties change slightly down the group.

DID YOU KNOW?

Francium was the last element discovered that was natural rather than made in the laboratory. Only about 30 grams of francium exist in the Earth's crust at any one time.

KEY POINTS

1 The Group I elements show a trend in physical properties such as hardness, melting points and density down the group.

2 Group I elements all react with water to produce hydrogen and a solution of the alkali metal hydroxide.

3 The chemical reactivity of the alkali metals increases down the group.

12.3 The Group VII elements

LEARNING OUTCOMES

- Describe the trends in physical properties of the halogens
- Describe how the halogens react with halide ions
- Predict the properties of other elements in Group VII

Chlorine is a gas, bromine is a liquid and iodine is a solid at room temperature and pressure

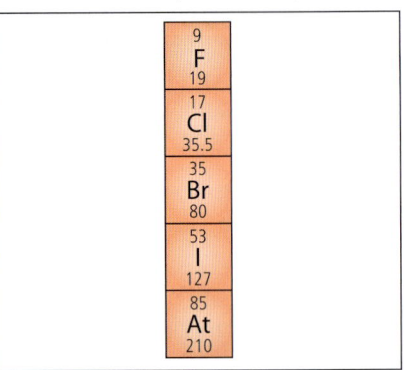

Figure 12.3.1 The Group VII elements

DID YOU KNOW?

About 2000 kg of poisonous mercury pollutes the atmosphere and oceans every year as a result of making chlorine using out-of-date electrolysis methods.

The Group VII elements are called the **halogens**. They are poisonous non-metals that have low melting and boiling points. They all exist as **diatomic** molecules – they have two atoms in each molecule. Chlorine, bromine and iodine are the only halogens that you will see in the school laboratory, so we will concentrate on these three.

Trends in physical properties

The halogens show trends in their physical properties:

halogen	electronic structure	melting point/°C	boiling point/°C	state at r.t.p.	colour
chlorine	2,8,7	−101	−35	gas	green
bromine	2,8,18,7	−7	+59	liquid	reddish-brown
iodine	2,8,18,18,7	+114	+184	solid	greyish-black

You can see from the table that:

- the melting and boiling points of the halogens increase down the group. This is the opposite trend to the Group I metals.
- as a result of the trend in melting and boiling points, the state of the halogens at room temperature changes from gas to liquid to solid down the group.
- the colour gets darker down the group.

You can predict the properties of other halogens by observing the trends down the group. For example, fluorine will be lighter in colour than chlorine and will have a lower boiling point. Fluorine is a pale yellow gas that boils at −188 °C. We can also predict that astatine will be a black solid with a higher boiling point than iodine.

Trends in chemical reactivity

Reactions with Group I elements

The salts formed when metals react with halogens are called halides. Chlorides, bromides and iodides are all halides. Chlorine, bromine and iodine all react with sodium to form halides. For example:

$$2Na(s) + Cl_2(g) \xrightarrow{heat} 2NaCl(s)$$

sodium + chlorine → sodium chloride

The reactivity of the halogens with sodium decreases down the group. This reactivity trend is opposite to the order of reactivity of the Group I elements.

Reactions of halogens with halide ions

When an aqueous solution of chlorine reacts with aqueous potassium bromide, potassium chloride and bromine are formed:

$$Cl_2(aq) + 2KBr(aq) \longrightarrow 2KCl(aq) + Br_2(aq)$$

chlorine + potassium bromide → potassium chloride + bromine

We call this a **displacement** reaction because one type of atom has replaced another. In this case chlorine has replaced the bromine in the potassium bromide. We say that chlorine has *displaced* the bromine. We can use displacement reactions to show that the trend in reactivity of the halogens decreases down the group.

PRACTICAL

Reacting halogens with halides

In this experiment we use aqueous solutions of halogens and halides. We add the halogen to the halide and observe any change of colour. We repeat this with different combinations of halogens and halides.

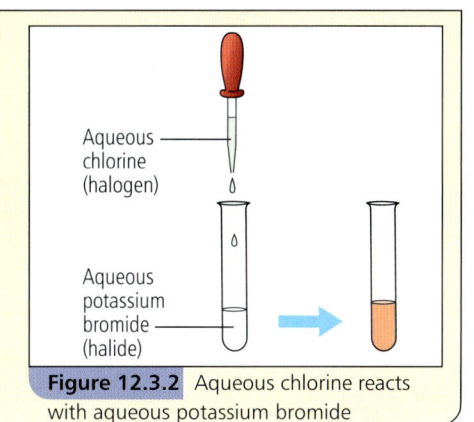

Figure 12.3.2 Aqueous chlorine reacts with aqueous potassium bromide

The table shows the colour changes when halogens are added to different halides. A cross shows that there is no colour change. In aqueous solution chlorine is very light green, bromine is orange and iodine is brown.

halogen	halide		
	potassium chloride	potassium bromide	potassium iodide
chlorine	x	turns orange	turns brown
bromine	x	x	turns brown
iodine	x	x	x

In these reactions the colour changes show that chlorine displaces bromine (orange) from potassium bromide, and iodine (brown) from potassium iodide. Bromine has only displaced iodine from potassium iodide. Iodine has not reacted at all. So the more reactive halogen displaces the less reactive halogen from a solution of its halide. Chlorine is the most reactive and iodine is the least reactive of these three halogens. So, aqueous iodine will not react with aqueous potassium bromide. This is because iodine is less reactive than bromine. The equations for all the displacement reactions are similar to the one at the bottom of p 156.

Supplement

These displacement reactions are examples of **redox reactions**. We can see this more clearly if we write ionic equations for the reactions. So for the reaction:

$$Cl_2(aq) + 2KBr(aq) \longrightarrow 2KCl(aq) + Br_2(aq)$$

the ionic equation is:

$$Cl_2(aq) + 2Br^-(aq) \longrightarrow 2Cl^-(aq) + Br_2(aq)$$

Chlorine has gained electrons to form chloride ions – this is reduction.

Bromide ions have lost electrons to form bromine – this is oxidation.

STUDY TIP

Make sure that you can distinguish between the halogens (elements) and halides (compounds). It is a common error to write chlorine ions instead of chloride ions.

KEY POINTS

KEY POINTS

1. Halides are salts formed when metals react with halogens.

2. The reactivity of halogens decreases down the group.

3. A more reactive halogen displaces a less reactive halogen from a solution of its halide.

SUMMARY QUESTIONS

1. Copy and complete using the words below:

   **chlorine   decreases
   diatomic   halogens
   iodine   liquid**

   We call the Group VII elements _____. All halogens have _____ molecules. Their reactivity _____ down the group. At r.t.p._____ is a green gas, bromine is a reddish-brown _____ and _____ is a greyish-black solid.

2. Explain why aqueous potassium chloride does not react with bromine.

3. **Supplement** Write an ionic equation for the reaction of chlorine with aqueous potassium iodide.

12.4 The noble gases and more

LEARNING OUTCOMES

- Describe the noble gases as being unreactive, monatomic gases and explain this in terms of electronic structure
- State the uses of the noble gases in providing an inert atmosphere
- Identify trends in groups other than I, VII and VIII given information about the elements concerned

The noble gases

The gases in Group VIII or 0 are called the **noble gases**. All these gases are unreactive because they have a full outer shell of electrons. They do not need to gain, lose or share electrons to form compounds. This is why they exist as single atoms – they are **monatomic**.

Many of the noble gases are found in the air. Argon forms about 1% of the air. The others are present in very small amounts. We usually obtain the noble gases by the fractional distillation of liquid air. We can extract helium from natural gas where it occurs as an impurity.

Properties of the noble gases

Physical properties: The noble gases show definite trends in their physical properties.

The density of the noble gases increases down the group. Helium is much lighter than air and neon is a little lighter than air. The other noble gases are heavier than air.

Chemical properties: the noble gases are colourless non-metals. They are monatomic (exist as single atoms) and are unreactive chemically. This is because their electronic structure with eight outer shell electrons (or two in helium) is energetically stable. Electrons in their outer shell cannot be easily removed, added or shared with other atoms to form compounds.

Uses of the noble gases

Many of the uses of the noble gases rely on the fact that they are unreactive.

Helium is used in weather balloons and airships because it is lighter than air. It is safer to use than hydrogen because it is inert. Hydrogen is highly flammable. A mixture of helium and oxygen is used by deep-sea divers.

Neon is used in advertising signs because it glows red when a high voltage is passed through it.

Argon is used as an inert atmosphere in the extraction of titanium to stop the metal oxidising as it is extracted. It is also used to provide an inert atmosphere during welding. We also use argon in electric light bulbs containing a tungsten filament to prevent the filament burning.

Krypton is used in lasers for eye surgery and in bulbs for car headlamps.

Xenon is used in lamps where a very bright light is required. Examples include in lamps for lighthouses and hospital operating theatres. It is also used in lasers.

Figure 12.4.1 All the noble gases are found in Group VIII of the Periodic Table

STUDY TIP

It is better to write that the noble gases are unreactive 'because the electronic configuration is energetically stable' than to write 'the outer shells are full or complete, or have eight electrons'.

Trends in other groups

We can identify trends in physical and chemical properties in most groups in the Periodic Table. You do not have to remember these trends but you should be able to identify them when given suitable information.

In Groups III to VI these trends are often associated with a change from non-metals to metals down the group. Very often these trends are general – sometimes one element may spoil a regular pattern.

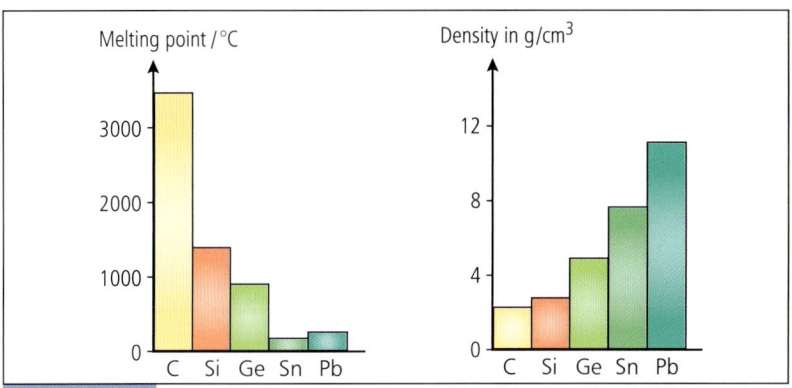

Figure 12.4.2 Trends in the melting point and density of Group IV elements

Sometimes the differences in the chemical properties within a group can be very marked. In Group VI the non-metal, oxygen, at the top of the group has totally different properties from the metal polonium at the bottom of the group.

Group II metals have some trends similar to Group I:

- they show increased reactivity with oxygen down the group.
- they show increased reactivity with water down the group.

But they do not show a very regular trend in density and melting points. There are other trends though. For example, the solubility of the hydroxides increases down Group II.

Helium is used to fill balloons because it is less dense than air

DID YOU KNOW?

Although it is often said that the noble gases are totally unreactive, some compounds of argon and xenon have been made.

SUMMARY QUESTIONS

1 Copy and complete using the words below:

   electrons   energy   Group   outer   noble   unreactive

   The elements in _____ VIII of the Periodic Table are called the _____ gases. They are generally _____ because it takes too much _____ to add or remove _____ from the _____ shell.

2 Explain why:
   a helium is used in weather balloons
   b argon is used in tungsten filament light bulbs.

3 Describe the trend in the density of the noble gases down the group.

KEY POINTS

1 The noble gases are unreactive because their electronic structure is energetically stable.

2 The noble gases show trends in physical properties down the group.

3 Many of the uses of the noble gases depend on their unreactive nature.

12.5 Transition elements

LEARNING OUTCOMES

- Describe transition elements as metals having high densities, high melting points, forming coloured compounds and acting (as elements and compounds) as catalysts
- Know that transition elements have variable oxidation states

Supplement

The **transition elements** form a block of about 30 elements in the middle of the Periodic Table. They are all metals with typical metallic properties. They all conduct heat and electricity, are **malleable** and **ductile**, and are shiny and sonorous.

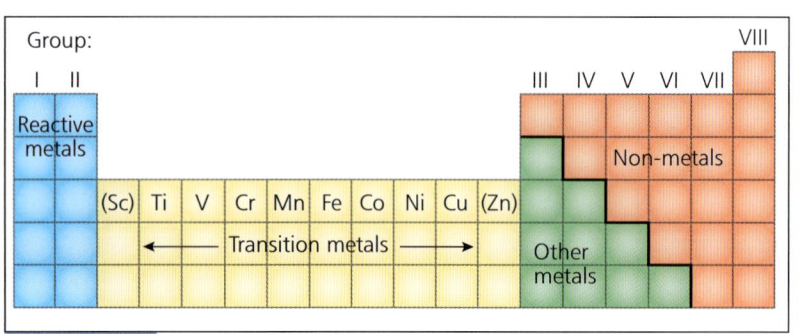

Figure 12.5.1 The position of the transition metals in the Periodic Table

Many of the properties of transition elements are similar but they are very different from the metals in Groups I, II and III. We can see this by comparing some of their physical and chemical properties.

This jet engine has strong turbine blades made from transition metal alloys that can withstand high temperatures

STUDY TIP

It is a common error to suggest that transition elements are highly coloured. It is the <u>compounds</u> of transition elements which have a range of colours.

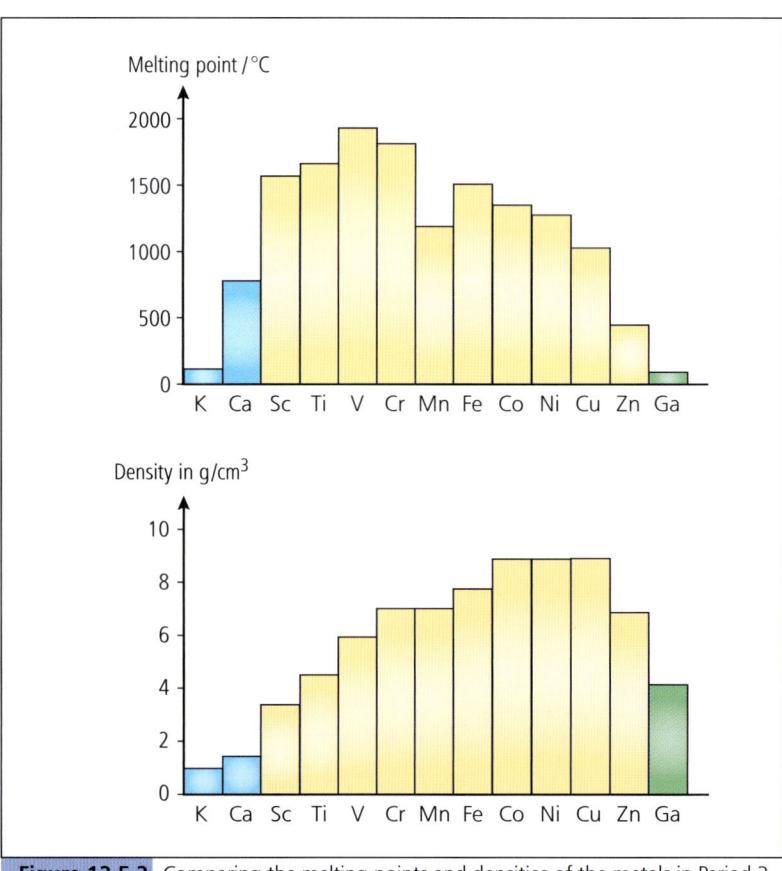

Figure 12.5.2 Comparing the melting points and densities of the metals in Period 3

Physical properties

We use the following properties to distinguish transition metals from metals in Groups I, II and III:

Transition elements:

- have very high melting and boiling points. For example, the melting point of chromium is 1857°C. The melting point of potassium in Group I is only 63°C.
- have very high densities. Compare the densities of chromium and potassium: chromium 7.2 g/cm³; potassium 0.86 g/cm³.
- are stronger and harder than Group I metals.

Chemical properties

Many of the chemical properties of the transition elements set them apart from other metals:

- They form coloured compounds. For example iron(II) salts are often light green in colour but iron(III) salts are yellow or brown. Salts of Group I and II metals on the other hand are usually colourless.
- They form complex ions. These are ions that are complicated in structure. In Topic 11.5 we identified copper(II) ions in solution by adding ammonia. When we add excess ammonia we get a dark blue solution. This colour is due to a complex ion. Ammonia forms bonds with the copper ions.
- Many transition elements and transition element oxides are good **catalysts**. Iron is a catalyst for the Haber process (see Topic 16.2). Vanadium(V) oxide is the catalyst used in the manufacture of sulfuric acid.
- They are less reactive than metals from other groups. They do not react with cold water, although some react with steam.
- They have more than one oxidation state – they have variable valency. For example, iron can form iron(II) ions, Fe^{2+}, or iron(III) ions, Fe^{3+}. Many transition elements have a very wide range of oxidation states. For example, manganese can exist in positive oxidation states ranging from +1 to +7. This does not mean that you can get ions with a 7+ charge. You cannot get ions with more than a 3+ charge.

DID YOU KNOW?

Coins made from transition elements are self-sterilising because the transition element ions poison any bacteria that may stick to their surface.

STUDY TIP

Remember that oxidation state does not always refer to the charge on the ions. For example, in potassium manganate(VII), $KMnO_4$, the oxidation state of manganese is +7 but the manganese ion with the highest charge is Mn^{2+}.

SUMMARY QUESTIONS

1 Copy and complete using the words below:

block   catalysts   coloured   densities   high   middle

The transition elements form a ____ of about 30 metals in the ____ of the Periodic Table. Most transition elements have ____ melting and boiling points and high ____ compared with the metals in Groups I, II and III. Transition elements form ____ compounds and are good ____.

2 State two differences in physical properties that distinguish transition elements from Group I elements.

3 Why do most oil paints used by artists contain transition element compounds?

KEY POINTS

1 Transition elements have high melting and boiling points and high densities.

2 Transition elements have several different oxidation states in their compounds.

3 Transition elements and their compounds often act as catalysts

4 Compounds of transition elements are highly coloured.

SUMMARY QUESTIONS

1 Copy and complete using words from the list.

alkali   coloured   darker   high   less melting   middle   oxidation   soft

The _____ metals are a group of _____ metals that decrease in _____ point down the group. Halogens are non-metals that get _____ in colour and _____ reactive down the group. The transition elements are a block of 30 elements in the _____ of the Periodic Table. They are metals with very _____ melting points and densities. They form _____ compounds in which they have a variety of _____ states.

2 Match the elements on the left with the phrases on the right.

chlorine	an unreactive gas
bromine	a metal with a very high melting point
lithium	the least reactive of the alkali metals
neon	a green poisonous gas
potassium	a reddish-brown liquid
iron	a metal that catches fire when it reacts with water

3 State four differences between an alkali metal and a transition element such as nickel.

4 List the trends in physical properties of:
   (a) the Group I metals   (b) the halogens.

5 Describe the differences between the reactions of lithium and potassium with water.

6 Write word equations for the reaction of:
   (a) sodium with water
   (b) sodium with chlorine
   (c) aqueous chlorine with aqueous potassium iodide.

Supplement

7 Write symbol equations for the reaction of:
   (a) lithium with water
   (b) aqueous bromine with aqueous sodium iodide.

8 Describe and explain how metallic character changes across a period.

9 Explain why chlorine reacts with aqueous potassium bromide but bromine does not react with aqueous potassium chloride.

PRACTICE QUESTIONS

1 Which one of these statements about the reaction of the Group I elements with water is true?

A A solution of halide ions is formed.

B An acidic solution is formed.

C Oxygen is given off.

D An alkaline solution is formed.

(Paper 1)

2 All transition elements in their compounds:

A are magnetic

B have variable oxidation states

C are white in colour

D form ions of the type M^- and M^{2-}.

(Paper 2)

3 Some properties of the Group I elements are shown in the table.

element	melting point / °C	boiling point / °C	reaction with water
lithium	180	1330	fairly reactive
sodium	98		reactive
potassium	64	760	very reactive
rubidium		690	

   (a) Suggest values for (i) the melting point of rubidium (ii) the boiling point of sodium. [2]

   (b) How does the reactivity of rubidium compare with the reactivity of potassium? [1]

   (c) Describe three observations that you can make when potassium reacts with water. [3]

   (d) The solution formed when potassium reacts with water is alkaline.
       (i) How can you show that the solution is alkaline? [2]
       (ii) Write a word equation for the reaction of potassium with water. [2]

(Paper 3)

4 The diagram shows some elements in the Periodic Table.

Li	Be	B	C	N	O	F	Ne
Na	Mg					Cl	Ar

(a) What determines the order of the elements in the Periodic Table? *[1]*

(b) To which period does chlorine belong? *[1]*

(c) Choose from the elements shown in the table above to answer these questions:
  (i) Which elements are halogens? *[1]*
  (ii) Which elements are noble gases? *[1]*
  (iii) Which elements react with cold water to form an alkaline solution? *[1]*
  (iv) Which element has three electrons in its outer shell? *[1]*
  (v) Which two elements react together to form an acidic oxide? *[1]*

(d) State one use for (i) Ar (ii) Ne. *[2]*

(Paper 3)

5 Argon is next to chlorine in the Periodic Table.

(a) Use your knowledge of the structures of argon and chlorine to explain why they are both gases at room temperature. *[2]*

(b) Explain why argon is unreactive. *[2]*

(c) Chlorine is a diatomic molecule. What do you understand by the term *diatomic*? *[1]*

(d) Chlorine reacts with hydrogen to form hydrogen chloride. Copy and complete the equation for this reaction:

___ + $H_2 \longrightarrow$ ___ HCl *[2]*

(e) Chlorine, bromine and iodine react with molten sodium to form sodium halides.
  (i) Suggest the order of reactivity of the halogens with sodium. Give a reason for your answer. *[2]*
  (ii) Write a symbol equation for the reaction of sodium with bromine to form sodium bromide, NaBr. *[2]*

(Paper 3)

6 In Period 3, the elements are arranged in order of increasing atomic number.

(a) Describe how the electronic structure of these elements changes across Period 3. *[3]*

(b) Describe how metallic and non-metallic character is related to the number of outer shell electrons. *[2]*

(c) The halogens are a group of elements with similar properties. Describe how the colour and reactivity of the halogens changes down the group. *[2]*

(d) Bromine reacts with aqueous potassium iodide but not with aqueous potassium chloride.
  (i) Explain why bromine does not react with aqueous potassium chloride. *[1]*
  (ii) Write a balanced equation for the reaction of bromine with potassium iodide. Include state symbols. *[3]*
  (iii) Chlorine reacts with cold sodium hydroxide to form sodium(I) chlorate, sodium chloride and water. The formula for the chlorate(I) ion is ClO⁻.

  $Cl_2(aq) + 2NaOH(aq) \longrightarrow$
  $NaCl(aq) + NaClO(aq) + H_2O(l)$

  Write an ionic equation for this reaction. *[1]*

(Paper 4)

7 Strontium is below calcium in Group II of the Periodic Table.

(a) Use ideas about atomic structure to explain:
  (i) why strontium is more reactive than calcium *[2]*
  (ii) why strontium is less reactive than rubidium. *[2]*

(b) Calcium reacts with water in a similar way to Group I elements.
  (i) Suggest two observations that you can make when calcium reacts with water. *[2]*
  (ii) Write a balanced equation for the reaction of calcium with water. *[2]*

(c) Suggest an element in Group II that is more reactive than strontium. *[1]*

(Paper 4)

13 Metals and reactivity

13.1 Alloys

LEARNING OUTCOMES

- Describe the general physical and chemical properties of metals
- Explain why metals are often used in the form of alloys
- Identify representations of alloys from diagrams of their structure

In Topic 2.5 we learned that metals are:

- good conductors of electricity and heat
- malleable – they can be hammered into different shapes
- ductile – they can be drawn into wires
- shiny.

A metal such as iron is rarely used on its own because it rusts easily. Pure copper is not very strong so cannot be used for parts of machines that are constantly in motion. We can change the properties of a metal to make it harder or more resistant to corrosion. We do this by mixing it with another metal or with a non-metal.

A mixture of two or more metals, or one or more metals with a non-metal, is called an **alloy**. An alloy is not just a mixture of metal crystals. The atoms of the second metal form part of the crystal lattice.

A mixture of metal crystals   An alloy

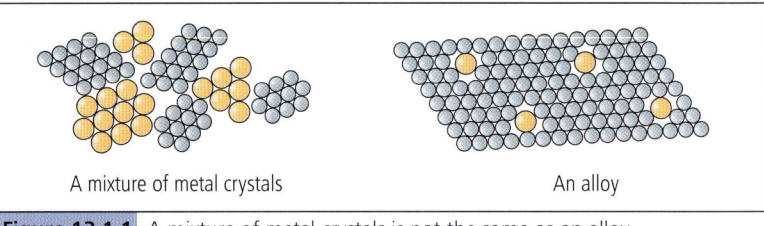

Figure 13.1.1 A mixture of metal crystals is not the same as an alloy

STUDY TIP

It is a common error to think that all metals are hard and have very high melting points. Remember that Group 1 metals are soft and have low melting points.

PRACTICAL

Tin, lead and solder

Solder is an alloy of lead and tin. It is used to join wires in electrical circuits. You put small pieces of tin, lead and solder on the steel 'tin' lid and heat the centre. You record the time taken for each metal to melt. The solder melts long before the tin and lead. This shows that the alloy has different properties from the tin and lead alone.

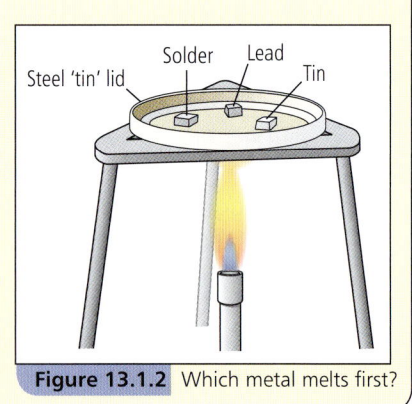

Figure 13.1.2 Which metal melts first?

Explaining the difference

The atoms in a pure metal are arranged in regular layers. When a force is applied, the layers slide over each other. This explains why metals are malleable and ductile. When a metal is alloyed with a second metal, the different sized metal atoms make the arrangement of the lattice less regular. We say that they disrupt the crystal lattice. This stops the layers of metal atoms sliding easily over each other when a force is applied. This is why an alloy is stronger and harder than a pure metal.

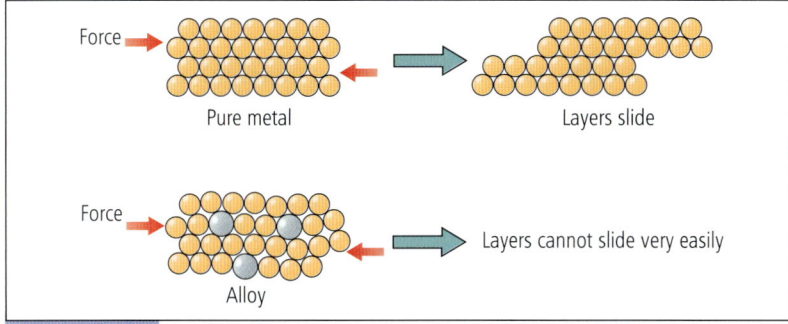

Figure 13.1.3 Alloys are stronger than pure metals because the layers cannot slide easily

Uses of alloys

Alloys have many uses. Some of these are given in the table.

alloy	properties	uses
brass (copper + zinc)	stronger than copper but still malleable	musical instruments, ornaments
bronze (copper + tin)	very hard	some moving parts of machines, statues, bells
stainless steel (iron + chromium + nickel)	does not rust like iron	car parts, cutlery, parts of chemical factories, surgical instruments

In recent years a number of alloys with 'memory' properties have been made. One of these, an alloy of nickel and titanium is called 'nitinol'. If a piece of nitinol wire is manufactured in an 'S' shape it remembers that shape. If you straighten it out, it stays straight. But when it is put into some hot water it changes back to its 'S' shape. These alloys are called 'shape memory alloys'. They are useful for making spectacle frames and dental braces.

DID YOU KNOW?

'Smart' alloys are able to 'remember' their shape. Springs made from these alloys 'remember' to open out at about 90°C but close up again when cooled. They can be used to activate fire sprinklers.

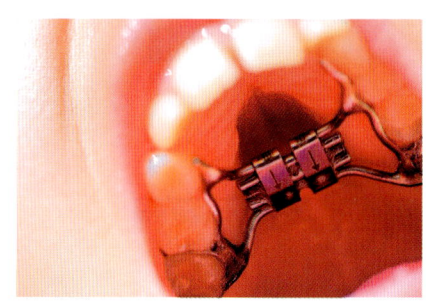

'Smart' alloys are used in some dental braces. As the alloy warms up, it pulls the teeth into the correct position.

SUMMARY QUESTIONS

1 Copy and complete using the words below:

alloy   mixture   layers   non-metal   slide   stronger

An alloy is a _____ of metals or a mixture of metals with a _____. By making a metal into an _____ it becomes _____ and harder. This is because the _____ of metal atoms in the alloy cannot _____ over one another very easily.

2 Explain why an alloy of aluminium and manganese is stronger than pure aluminium.

3 Draw a diagram to show the arrangement of the atoms in:
   a a pure metal
   b an alloy.

KEY POINTS

1 Alloys are mixtures of metal atoms with other metal atoms or non-metal atoms.

2 The properties of a metal are changed by making it into an alloy.

3 Metals are made into alloys to improve their strength, hardness or resistance to corrosion.

13.2 The metal reactivity series

LEARNING OUTCOMES

- Place metals in a reactivity series by referring to their reactions with water, steam and hydrochloric acid
- Deduce an order of reactivity from information given

Some metals are very reactive. The Group I meals react very rapidly with water. Other metals are less reactive. The transition elements either do not react with water or react only with steam. We can put the metals in order of their reactivity by investigating how well they react with oxygen, water or dilute hydrochloric acid.

Reacting metals with oxygen

The list below shows what happens when different metals are heated in air. The metals react with the oxygen in the air to form metal oxides.

- **copper**: does not burn but turns black on its surface
- **iron**: burns only when in powder form or as iron wool
- **gold**: does not burn at all, even as a powder
- **magnesium**: burns rapidly with a bright white light

From this list of reactions we can put these metals in order of reactivity:

magnesium → iron → copper → gold
(most reactive → least reactive)

Reacting metals with water or steam

If a metal does not react with cold water it may react with steam.

DEMONSTRATION

Reacting iron wool with steam

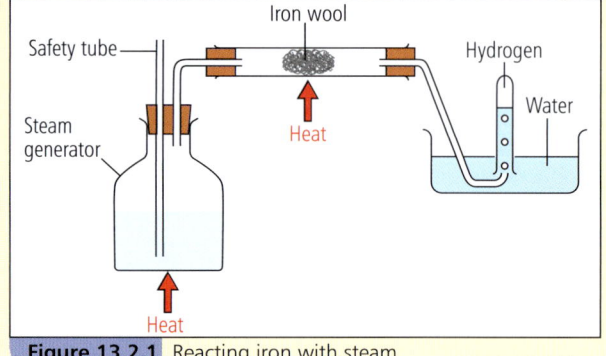

Figure 13.2.1 Reacting iron with steam

We pass steam over red-hot iron wool. The iron turns black. We collect the gas in a test tube. The gas pops with a lighted splint. The iron has reacted and formed iron oxide and hydrogen.

Figure 13.2.2 The metal reactivity series showing the position of hydrogen

More reactive with water:
K, Na, Ca, Mg, Zn, Fe

H

Do not react:
Cu, Ag

DID YOU KNOW?

An isotope of caesium, a very reactive metal, is used in atomic clocks to measure time with great accuracy.

The table shows how different metals react with water or steam.

calcium	reacts rapidly with cold water
copper	no reaction with cold water or steam
magnesium	reacts very slowly with cold water but reacts rapidly with steam
sodium	reacts violently with cold water
zinc	reacts only when powdered and heated strongly in steam

From the information in the table we can put these metals in order of their reactivity:

sodium   calcium   magnesium   zinc   copper

most reactive ⟶ least reactive

If a metal reacts with cold water, a metal hydroxide and hydrogen are formed:

$$Ca(s) + 2H_2O(l) \longrightarrow Ca(OH)_2(aq) + H_2(g)$$
calcium + water ⟶ calcium hydroxide + hydrogen

If a metal reacts only with steam, a metal oxide is formed:

$$3Fe(s) + 4H_2O(g) \longrightarrow Fe_3O_4(s) + 4H_2(g)$$
iron + steam ⟶ iron oxide + hydrogen

Reaction with dilute acid

The table shows how different metals react with dilute hydrochloric acid.

sodium	very violent – explosive	most reactive
calcium	very rapid – lots of hydrogen bubbles produced	↑
magnesium	rapid – bubbles of hydrogen produced steadily	
zinc	slow – bubbles of hydrogen produced slowly	
copper	no reaction with dilute or concentrated acid	least reactive

The metal reactivity series

We can use the reactions of elements with oxygen, water and hydrochloric acid to build up part of the **reactivity series**. By reacting metals with different solutions of metals ions we can extend this reactivity series. If you look at the reactivity series shown in Figure 13.2.2 you will notice that we have included hydrogen in the reactivity series. Metals below hydrogen do not react with cold water or steam. They do not release hydrogen from hydrochloric acid either. So copper, silver and gold are very unreactive.

> **STUDY TIP**
>
> Remember that metals that react with cold water form metal hydroxides. When a metal is heated in steam, an oxide is formed.

We can put magnesium, zinc and iron in order of reactivity by comparing how rapidly they react with an acid:
1. Mg
2. Zn
3. Fe

SUMMARY QUESTIONS

1 Copy and complete using the words below:

   cold   hydrogen   hydroxide   iron   potassium   steam

   Sodium, ____ and calcium react with ____ water to form a metal ____ and hydrogen. Less reactive meals such as ____ and zinc do not react with cold water but they do react with ____. Metals below ____ in the reactivity series do not react with water or steam.

2 Platinum is never found in nature combined with oxygen. What does this tell you about the reactivity of platinum?

3 Tin is between iron and lead in the reactivity series. Suggest how tin will react with:
   a   cold water   b   steam.

KEY POINTS

1 Metals can be arranged in a reactivity series by comparing how easily they react with water, steam and hydrochloric acid.

2 Only metals above hydrogen in the reactivity series react with hydrochloric acid.

3 Only metals above hydrogen in the reactivity series will react with water or steam.

13.3 More about metal reactivity

LEARNING OUTCOMES

- Describe the reactivity series related to the tendency of a metal to form its positive ion.

In Topic 12.3 we saw that a more reactive halogen will replace a less reactive halogen in a metal halide. We can think of this as a competition to see which halogen combines the best with the metal. We call this type of redox reaction a **displacement** reaction. Can metals compete in a similar way?

DEMONSTRATION

The 'thermit' reaction

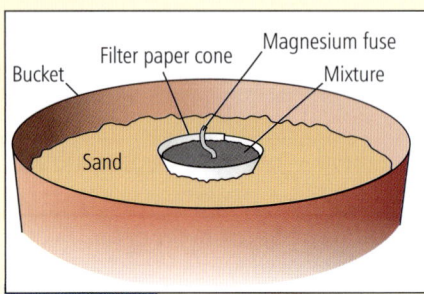

Figure 13.3.1 Setting up the 'thermit' reaction

A mixture of aluminium powder and iron(III) oxide is put into a cone of filter paper. This is placed in a bucket of sand. The magnesium fuse is lit. A vigorous reaction occurs with flames, light and smoke. The result is a lump of iron where the mixture was.

In the 'thermit' reaction the aluminium displaces the iron from the iron(III) oxide:

$$Fe_2O_3 + 2Al \rightarrow 2Fe + Al_2O_3$$
iron(III) oxide + aluminium → iron + aluminium oxide

We can carry out similar experiments using metals and solutions of metal ions. When we add excess zinc to a solution of copper(II) sulfate, the zinc gets coated with copper and the solution turns colourless.

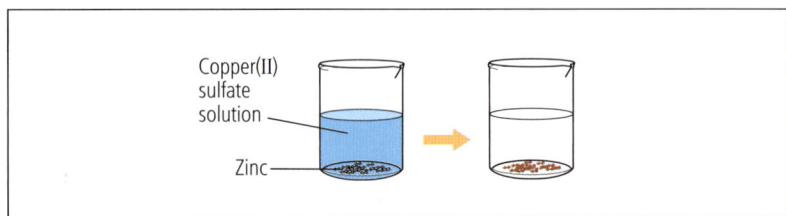

Figure 13.3.2 The more reactive metal, zinc, displaces the less reactive copper from copper(II) sulfate solution

Zinc is higher in the reactivity series than copper. So it displaces copper from the copper(II) sulfate solution. The solution turns colourless because colourless zinc sulfate is formed:

$$Zn(s) + CuSO_4(aq) \rightarrow ZnSO_4(aq) + Cu(s)$$

DID YOU KNOW?

When a small piece of sodium reacts with water, about 100 000 000 000 000 000 sodium atoms change to sodium ions every second!

By carrying out experiments using different combinations of metals and solutions of metal salts, we can arrange all the metals in a metal reactivity series. A more reactive metal will displace a less reactive metal from a solution of its salt. We sometimes call this reactivity series the electrochemical series.

Because these are redox reactions, we can write half-equations for oxidation and reduction. For example:

oxidation of zinc to zinc ions:

$$Zn(s) \longrightarrow Zn^{2+}(aq) + 2e^-$$

reduction of copper ions to copper:

$$Cu^{2+}(aq) + 2e^- \longrightarrow Cu(s)$$

Explaining reactivity

In the half-reactions above, we can see that each atom of the more reactive metal loses electrons and each ion of the less reactive metal gains electrons. The more reactive a metal is, the more easily it loses its outer shell (valency) electrons. It is easier to lose electrons from the outer (valency) shell if:

- the valency electrons are further away from the pull of the nucleus. Remember that negative electrons are attracted to positive protons.
- there are more electron shells between the nucleus and the valency electrons. These shells shield the valency electrons from the charge of the nucleus.
- there are fewer protons in the nucleus (less nuclear charge) to pull the electrons towards them.

Potassium is more reactive than sodium because its valency electrons are further from the nucleus and there are more shells between the valency electrons and the nucleus. This outweighs the effect of the increased number of protons in the nucleus of potassium. So potassium can lose its valency electron more easily than can sodium.

Magnesium is less reactive than sodium because even though the valency electrons are in the same shell, magnesium has a greater nuclear charge. So magnesium will not lose its valency electrons as easily as sodium.

The copper has reacted with silver nitrate and crystals of silver have formed on its surface

SUMMARY QUESTIONS

1 Copy and complete using the words below:

   displaces   less   more   solution   valency

   A more reactive metal _____ a _____ reactive metal from a _____ of its salt. This is because the _____ reactive metal loses its _____ electrons more easily.

2 Silver is less reactive than copper but copper is less reactive than magnesium. Write symbol equations for:
   a the reaction of copper(II) sulfate with magnesium
   b the reaction of silver nitrate with copper.

3 Explain why copper will not react with iron(II) sulfate.

KEY POINTS

1 A more reactive metal displaces a less reactive metal from a solution of its salt.

2 A more reactive metal loses its valency electrons more easily than a less reactive metal.

3 The ease with which a metal loses its valency electrons to form ions depends on the distance of the valency electrons from the nucleus, the nuclear charge and the number of electrons shells.

13.4 From metal oxides to metals

LEARNING OUTCOMES

- Describe the use of carbon as a reducing agent for some metal oxides
- Explain the use of carbon as a reducing agent
- Explain the apparent unreactivity of aluminium

Competing for oxygen

When you heat powdered iron with copper(II) oxide, CuO, the iron displaces the copper:

$$Fe(s) + CuO(s) \xrightarrow{heat} FeO(s) + Cu(s)$$

The iron competes better to 'hold onto' the oxygen. This is because iron is higher in the reactivity series than copper. The iron is oxidised to iron(II) oxide and the copper(II) oxide is reduced to copper. The more reactive metal, in this case iron, is the reductant. It removes the oxygen from the less reactive metal's oxide.

Reducing metal oxides with carbon

We can also carry out reduction using carbon as a reductant (reducing agent). For example: carbon is more reactive than copper. So carbon removes the oxygen from copper(II) oxide when heated.

$$2CuO(s) + C(s) \xrightarrow{heat} 2Cu(s) + CO_2(g)$$

copper(II) oxide + carbon $\longrightarrow$ copper + carbon dioxide

PRACTICAL

Reducing copper(II) oxide with carbon

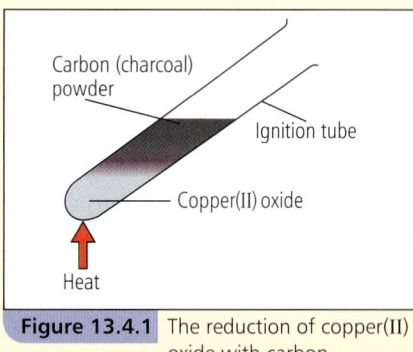

Figure 13.4.1 The reduction of copper(II) oxide with carbon

You put a layer of charcoal powder (carbon) over a layer of copper(II) oxide then heat the tube strongly. When the reaction is over you can see some pinkish-brown copper metal where the two layers of powder meet.

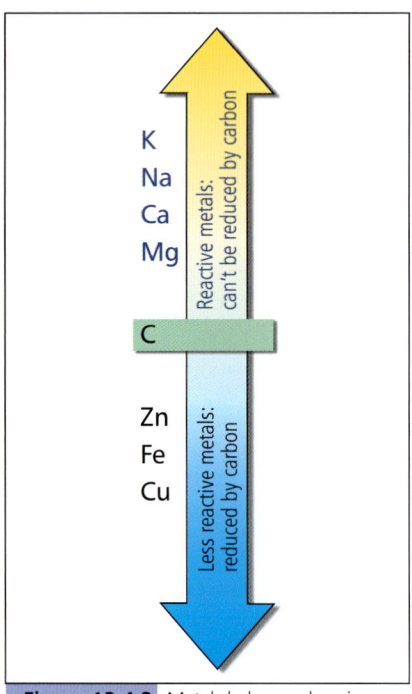

Figure 13.4.2 Metals below carbon in the reactivity series can be extracted by heating with carbon. Metals above carbon are extracted by electrolysis.

K
Na
Ca
Mg
— Reactive metals: can't be reduced by carbon
C
Zn
Fe
Cu
— Less reactive metals: reduced by carbon

So which metals can carbon reduce? Look at the position of carbon in the reactivity series (Figure 13.4.2). Only the oxides of the metals below carbon can be reduced to the metal by heating with carbon. Metals more reactive than carbon have to be extracted by electrolysis.

Explaining the use of reducing agents

We can explain the use of reactive metals and carbon to reduce metal oxides in terms of movement of electrons. A more reactive metal loses its outer shell electrons and combines with oxygen more easily than a less reactive metal. So a more reactive metal will be able to remove the oxygen from the oxide of a less reactive metal. The more reactive metal is a better reducing agent.

If a metal is below carbon in the reactivity series the oxygen in its oxide will form covalent bonds more easily with carbon than with the metal. The carbon is the reducing agent.

Why does aluminium seem unreactive?

Aluminium is high in the reactivity series but it does not seem to react with water or acids. It will react with acids only when it is freshly made. This is because, when the surface of freshly made aluminium is left in the air, a thin layer of aluminium oxide quickly forms on its surface:

$$4Al(s) + 3O_2(g) \longrightarrow 2Al_2O_3(s)$$

This layer is only about 0.0002 cm thick, but this is enough to make the metal resistant to corrosion. The tough oxide layer sticks to the surface of the aluminium very strongly and does not flake off. The oxide layer is unreactive.

DID YOU KNOW?
The earliest evidence for the extraction of lead comes from Turkey. Beads of lead have been found that are thought to have been extracted 8500 years ago.

STUDY TIP
Remember that aluminium is a reactive metal. It must be reactive if it forms an oxide layer on its surface so quickly.

Charcoal kilns like these have been used for centuries to provide the carbon for metal extraction

SUMMARY QUESTIONS

1 Copy and complete using the words below:

below   heated   metals   oxygen   reduced   reducing

Metal oxides _____ carbon in the reactivity series are _____ to _____ when they are _____ with carbon. In this reaction carbon is the _____ agent because it removes the _____ from the metal oxide.

2 Write balanced equations for:

   a the reaction of zinc oxide, ZnO, with carbon to form zinc and carbon dioxide

   b the reaction of magnesium with copper(II) oxide to form magnesium oxide and copper.

3 Explain why freshly made aluminium reacts with hydrochloric acid but old aluminium does not react.

KEY POINTS

1 Metal oxides below carbon in the reactivity series are reduced by carbon when heated.

2 When a more reactive metal is heated with the oxide of a less reactive metal, the more reactive metal acts as a reducing agent.

3 The apparent lack of reactivity of aluminium is due to an unreactive oxide layer that forms on its surface.

13.5 Thermal decomposition

LEARNING OUTCOMES

- Describe the action of heat on selected hydroxides and nitrates
- Link the thermal decomposition of nitrates to the reactivity of the metals

When we heat some carbonates, nitrates and hydroxides, they break down to form two or more different products. We call this type of reaction **thermal decomposition**. For example, the equation:

$$CaCO_3(s) \xrightarrow{heat} CaO(s) + CO_2(g)$$

represents the thermal decomposition of calcium carbonate.

Thermal decomposition of metal hydroxides

Most metal hydroxides decompose when heated. A metal oxide and water are formed. For example:

$$Zn(OH)_2(s) \xrightarrow{heat} ZnO(s) + H_2O(g)$$

zinc hydroxide → zinc oxide + water

All Group II hydroxides decompose in a similar way. Most alkali metal hydroxides, however, do not decompose. They are stable to heat. There is one exception to this: lithium hydroxide.

$$2LiOH(s) \xrightarrow{heat} Li_2O(s) + H_2O(g)$$

lithium hydroxide → lithium oxide + water

Lithium is the least reactive of the alkali metals. So does thermal decomposition depend on the reactivity of the metal? We shall look at the decomposition of nitrates and carbonates to answer this question.

Thermal decomposition of nitrates

All nitrates decompose when heated. But there are some differences in how they decompose. The alkali metal nitrates decompose to form a **nitrite** and oxygen. You can see that the nitrite ion is similar to the nitrate ion but has one fewer oxygen atom:

$$2KNO_3(s) \xrightarrow{heat} 2KNO_2(s) + O_2(g)$$

potassium nitrate → potassium nitrite + oxygen

Nitrates of other metals decompose on heating to form an oxide, nitrogen dioxide and oxygen. For example:

$$2Mg(NO_3)_2(s) \xrightarrow{heat} 2MgO(s) + 4NO_2(g) + O_2(g)$$

magnesium nitrate → magnesium oxide + nitrogen dioxide + oxygen

Lithium nitrate, a compound of the least reactive alkali metal, decomposes in this way too.

Nitrates can decompose explosively to cause a large amount of damage, as shown by the remains of this fertiliser factory

STUDY TIP

You need to remember the products from the thermal decomposition of nitrates. If you don't know these, you won't be able to write equations for thermal decomposition.

Nitrates of very unreactive metals, such as silver, decompose to form the metal when they are heated:

$$2AgNO_3(s) \xrightarrow{heat} 2Ag(s) + 2NO_2(s) + O_2(g)$$

silver nitrate → silver + nitrogen dioxide + oxygen

DEMONSTRATION

The decomposition of Group II nitrates

This experiment is demonstrated using a fume cupboard because nitrogen dioxide gas is poisonous.

We start a stopclock when we begin to heat the magnesium nitrate. The nitrate is heated until we see dark brown fumes of nitrogen dioxide in the tube. We record the time taken to see these fumes. Then we repeat the experiment using other nitrates. In this experiment we must keep the amount of nitrate and the rate of heating the same.

The longer it takes for the nitrate to decompose, the more stable the nitrate is.

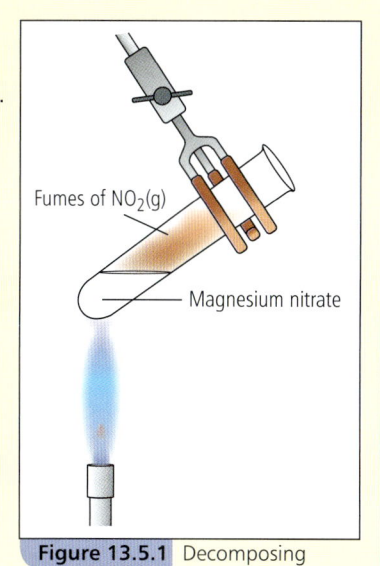

Figure 13.5.1 Decomposing magnesium nitrate

Magnesium nitrate decomposes at 402 °C but barium nitrate decomposes at 865 °C. So it appears that barium nitrate is more stable than magnesium nitrate. Barium is more reactive than magnesium. So the more reactive the metal, the more stable to thermal decomposition its compound is. This is well demonstrated if we look at the temperatures at which carbonates decompose when heated.

Thermal decomposition of carbonates

The table shows the temperatures at which the Group II carbonates decompose.

Group II carbonate	magnesium carbonate	calcium carbonate	strontium carbonate	barium carbonate
decomposition temperature/°C	540	900	1280	1360

The reactivity of the Group II metals increases down the group. You can see from the table that as you go down the group it gets more difficult to decompose the carbonates.

So, the more reactive the metal, the more stable its nitrate, carbonate or hydroxide is.

KEY POINTS

1. Thermal decomposition is the breakdown of a compound into two or more different products by heat.
2. The more reactive the metal, the more stable its nitrate, hydroxide or carbonate.
3. Metal hydroxides decompose to oxides and water when heated.
4. Most nitrates decompose to either nitrites and oxygen or to oxides, nitrogen dioxide and oxygen when heated.

SUMMARY QUESTIONS

1. Copy and complete using the words below:

   **alkali   dioxide   heated
   less   nitrite   oxygen**

   Nitrates of the _____ metals decompose when _____ to form the metal _____ and oxygen. Nitrates of _____ reactive metals form metal oxides, nitrogen _____ and _____ when heated.

2. Write equations for the thermal decomposition of:

   a copper(II) hydroxide, $Cu(OH)_2$

   b calcium nitrate, $Ca(NO_3)_2$.

3. Suggest why lithium nitrate decomposes in a similar way to a Group II nitrate rather than a Group I nitrate.

SUMMARY QUESTIONS

1 Copy and complete using the words from the list below.

   **alkali   hydroxides   iron   oxide
   oxygen   potassium   powder
   sodium   water**

   Metals can be put in order of reactivity using their reactivity with _____ as a guide. Alkali metals such as _____ and _____ react rapidly with water as well as _____. The _____ metals react with cold water to form alkali metal _____. Less reactive metals such as _____ will only react with oxygen if they are in wire or _____ form. Iron reacts with steam to form iron _____.

2 Use the following reactivity series to answer the questions below.

   calcium   magnesium   zinc   iron   lead   copper

   most reactive ⟶ *least reactive*

   (a) Which metals in the list will react with dilute hydrochloric acid?
   (b) Which metals in the list will react with cold water?
   (c) Which metals in the list will react with steam?
   (d) Where would **(i)** potassium and **(ii)** silver come in this list?

3 Match the metals on the left with the phrases on the right.

sodium	a reactive metal that burns with a bright light to form a metal oxide
copper	a metal that reacts with water to form an alkaline solution
iron	a grey, fairly unreactive metal
lead	a pinkish-brown metal that is unreactive
magnesium	a metal that reacts with steam but not with cold water

4 Describe how the reactivity series depends on the ease with which a metal forms a positive ion.

5 Aluminium is a metal high in the reactivity series. Explain why aluminium apparently does not react with dilute hydrochloric acid.

PRACTICE QUESTIONS

1 Some information about the reaction of three metals with hydrochloric acid is given below.
   - Metal P: dissolves slowly and a few bubbles are formed.
   - Metal Q: dissolves rapidly and bubbles are formed very rapidly.
   - Metal R: dissolves rapidly and bubbles are formed rapidly.

   The order of reactivity of these metals, starting with the most reactive, is:

   A   PQR   B   RQP
   C   QRP   D   PRQ

   (Paper 1)

2 The order of reactivity of four metals is shown below:

   barium > calcium > magnesium > copper
   most reactive ⟶ *least reactive*

   Which one of these statements about the decomposition of metal compounds is correct?

   A Magnesium carbonate decomposes at a lower temperature than copper carbonate.

   B Calcium carbonate decomposes at a higher temperature than barium carbonate.

   C Barium carbonate decomposes at a lower temperature than magnesium carbonate.

   D Calcium carbonate decomposes at a higher temperature than magnesium carbonate.

   (Paper 2)

3 Some of the elements in the reactivity series are shown below:

   sodium   calcium   magnesium   zinc   iron   lead   copper
   most reactive ⟶ *least reactive*

   (a) Which of these elements will react with cold water? [1]
   (b) Iron reacts with steam to form iron(III) oxide.
      (i) Name one other element in the list that reacts with steam but does not react with cold water.
      (ii) Copy and balance the following equation:
      $$\_\_Fe + \_\_H_2O \longrightarrow Fe_3O_4 + \_\_H_2$$
      [3]

(c) Magnesium reacts with hydrochloric acid. Magnesium chloride and hydrogen are formed.
  (i) Write a word equation for this reaction. [1]
  (ii) Write a symbol equation for this reaction. [2]
  (iii) How can you test for the hydrogen given off in this reaction? [2]
(d) Magnesium reacts with black copper(II) oxide when heated.
  (i) Describe what you observe during this reaction. [2]
  (ii) Write a word equation for this reaction. [1]
  (iii) Which reactant is reduced in this reaction? [1]

(Paper 3)

4 Brass is an alloy of zinc and copper.
  (a) What do you understand by the term *alloy*? [1]
  (b) Zinc is more reactive than copper.
    (i) State two observations that you can make when zinc reacts with aqueous copper(II) sulfate. [2]
    (ii) Write a word equation for this reaction. [1]
    (iii) What type of chemical reaction is this? [1]
  (c) Zinc reacts with steam but not with cold water. Potassium reacts with cold water.
    (i) Suggest why zinc does not react with cold water but potassium does. [1]
    (ii) Copy and balance the equation for the reaction of potassium with water:
    __K + __H_2O ⟶ __KOH + H_2 [3]

(Paper 3)

5 Zinc powder reacts with copper(II) oxide on heating:

$$Zn + CuO \longrightarrow ZnO + Cu$$

  (a) Which is the reductant in this reaction? Explain your answer. [2]
  (b) Describe the direction of electron transfer in this reaction. [2]

  (c) Explain why the reverse reaction does not occur. [1]
  (d) Both magnesium and lead are above copper in the reactivity series. Explain, in terms of ease of formation of ions, why magnesium is able to remove oxygen from copper(II) oxide more readily than lead. [2]
  (e) Copper(II) nitrate, $Cu(NO_3)_2$, decomposes in a similar way to magnesium nitrate. Write a balanced equation for the thermal decomposition of copper(II) nitrate, including state symbols. [3]

(Paper 4)

6 Sodium and magnesium are both reactive metals.
  (a) (i) Describe the differences between the reactions of sodium and magnesium with water. [2]
    (ii) Write a balanced equation to show the reaction of magnesium with steam. Include state symbols. [3]
  (b) Sodium nitrate and magnesium nitrate behave differently on thermal decomposition.
    (i) What do you understand by the term *thermal decomposition*? [1]
    (ii) Write a balanced equation to show the thermal decomposition of sodium nitrate. [2]
    (iii) Write a word equation for the thermal decomposition of magnesium nitrate. [2]
  (c) Magnesium is more reactive than aluminium.
    (i) Describe how you could use magnesium to extract aluminium from molten aluminium oxide. [1]
    (ii) Suggest why this method not used to extract aluminium industrially. [1]
    (iii) Explain why aluminium containers can be used to store acidic foods even though it is a reactive metal. [2]

(Paper 4)

14 Metal extraction

14.1 Metals from their ores

LEARNING OUTCOMES

- Describe bauxite as an ore of aluminium
- Describe how the ease of obtaining metals from their ores depends on the position of the metal in the reactivity series
- Describe the extraction of zinc from zinc blende

Most metals found in the Earth's crust are present as compounds in the rocks. A rock from which a metal can be extracted is called an **ore**. Most ores are oxides or sulfides. Some important ores are **hematite** (iron ore), **bauxite** (aluminium ore) and **zinc blende** (zinc ore).

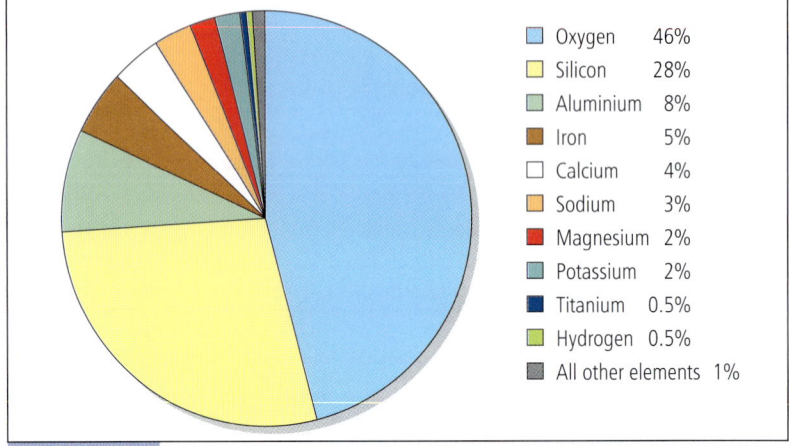

- Oxygen 46%
- Silicon 28%
- Aluminium 8%
- Iron 5%
- Calcium 4%
- Sodium 3%
- Magnesium 2%
- Potassium 2%
- Titanium 0.5%
- Hydrogen 0.5%
- All other elements 1%

Figure 14.1.1 The Earth's crust is made up of many different elements

Metal extraction and the reactivity series

The way we extract a metal from its ore depends on the position of the metal in the **reactivity series**. Carbon is used to reduce oxides of metals below it in the reactivity series. For example: oxides of zinc, lead and iron can be reduced by carbon. The carbon is usually used in the form of coke. This is coal from which some impurities have been removed.

$$PbO(s) + C(s) \longrightarrow Pb(s) + CO(g)$$
lead(II) oxide   carbon(coke)   lead   carbon monoxide

Carbon monoxide is also a good reducing agent. This is formed when carbon undergoes incomplete combustion:

$$2C(s) + O_2(g) \longrightarrow 2CO(g)$$
carbon + oxygen → carbon monoxide

Carbon dioxide is often produced in furnaces used for the extraction of metals:

$$ZnO(s) + CO(g) \longrightarrow Zn(s) + CO_2(g)$$

Metals above carbon in the reactivity series cannot be extracted from their oxides by heating with carbon. This is because the metal bonds to oxygen too strongly and the carbon is not reactive enough to remove it. So we have to use **electrolysis** to extract metals such as aluminium, magnesium and calcium.

It is possible to use electrolysis to extract metals less reactive than carbon. This is not done because much more energy is needed to carry out electrolysis compared with extraction using carbon.

DID YOU KNOW?

About 30% of the pure zinc produced in the world comes from recycling the metal.

STUDY TIP

You should be prepared to label a diagram and write relevant equations.

Extracting zinc

Supplement

The raw materials used in the extraction of zinc are zinc blende, coke (carbon) and air. The main ore of zinc is zinc blende which is zinc sulfide, ZnS. The ore is first crushed and treated to remove waste rock and other impurities. The zinc blende is then roasted (strongly heated) in air to form zinc oxide:

$$2ZnS(s) + 3O_2(g) \longrightarrow 2ZnO(s) + 2SO_2(g)$$

The zinc oxide is then heated with coke (carbon) in a **blast furnace**.

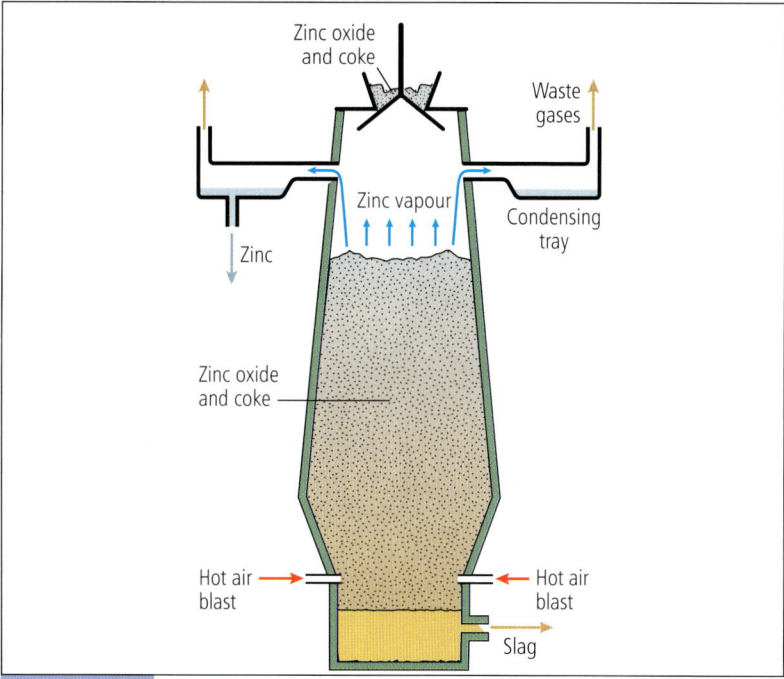

Figure 14.1.2 A blast furnace used for extracting zinc

A blast of air is blown into the bottom of the furnace. The excess carbon reacts with oxygen in the air to form carbon monoxide:

$$2C(s) + O_2(g) \longrightarrow 2CO(g)$$

Higher up the furnace, carbon monoxide reduces zinc oxide to zinc:

$$ZnO(s) + CO(g) \longrightarrow Zn(g) + CO_2(g)$$

The carbon dioxide formed can react with more carbon to reform carbon monoxide:

$$CO_2(g) + C(s) \longrightarrow 2CO(g)$$

Some zinc oxide may also react directly with the carbon:

$$ZnO(s) + C(s) \longrightarrow Zn(g) + CO(g)$$

The temperature in the furnace is higher than the boiling point of zinc. So the zinc vapour is carried up through the furnace by the stream of carbon monoxide and carbon dioxide. The vapour condenses in trays at the top of the furnace together with lead which is extracted at the same time. The zinc is then purified by distillation.

This method produces only about 20% of the world's zinc. Electrolysis of zinc sulfate is now preferred because this produces much purer zinc.

A zinc-smelting works

SUMMARY QUESTIONS

1 Copy and complete using words from the list below:

   bauxite   carbon   extract   iron   ores   zinc

   Rocks from which we can _____ metals are called _____. The main ore of aluminium is _____. Hematite is one of the main ores of _____. We can use _____ to reduce metal oxides such as iron oxide and _____ oxide.

2 Explain why you cannot use carbon to extract magnesium from magnesium oxide.

3 Draw a flow diagram to show the main stages in the extraction of zinc using carbon.

KEY POINTS

1 Metal oxides below carbon in the reactivity series are reduced by carbon when heated.

2 Metals more reactive than carbon are extracted by electrolysis.

3 Zinc is extracted from zinc oxide in a furnace using carbon as a reducing agent.

14.2 Extracting iron

LEARNING OUTCOMES

- State that iron can be extracted from hematite ore
- Describe the essential reactions in the blast furnace for the production of iron

Iron is the second most common metal in the Earth's crust. The main ore of iron, **hematite**, usually contains more than 60% iron. Hematite is largely iron(III) oxide. We extract iron by reduction of the iron(III) oxide with carbon.

The raw materials for making iron are hematite, coke, limestone and air. The hematite, coke and limestone are added at the top of the furnace. A strong current of hot air is blown in at the bottom of the furnace. This is why it is called a **blast furnace**. The temperature of the hot air is between 550 °C and 850 °C. This is high enough to react with the coke.

The main reducing agent in the blast furnace is carbon monoxide, but in some parts of the blast furnace carbon also reduces the iron(III) oxide. The temperature in the blast furnace ranges from 1500 °C at the bottom where the air enters to 250 °C at the top.

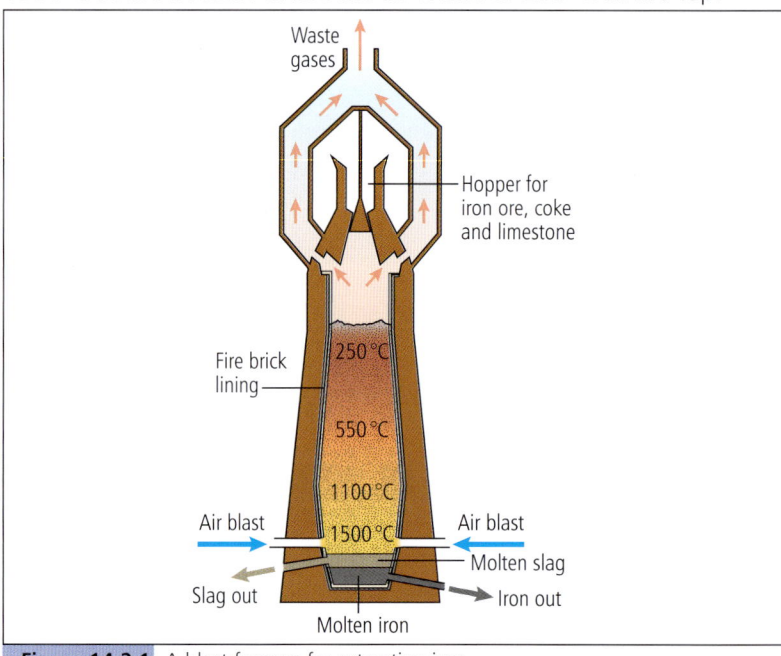

Figure 14.2.1 A blast furnace for extracting iron

STUDY TIP

You should be prepared to answer questions related to a diagram of the blast furnace and the reactions involved.

The chemical reactions in a blast furnace

Extracting the iron

The chemical reactions below result in the production of iron from iron(III) oxide:

- At the bottom of the furnace the coke burns in the hot air blast to form carbon dioxide. This reaction is exothermic. The heat released helps heat the furnace.

$$C(s) + O_2(g) \longrightarrow CO_2(g)$$

- The carbon dioxide reacts with the coke to form carbon monoxide:

$$CO_2(g) + C(s) \longrightarrow 2CO(g)$$

- The carbon monoxide reduces the iron(III) oxide to iron:

$$Fe_2O_3(s) + 3CO(g) \rightarrow 2Fe(l) + 3CO_2(g)$$

Iron(III) oxide + carbon monoxide → iron + carbon dioxide

Most of the iron is produced in this way. The iron flows to the bottom of the furnace and is removed from time to time as a liquid. It flows into moulds and is left to solidify.

Other reactions may take place in the furnace. In the hotter parts of the furnace, carbon reduces iron(III) oxide directly:

$$Fe_2O_3(s) + 3C(g) \rightarrow 2Fe(l) + 3CO(g)$$

The hot waste gases exiting from the top of the furnace are used to heat the air going into the furnace thus reducing energy costs.

Why do we add limestone?

Hematite contains sand (silicon(IV) oxide) as a major impurity. The limestone (calcium carbonate) helps remove most of the impurities in the following way:

- The heat from the furnace decomposes the limestone:

$$CaCO_3(s) \rightarrow CaO(s) + CO_2(g)$$

calcium carbonate (limestone) → calcium oxide + carbon dioxide

- The calcium oxide reacts with the silicon(IV) oxide to form a 'slag' of calcium silicate:

$$CaO(s) + SiO_2(s) \rightarrow CaSiO_3(l)$$

calcium oxide + silicon(IV) oxide → calcium silicate (slag)

- The liquid slag runs down and forms a layer on top of the liquid iron because it has a lower density than iron. The slag is run off. The solid slag is used as a building material, particularly in road building.

> **DID YOU KNOW?**
>
> More than 2500 years ago people near Lake Victoria in Africa extracted iron from iron oxides at a temperature of 1400 °C.

SUMMARY QUESTIONS

1 Copy and complete using the words below:

air   blast   coke   hematite   oxide   reduces

We extract iron from iron ore in a _____ furnace. The most common ore of iron is _____. The other raw materials used are _____, limestone and _____. Inside the blast furnace, carbon monoxide _____ the iron(III) _____ to iron.

2 Write word equations for these reactions which take place in the blast furnace during the extraction of iron:
   a The reaction of iron(III) oxide with carbon monoxide
   b The reaction of calcium oxide with silicon(IV) oxide.

3 Explain why limestone is added to the blast furnace.

KEY POINTS

1 The raw materials used in the extraction of iron in the blast furnace are iron ore, coke, limestone and air.

2 In the blast furnace, carbon monoxide reduces iron(III) oxide to iron.

3 The thermal decomposition of limestone produces calcium oxide which reacts with silicon(IV) oxide impurities in the iron ore to form 'slag'.

14.3 Iron into steel

LEARNING OUTCOMES

- Describe how iron is converted into steel
- Understand the role of basic oxides and oxygen in steelmaking
- Describe the idea of changing the properties of iron by controlled use of additives

Supplement

The iron produced in the blast furnace is only about 95% pure. The impurities are mainly carbon but also include sulfur, silicon and phosphorus. The impurities make the iron very brittle – it breaks easily. If all the impurities are removed, the iron becomes very soft. In this condition, it is easily shaped but it is too soft for many uses. Pure iron also rusts very easily.

To make the iron strong, only some of the impurities are removed to produce various types of **steel**. Steel is an **alloy** of iron with carbon and/or with other metals.

Steelmaking

We make steel using a basic oxygen converter (See Figure 14.3.1). This is often just called a steelmaking furnace. The converter is a very large bucket which can be tipped at an angle.

The impurities are removed from the iron, and steel is made in the following way:

- The converter is tipped to one side and molten iron and scrap iron are poured in.
- The converter is put back into a vertical position. A water-cooled tube called an oxygen lance is lowered into the converter.
- Oxygen and powdered calcium oxide are blown onto the surface of the molten iron through the lance.
- The oxygen oxidises carbon, sulfur, silicon and phosphorus to their oxides. For example:

$$Si + O_2 \longrightarrow SiO_2$$
silicon → silicon(IV) oxide

$$4P + 5O_2 \longrightarrow 2P_2O_5$$
phosphorus + oxygen → phosphorus(V) oxide

The carbon dioxide and sulfur dioxide escape from the converter because they are gases.

These reactions are very exothermic. The heat released in these oxidation reactions keeps the iron molten.

- Silicon and phosphorus oxides are solids. They are acidic oxides. So these react with the powdered calcium oxide which is basic. A **slag** is formed. For example:

$$CaO(s) + SiO_2(s) \longrightarrow CaSiO_3(l)$$
calcium oxide + silicon(IV) oxide → calcium silicate (slag)

The slag floats on the surface of the molten iron and is removed.

- The amount of carbon in the steel is controlled by the amount of oxygen blown into the impure iron. The longer the oxygen blast the more carbon is removed.

DID YOU KNOW?

A steel weapon called the Falcata was made nearly 2500 years ago in the area which is now Spain and Portugal.

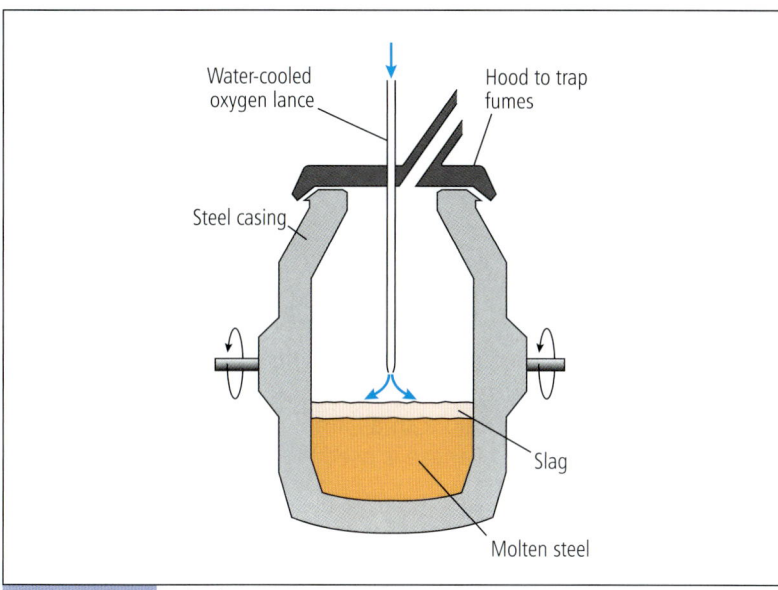

Figure 14.3.1 A basic oxygen converter

Labels on figure: Water-cooled oxygen lance; Hood to trap fumes; Steel casing; Slag; Molten steel

Pouring molten steel into moulds can be quite spectacular

A modern converter can make up to 350 tonnes of medium quality steel in 40 minutes. High quality steel is usually made using an electric furnace. In this process the high temperatures needed to keep the iron molten are produced by an electric current.

> **Supplement**
>
> After the required amount of carbon has bene removed, other metals such as chromium or manganese are added in controlled amounts to the molten iron to make particular alloys of steel with specific properties. The addition of chromium or nickel makes the steel hard and more resistant to corrosion and heat. The addition of manganese to steel makes the steel stronger.

STUDY TIP

Do not confuse steelmaking with the blast furnace. In steelmaking the impurities are removed from the impure iron we get from the blast furnace. In the blast furnace the impure iron is extracted from the iron ore.

SUMMARY QUESTIONS

1 Copy and complete using the words below:

   **basic   carbon   converter   oxide   oxidises
   silicon   slag   surface**

   The iron from the blast furnace contains about 5% _____ and other impurities. These impurities are removed in a _____ oxygen _____. Oxygen is blown onto the _____ of the iron. This _____ the impurities. Calcium _____ is added to remove oxides of _____ and phosphorus as _____.

2 Explain why calcium oxide is added to the basic oxygen converter.

3 Explain how we get different types of steel by using the basic oxygen converter.

KEY POINTS

1 The iron from the blast furnace contains 5% impurities, most of which is carbon.

2 Impurities are removed from iron using a basic oxygen converter.

3 Pure iron is too weak and soft for it to be very useful.

4 The percentage of carbon in steel is controlled by the amount of oxygen blown into the steel.

5 The addition of other metals to iron results in the formation of steel alloys. *(Supplement)*

14.4 Uses of metals

LEARNING OUTCOMES

- Name the use of aluminium in the manufacture of aircraft and food containers
- Name the uses of mild steel and stainless steel
- Name the uses of copper related to its properties
- Discuss the advantages and disadvantages of recycling metals
- Explain the uses of zinc for galvanising and for making brass

Supplement

DID YOU KNOW?

Although coins have been made of metal for many centuries, in China 2800 years ago model farming tools were used for money.

STUDY TIP

Make a list of all the substances in the syllabus whose uses you need to know. Divide the page into two with the names down one side and the uses on the other. Then test yourself.

Steel alloys

Pure iron is too soft and weak to be very useful but the iron from the blast furnace has too much carbon in it to make it useful. It is too brittle to be used for constructions such as bridges and steel frames for buildings.

In Topic 14.3 we described how the amount of carbon in steel is controlled. **Steel** is an **alloy** of iron with carbon or with carbon and other metals. All steels contain a small amount of carbon. There are many types of steel. Each of these is used for a particular purpose:

- Mild steel is low carbon steel. It contains about 0.25% carbon. It is soft, malleable and can be drawn into wires easily. We use it to make car bodies and parts of machinery where it will not be worn away. It is also used for buildings and general engineering purposes.
- High carbon steels contain between 0.5% and 1.4% carbon. As the percentage of carbon in the steel increases, it becomes more brittle. But it also becomes harder. These steels are used for tools such as hammers and chisels.
- Low alloy steels contain between 1% and 5% of other metals such as nickel, chromium, manganese and titanium. They are hard and do not stretch much. Nickel steels are used for bridges where strength is needed and for bicycle chains. Tungsten steel is used for high-speed tools because it does not change shape at high temperatures.
- Stainless steels are high alloy steels. They may contain up to 20% chromium. Many stainless steels contain 70% iron, 20% chromium and 10% nickel. They are very strong and resist corrosion. So they are used in the construction of pipes and towers in chemical factories. Stainless steel is also used to make cutlery – knives, forks and spoons. Surgical instruments are also made from stainless steel.

Uses of other metals and alloys

Aluminium

Aluminium is used for making aircraft bodies because it is lightweight – it has a low density. It is also quite strong. Most aircraft are made from aluminium alloys containing about 90% aluminium and smaller amounts of zinc and copper.

Some food containers and cooking foil are made from aluminium. This is because there is an unreactive oxide layer on its surface which does not flake off (see Topic 13.4). This oxide layer does not react with the acids that are present in many foods.

Copper

Copper is used for electrical wiring because of its high electrical conductivity. It is one of the most malleable and ductile metals so it can easily be shaped and drawn into wires. It is also used for the base of cooking pans because it is an excellent conductor of heat.

Supplement

Zinc

Zinc is used to **galvanise** iron or mild steel to prevent rusting (see Topic 15.7). To galvanise a steel object, we dip the object into liquid zinc. The zinc forms a coating on the surface of the steel. About a third of the zinc produced in the world is used to galvanise steel. Galvanised steel is used for roofing because it is weather resistant.

Brass

Brass is an alloy of 70% copper and 30% zinc. It is stronger than copper and does not corrode. Although it is strong, it can still be easily beaten into shape. Its gold colour makes it attractive. So it is used to make musical instruments, door handles, ornaments and screws.

Recycling metals

Recycling is the processing of used materials to make new products. Some advantages and disadvantages of recycling metals, such as aluminium, iron and steel, are shown in the table.

advantages	disadvantages
conserves metal ores and other raw materials (natural resources) used	collecting and storing metals may be costly
saves energy because less fuel is used (extracting and purifying metals uses a lot of energy)	takes time and energy to collect the waste metals …
reduces pollution arising from extracting and purifying materials	… so there may be more lorries and noise on the roads of towns and villages
saves land that may be used for extracting ores	takes time and money to sort out the metals from mixtures of metals
reduces waste and problems of disposal of unwanted material, e.g. less landfill	

KEY POINTS

1. The properties of iron are changed by adding controlled amounts of other metals or carbon to make steels.
2. Mild steel is used in buildings and for making car bodies.
3. Stainless steel is used in chemical factories and in making cutlery.
4. Recycling metals has some advantages and disadvantages.
5. Steel is often covered with a layer of zinc to protect it from rusting. This is called galvanising.

SUMMARY QUESTIONS

1. Copy and complete using the words below:

   **car   carbon   cutlery
   different   mild   stainless**

   We use _____ types of steel for different jobs. Low _____ steel, often called _____ steel, is used to make _____ bodies and machinery. We use _____ steel to make _____ and parts of chemical factories.

2. Explain why aluminium is used to make food containers.

3. Describe the advantages of using steel alloys rather than pure iron.

Aircraft bodies are made from alloys of aluminium because these alloys are strong yet lightweight

SUMMARY QUESTIONS

1. Match each metal on the left with its ore on the right.

iron	rock salt
aluminium	zinc blende
zinc	bauxite
sodium	hematite

2. Draw a diagram of a blast furnace. On your diagram show:
   - where the solid raw materials are loaded into the furnace
   - where air enters the furnace
   - where the iron and slag collect.

3. Copy and complete using the words from the list below.

   air   blast   calcium   coke   decomposes   impurities   monoxide   slag

   Iron is extracted in a _____ furnace from iron ore using carbon _____ as a reducing agent. The carbon monoxide is formed when _____ burns in a blast of _____. The limestone added to the blast furnace _____ to form _____ oxide. Calcium oxide combines with the _____ in the iron ore to form _____.

4. Match each metal on the left with its use on the right.

aluminium	galvanising iron roofs
mild steel	electrical wiring in the home
stainless steel	aircraft bodies
copper	cutlery
zinc	car bodies

5. Two methods for extracting metals are (i) heating with carbon and (ii) electrolysis. Which of these methods is best used to extract each of the following metals?
   (a) sodium
   (b) lead
   (c) calcium
   (d) iron
   (e) aluminium

PRACTICE QUESTIONS

1. Which one of these statements about the extraction of iron in a blast furnace is true?
   A Limestone is added to combine with excess carbon dioxide.
   B A slag of iron(III) oxide forms at the bottom of the furnace.
   C Hot air is blown in at the top of the furnace.
   D Carbon monoxide reduces iron(III) oxide to iron.

   (Paper 1)

2. Calcium oxide is added to a steelmaking furnace to:
   A oxidise basic oxides
   B oxidise carbon to carbon dioxide
   C react with acidic impurities
   D react with silicon to form silicon dioxide.

   (Paper 1)

3. (a) Describe three advantages of recycling metals. For each advantage give an explanation of why it is advantage. [6]
   (b) Give three examples of energy costs involved in recycling metals. [3]

   (Paper 3)

4. Iron is extracted in a blast furnace.
   (a) Name the four raw materials used to extract iron. [4]
   (b) The diagram shows a blast furnace.

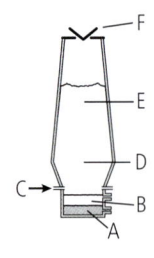

6. Put the following phrases about the extraction of zinc in the correct order.
   A zinc vapour rises up the furnace
   B zinc oxide is heated with coke to produce zinc
   C zinc sulfide is roasted in air to produce zinc oxide
   D the zinc condenses in trays

Which letter in the diagram shows:
  (i) where the solid raw materials are put into the furnace [1]
  (ii) where the furnace is hottest [1]
  (iii) where the slag is collected? [1]
(c) In the blast furnace, carbon monoxide reacts with iron(III) oxide.
  (i) Write a word equation for this reaction. [1]
  (ii) The carbon monoxide is formed by carbon dioxide reacting with excess carbon. Write a symbol equation for this reaction. [2]

(Paper 3)

5 Iron is converted to steel by blowing oxygen into molten iron in a basic oxygen converter.
  (a) Name three impurities present in the iron extracted from the blast furnace. [3]
  (b) Explain the function of the oxygen 'blast' in steelmaking. [1]
  (c) Powdered calcium oxide is added at the same time as the oxygen. Why is calcium oxide added? [2]
  (d) Stainless steel is used for many purposes.
    (i) State one use of stainless steel. [1]
    (ii) What is the difference between stainless steel and iron? [2]
    (iii) Stainless steel is an alloy. What do you understand by the term *alloy*? [1]

(Paper 3)

6 Some of the metals in the reactivity series are shown below:

sodium   calcium   zinc   iron   copper

most reactive ⟶ least reactive

  (a) (i) State the names of two metals from this list that can be extracted by heating with carbon. [2]
    (ii) State the names of two metals from this list that are extracted using electrolysis. [2]
  (b) Aluminium can be extracted by electrolysis. State two uses of aluminium. [2]

  (c) Carbon is used to extract tin from tin(IV) oxide, SnO_2.
    (i) What condition is needed for this extraction? [1]
    (ii) The products of the reaction are tin and carbon monoxide. Write a symbol equation for the extraction of tin. [2]

(Paper 3)

7 Zinc is extracted from its sulfide ore.
  (a) State the name of the sulfide ore of zinc. [1]
  (b) The sulfide ore is first converted to zinc oxide. Write a balanced equation for this reaction. [2]
  (c) The zinc oxide is reduced in a blast furnace. Describe in outline how zinc is extracted from zinc oxide. [4]
  (d) At high temperatures, zinc oxide reacts with carbon. Explain why this is a redox reaction in terms of electron transfer. [2]
  (e) (i) Give one use of zinc. [1]
    (ii) Explain why zinc is used for the purpose you stated in part (i). [1]

(Paper 4)

8 Steel is an alloy of iron with controlled amounts of carbon and other metals added.
  (a) Explain how the amount of carbon in the steel is controlled. [2]
  (b) One of the impurities in the iron from the blast furnace is phosphorus. Explain how phosphorus is removed from the iron when steel is made in the basic oxygen converter. [3]
  (c) State two uses of mild steel. [1]
  (d) (i) Draw the structure of a typical alloy. [2]
    (ii) Explain why steel is harder than pure iron. [2]
  (e) Brass is an alloy. State the names of the metals present in brass. [1]

(Paper 4)

15 Air and water

15.1 Water

LEARNING OUTCOMES

- Describe chemical tests for water
- Describe the treatment of the water supply in terms of filtration and chlorination
- Name some uses of water
- Describe the problems of an inadequate supply of water *(Supplement)*

Water: a valuable resource

Water is all around us in the sea, rivers and lakes as well as in the rocks below the ground and in the air as water vapour. But only 3% of this water is fresh water. And most of this fresh water is locked up as ice in the Arctic and Antarctic.

Water is used in the home, in industry and in agriculture:

- In the home we use it for drinking, cooking, washing and cleaning.
- In industry it is used as a solvent for many chemicals and as a coolant to stop industrial processes from getting too hot. It is also used as a cheap raw material for some chemical manufacturing processes. Water is also used to generate electrical power, either in hydroelectric power stations or by turning it into steam to drive turbines.

> **(Supplement)** We cannot survive without water for more than a few days – it is essential to life. Many places in the world are in short supply of water and this causes many problems: crops cannot grow, animals do not have enough drinking water and die, and there is an increased risk of disease.
>
> In agriculture, water is used on farms for irrigating (watering) crops and for animals to drink. A shortage of water for irrigation may lead to the death of crops and lack of food for populations relying on these crops.

Clean water is a valuable resource but it needs to be free of bacteria if we are to remain healthy

Testing for water

In Topic 9.1 we learned about a reversible reaction involving copper(II) sulfate. We can use this reaction to test for water. When we add water to white anhydrous copper(II) sulfate it turns blue.

$$CuSO_4(s) + 5H_2O(l) \longrightarrow CuSO_4 \cdot 5H_2O(s)$$

anhydrous copper(II) sulfate (white) + water ⟶ hydrated copper(II) sulfate (blue)

We can also use anhydrous cobalt chloride which changes from blue to pink when water is added:

$$CoCl_2(s) + 6H_2O(l) \longrightarrow CoCl_2 \cdot 6H_2O(s)$$

anhydrous cobalt(II) chloride (blue) + water ⟶ hydrated cobalt(II) chloride (pink)

Cobalt chloride test papers are often used as a convenient test for water. These tests show us that there is water present. But it might not be *pure* water.

Water purification

Water from rivers and lakes is not pure. It contains dirt, pieces of dead animals and plants, dissolved compounds and bacteria, many of which are harmful to health.

> **STUDY TIP**
>
> You do not need to know all the details about water treatment. Filtration and chlorination are the learning objectives on the syllabus.

Cholera and typhoid are just two of the many diseases caused by bacteria in untreated water. These diseases kill millions of people each year. So it is important that water is treated to remove the bacteria.

A modern water treatment plant removes insoluble materials and bacteria by filtration and **chlorination**.

Treating water is a complex process and there are many stages:

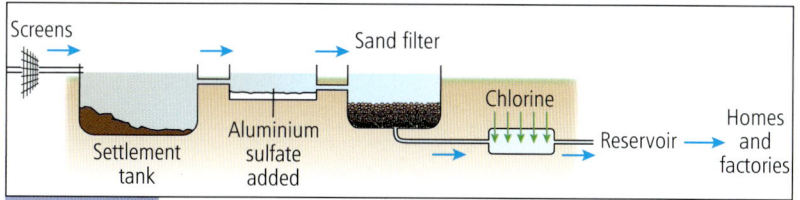

Figure 15.1.1 Treatment of water removes insoluble particles and kills bacteria. Not all the stages are shown.

- The water passes through metal screens which collect objects such as twigs and leaves.
- In the settlement tank, particles of solids such as soil particles settle to the bottom of the tank.
- Aluminium sulfate is added to make small particles in the water stick together. The particles then fall to the bottom of the tank.
- The water then passes through a filter made of sand and gravel or crushed coal. This removes any small insoluble particles that were not removed in the other tanks.
- Chlorine is added to kill bacteria.
- The pH of the water is adjusted and the water is run off and stored or goes directly to homes and factories.

Although treated water appears clear, it still contains dissolved salts. So it is not pure.

DID YOU KNOW?

In some places in the Middle East drinking water is made by 'reverse osmosis'. Water is pushed through a type of filter under high pressure.

SUMMARY QUESTIONS

1 Copy and complete using the words below:

   bacteria   chlorine   drinking   filter   insoluble

   We make water fit for _____ by passing the water through a sand and gravel _____. This removes any small _____ particles. We then add _____ to kill _____.

2 State two uses of water:
   a in the home
   b in industry.

3 Tap water is not pure. Explain why not.

KEY POINTS

1 Water is used in the home for drinking and washing, and in industry as a coolant and a solvent.

2 Anhydrous copper(II) sulfate is used to test for water.

3 Water is purified by filtration and chlorination.

4 An inadequate supply of water may result in crop failure and unsafe drinking water.

15.2 Air

LEARNING OUTCOMES

- State the composition of clean, dry air
- Describe the separation of nitrogen and oxygen from liquid air

Supplement

The gases in the air

The **atmosphere** around the Earth has several layers. The lowest layer contains most of the air around our planet. About one-fifth of air is oxygen. Oxygen is the reactive gas in the air. It is needed for combustion and rusting as well as for respiration (see Topic 15.6).

PRACTICAL

The percentage of oxygen in the air

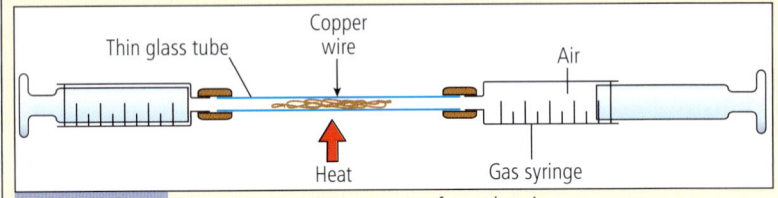

Figure 15.2.1 Hot copper removes oxygen from the air

1. We start by having 100 cm³ of air in one gas syringe and none in the other. The oxygen in the air reacts with the heated copper.
2. We pass the air over the copper by pushing the gas syringes back and forth.
3. We keep doing this until the volume of gas in the syringes remains constant. The total volume of gas in the syringes is 79 cm³. This is the approximate percentage of unreactive gases in the air – mainly nitrogen. So the percentage of oxygen is (100 − 79) 21%.

Nearly four-fifths of air is nitrogen. Nitrogen is an unreactive gas. But processes such as burning and rusting would be much quicker without nitrogen to dilute the oxygen.

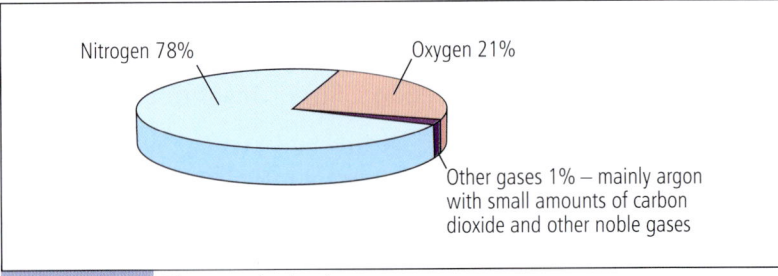

Figure 15.2.2 The gases in dry air

The rest of the air – about 1% of it – is largely made up of argon. But there are also tiny amounts of other gases. Carbon dioxide is the most important of these. However, only about 0.04% of the air is carbon dioxide. This gas has a great effect on our climate because it is a 'greenhouse gas' (see Topic 15.5).

Some of the first living things on Earth were bacteria which may have lived in colonies on rocks in the sea. They absorbed sunlight and released oxygen into the early atmosphere of the Earth.

DID YOU KNOW?

Most of the oxygen in the air was originally formed by bacteria which were carrying out photosynthesis millions of years ago.

In addition to these gases, the atmosphere contains water vapour. The amount of water vapour varies from place to place but it usually makes up between 1% and 4% of the atmosphere. Much of the air around us is polluted – it contains unwanted substances. The pollutants are dust particles and smoke as well as the gases sulfur dioxide and nitrogen oxides. Although these are present in small amounts, they have a big effect on our environment.

Separating the gases

Most of the gases in the air are useful. Oxygen is used in welding, steelmaking and to support respiration in hospitals. Because nitrogen is very unreactive, it is used to provide an inert atmosphere for food packaging, chemical processes and silicon chip production. The gases in air are separated by **fractional distillation** of liquid air:

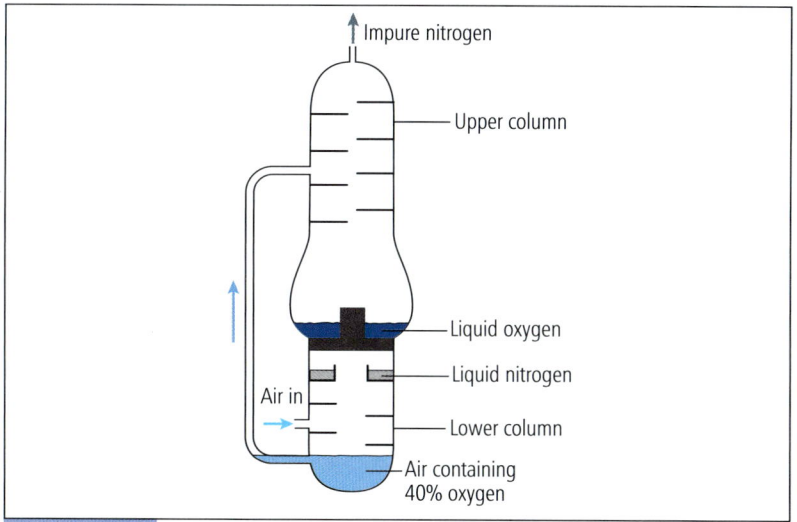

Figure 15.2.3 A double distillation column separates nitrogen from oxygen

- Water is removed from the air by passing it through a drying agent and carbon dioxide is removed by reacting it with sodium hydroxide solution.
- The air is cooled to –23 °C and compressed (squashed into a small space).
- The cold compressed air is allowed to expand into a larger space. When it expands, the air cools. After this is repeated several times the temperature drops to –200 °C. Most of the air is now liquid.
- Argon, krypton and xenon are removed from oxygen by further distillation. Helium and neon do not condense at –200 °C.
- When the liquid air is warmed, the nitrogen boils off first because it has a lower boiling point. Some of the nitrogen condenses at the top of the lower distillation column leaving a mixture of impure oxygen and nitrogen at the bottom. This mixture is then expanded into a gas.
- The oxygen–nitrogen gas mixture is then fed into the top column. The temperature in the top column is below the boiling point of oxygen but above the boiling point of nitrogen. So the oxygen condenses at the bottom and nitrogen gas is removed at the top.

STUDY TIP

The syllabus learning objectives ask you to describe the separation of gases from air only by fractional distillation

KEY POINTS

1 About four-fifths of air is nitrogen and about one-fifth is oxygen.

2 About 1% of air is argon with small amounts of other noble gases and carbon dioxide.

3 Nitrogen and oxygen are separated by the fractional distillation of liquid air.

SUMMARY QUESTIONS

1 Copy and complete using the words below:

**chemically   faster   fifths
nitrogen   rusting**

About four-____ of air is ____ and one-fifth is oxygen. Nitrogen is not very ____ reactive. Burning and ____ would be much ____ without nitrogen in the air.

2 What other gases apart from nitrogen and oxygen are present in unpolluted air?

3 Draw a flow chart to show how nitrogen is separated from the other gases in air.

15.3 Air pollution

LEARNING OUTCOMES

- Describe some sources of carbon monoxide, lead compounds and sulfur dioxide in the atmosphere
- State the adverse effects of each of these pollutants

We burn fossil fuels such as coal, oil and gas in power stations and in the home. We use hydrocarbons from the distillation of petroleum for transport – in cars, trains, ships and aircraft. The gases given off when we burn fuels pollute the atmosphere and cause harmful effects in the environment.

These pollutant gases include carbon monoxide, carbon dioxide, sulfur dioxide and oxides of nitrogen. When we burn fuels, tiny particles of solids also get into the air. These are called **particulates**. Lead dust and soot are examples of particulates. These pollutants have harmful effects on the atmosphere, seas and lakes, and on living things. This is why pollutants are of a global concern.

Carbon monoxide

When you burn a fuel in plenty of air, carbon dioxide and water are produced. This is complete **combustion**. But if there is not enough oxygen to burn the fuel completely, carbon monoxide and water are formed. We call this **incomplete** combustion.

Carbon monoxide is a poisonous gas. It combines with the red pigment haemoglobin in the cells of the blood that carry oxygen around the body. So if you get carbon monoxide poisoning, respiration in your cells will stop. Carbon monoxide is colourless and has no smell – so we can't easily tell if it is present or not. So it is important that gas boilers are serviced regularly to prevent incomplete combustion caused by not enough air reaching the fuel.

Lead compounds

Not so long ago, lead compounds were added to petrol to help it combust efficiently in car engines. Although lead is no longer used for this purpose in most grades of petrol, there is still a considerable amount of lead in the environment as particulates. This is because some lead compounds are not easily broken down in the environment. Lead compounds are also found in some paints and water pipes in old houses.

Lead is particularly hazardous if it gets into our food or water supply or if we breathe it in. Lead builds up in the body – our body does not get rid of it very easily. Even small quantities of lead in the body can cause damage to the nervous system, especially to the brain in children.

Sulfur dioxide and acid rain

Rain is naturally slightly acidic. This is because carbon dioxide from the air dissolves in rainwater and forms a solution of the weak acid, carbonic acid. However, if the pH of the rain falls below 5.0, the rain is called **acid rain**. How is acid rain formed?

Coal contains some sulfur impurities. Oil and natural gas also have some sulfur in them but much less. When these fossil fuels are burned the sulfur is oxidised to sulfur dioxide gas which escapes into the atmosphere:

These trees have died because of acid rain

STUDY TIP

It is a common error to suggest that sulfur rather than sulfur dioxide is responsible for acid rain. Comments such as 'sulfur dissolves in water to form acid rain' are incorrect.

$$S(s) + O_2(g) \longrightarrow SO_2(g)$$

Volcanoes are also a natural source of sulfur dioxide. They produce about a third of the sulfur dioxide that pollutes the atmosphere. The sulfur dioxide reacts with various compounds in the atmosphere to form sulfur trioxide. The sulfur trioxide then reacts with water vapour in the air to form a solution of sulfuric acid:

$$\underset{\text{sulfur trioxide}}{SO_3(g)} + H_2O(l) \longrightarrow \underset{\text{sulfuric acid}}{H_2SO_4(aq)}$$

When the acidic water vapour condenses, acid rain is formed.

Sulfur dioxide is not the only source of acid rain. Nitrogen oxides from car exhausts also play a part (see Topic 15.4).

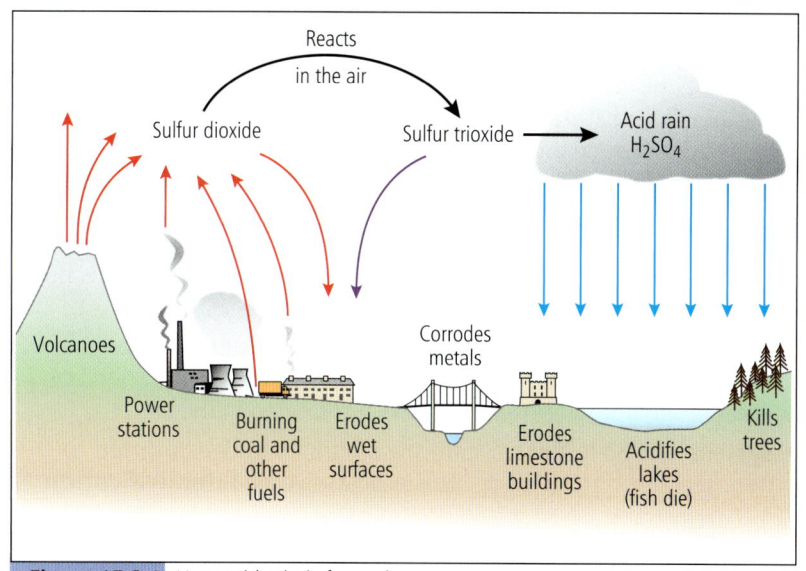

Figure 15.3.1 How acid rain is formed

Acid rain may fall in an area far away from the source of pollution. Sulfur dioxide and sulfur trioxide may be carried by winds as far as 200 km from the source. Acid rain can have a pH as low as 4. Acid rain has many negative effects on the environment:

- Trees, especially pine trees, have their leaves damaged by the acid. They can no longer carry out photosynthesis. The leaves fall off and the trees die.
- Lakes and rivers become acidic. If they are too acidic fish and other aquatic life may die.
- Soil may become too acidic to grow crop plants. The reaction of the acid with the soil may also cause some minerals to dissolve. Higher than normal concentrations of these minerals in soil water may be harmful to crops.
- Buildings made from carbonate rocks will be eroded. The acid reacts with the carbonate to release carbon dioxide and the surface of the building crumbles. Buildings made from limestone and marble are particularly badly affected.
- Metal structures such as bridges and iron railings corrode.

DID YOU KNOW?

The clouds around the planet Venus are made up of droplets of concentrated sulfuric acid!

KEY POINTS

1 Carbon monoxide is a poisonous gas formed from the incomplete combustion of compounds that contain carbon.

2 The source of acid rain is burning of fossil fuels.

3 Acid rain kills trees and acidifies lakes leading to the death of fish and acidifies soils leading to reduction in crop growth.

4 Lead pollution from old sources such as petrol, paint and water pipes can hinder brain development.

SUMMARY QUESTIONS

1 Copy and complete using the words below:

blood   hydrocarbon monoxide   oxygen poisonous

Carbon _____ is formed when there is not enough _____ to burn carbon or a _____ completely. Carbon monoxide is _____ because it stops _____ transporting oxygen.

2 Draw a flow chart to show how acid rain is formed.

3 Explain why lead pollution can still be a problem even though it is rarely used in petrol nowadays.

15.4 The nitrogen oxide problem

LEARNING OUTCOMES

- Describe some sources of nitrogen oxides in the atmosphere
- Describe the adverse effects of nitrogen oxides
- Describe and explain how catalytic converters remove oxides of nitrogen

Sources and effects

Nitrogen forms several oxides. They are all gases. We usually call these by their common names rather than by their correct chemical names. *Nitrous* oxide is N_2O (nitrogen(I) oxide), *nitric* oxide is NO (nitrogen(II) oxide) and nitrogen *dioxide* is NO_2 (nitrogen(IV) oxide). Oxides of nitrogen escape into the atmosphere from a number of sources:

- Most nitrogen oxides polluting the atmosphere are formed in car engines. Nitrogen is usually unreactive. But the high temperature and pressure inside an internal combustion engine causes nitrogen and oxygen to combine. A mixture of different nitrogen oxides is formed. This mixture comes out with the exhaust gases from the engine. This mixture is sometimes called Nox to show that several oxides of nitrogen are present.

    formation of nitric oxide: $\quad N_2(g) + O_2(g) \longrightarrow 2NO(g)$
    formation of nitrogen dioxide: $\quad N_2(g) + 2O_2(g) \longrightarrow 2NO_2(g)$

- From high-temperature furnaces – the temperature in these is high enough to allow nitrogen and oxygen to combine.
- In areas where there are a lot of thunderstorms, the electrical energy in the lightning causes the formation of large amounts of nitric oxide and nitrogen dioxide from nitrogen and oxygen in the air.
- Bacterial action in the soil can cause nitrates to be converted to nitrogen and nitrous oxide. This process is called **denitrification**.

In the atmosphere, a complicated series of reactions occurs. The various nitrogen oxides formed can be converted into one another by oxidation reactions or by photochemical reactions. Nitrous oxide is more stable than the other two oxides and remains in the atmosphere for longer.

Nitrogen oxides have several adverse effects on the environment:

- Nitrogen dioxide can cause acid rain. The nitrogen dioxide reacts with oxygen in the presence of rainwater to form a solution of nitric acid. Nitrous oxide and nitric oxide are not acidic, but nitric oxide can be easily converted into nitrogen dioxide in the atmosphere:

    $$2NO(g) + O_2(g) \longrightarrow 2NO_2(g)$$

- Nitrogen oxides from car exhausts can combine with hydrocarbons and other compounds in the atmosphere to form smog – a smoky fog. Smog often takes a long time to clear and often traps other particulates in it. This harmful mixture of gases and particulates causes irritation in the throat and nose.
- High levels of nitrogen oxides have been linked to certain medical conditions such as asthma.

Smog is a mixture of SMoke and fOG formed from car pollution

DID YOU KNOW?

Many large cities such as Mexico City have problems with the smog caused by nitrogen oxides from car exhausts reacting with hydrocarbons. Light catalysed smog can be very harmful to health.

Catalytic converters

We have seen how nitrogen oxides are formed in the car engine. Poisonous carbon monoxide is also formed in the engine by incomplete combustion of the fuel. A **catalytic converter** can be fitted to a car exhaust to remove these two poisonous gases.

The exhaust gases from the car engine are passed through a 'honeycomb' in a catalytic converter. The surfaces on the honeycomb are covered with a thin layer of catalyst made of platinum, rhodium or palladium or a mixture of these. The gases react on the surface of the catalyst.

Most catalytic converters have two compartments. In the first compartment the metals mainly catalyse the conversion of nitrogen oxides to nitrogen. In the second compartment carbon monoxide is converted to carbon dioxide:

$$2NO(g) \longrightarrow N_2(g) + O_2(g)$$
nitric oxide

$$2NO_2(g) \longrightarrow N_2(g) + 2O_2(g)$$
nitrogen dioxide

$$2CO(g) + O_2(g) \longrightarrow 2CO_2(g)$$

The reactions in a catalytic converter are **redox reactions**. But they may be more complicated than shown. Some of the carbon monoxide may react directly with the nitrogen oxides:

$$2NO(g) + 2CO(g) \longrightarrow N_2(g) + 2CO_2(g)$$
nitric oxide   carbon monoxide

$$2NO_2(g) + 4CO(g) \longrightarrow N_2(g) + 4CO_2(g)$$
nitrogen dioxide   carbon monoxide

Unburned hydrocarbons may reduce the nitrogen oxides too.

The gases leaving the car exhaust pipe now are not poisonous. However, carbon dioxide contributes to **global warming** (see Topic 15.5).

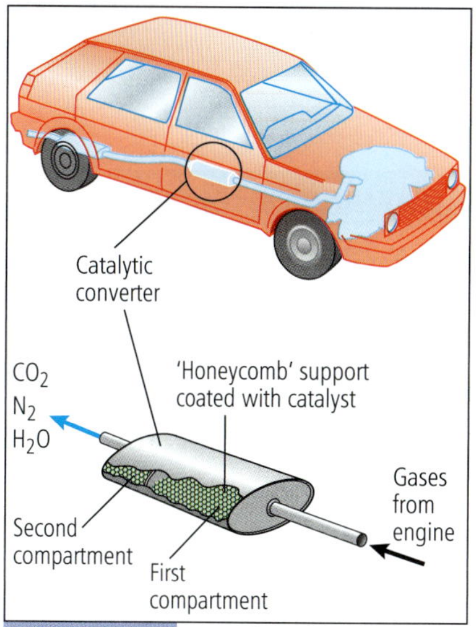

Figure 15.4.1 A catalytic converter

STUDY TIP

The best equations to remember about reactions in the catalytic converter are the reactions of nitrogen oxides with carbon monoxide to form nitrogen and carbon dioxide.

SUMMARY QUESTIONS

1 Explain how nitrogen oxides are produced:
   a in a car engine
   b during thunderstorms.

2 State three negative effects of nitrogen oxides on the environment.

3 Copy and complete using the words below:

   **converter   exhaust   hydrocarbons   monoxide
   nitrogen   oxidised   reduced**

   The _____ gases from a car engine include _____ oxides, carbon _____ and unburned _____. These gases are passed through a catalytic _____ where nitrogen oxides are _____ to nitrogen and carbon monoxide is _____ to carbon dioxide.

KEY POINTS

1 Nitrogen oxides are formed in car engines by nitrogen and oxygen reacting under high pressures and temperatures.

2 Nitrogen oxides can cause smog, breathing difficulties and acid rain.

3 Nitrogen oxides, carbon monoxide and unburned hydrocarbons can be removed from car engine exhausts by catalytic converters.

15.5 Global warming

LEARNING OUTCOMES

- State that carbon dioxide, methane and oxides of nitrogen are greenhouse gases and explain how they may contribute to climate change
- State the sources of methane

Burning fossil fuels may lead to increased global warming

DID YOU KNOW?

CFCs are still used for refrigeration and in spray cans although many countries have banned the use of these compounds. Like nitrous oxide, they stay in the atmosphere for a long time and are 'good' greenhouse gases. Their main effect is not as a greenhouse gas but in the depletion of the ozone layer which surrounds the Earth. Fortunately the use of CFCs has decreased rapidly over the last few years.

Greenhouse gases

A **greenhouse gas** is a gas that absorbs heat energy and stops heat escaping into space. The main greenhouse gases are carbon dioxide, methane, nitrous oxide and CFCs (chlorofluorocarbons). These gases are present in the atmosphere in tiny quantities compared with the amounts of nitrogen and oxygen. But even a small change in the concentration of these gases may create an effect that could change our climate. So what are the sources of these greenhouse gases?

- Carbon dioxide is the main greenhouse gas. It is naturally present in the atmosphere. But over the past century the amount of carbon dioxide has been increasing steadily and is now 0.038%. Much of this increase is due to burning fossil fuels in power stations and cars. Smaller amounts are produced by industrial processes, such as the thermal decomposition of calcium carbonate to make cement and lime.
- Methane is in the atmosphere at a much lower concentration than carbon dioxide. But it absorbs much more heat energy. The growing numbers of cows, pigs and sheep release a lot of methane by bacterial action in their digestive systems. The decomposition of vegetation, especially in swampy areas, also produces methane, as do rice paddy fields. Even termites (ant-like insects) produce methane.
- Nitrous oxide produced by bacteria in the soil is a greenhouse gas that remains in the atmosphere for a long time.

The greenhouse effect and global warming

The greenhouse effect has always been with us. Greenhouse gases prevent the Earth from cooling down too rapidly when we are not exposed to the Sun's rays. If we did not have greenhouse gases in the atmosphere, the Earth would be extremely cold.

The greenhouse effect works like this:

- Ultraviolet rays from the Sun have very short wavelengths. They get through the Earth's atmosphere easily and are not absorbed by carbon dioxide.
- The ultraviolet rays hit the Earth's surface.
- The Earth's surface absorbs the ultraviolet rays and it heats up.
- Heat energy is lost from the Earth's surface as infrared rays. This is long wavelength radiation.
- Infrared rays can be absorbed by carbon dioxide.
- Some of the heat is re-radiated back to the Earth and some escapes into space.

The more carbon dioxide there is in the atmosphere, the more heat is absorbed by this greenhouse gas. More heat is re-radiated back to Earth and less is lost into space. So the atmosphere heats up more. We have **global warming**.

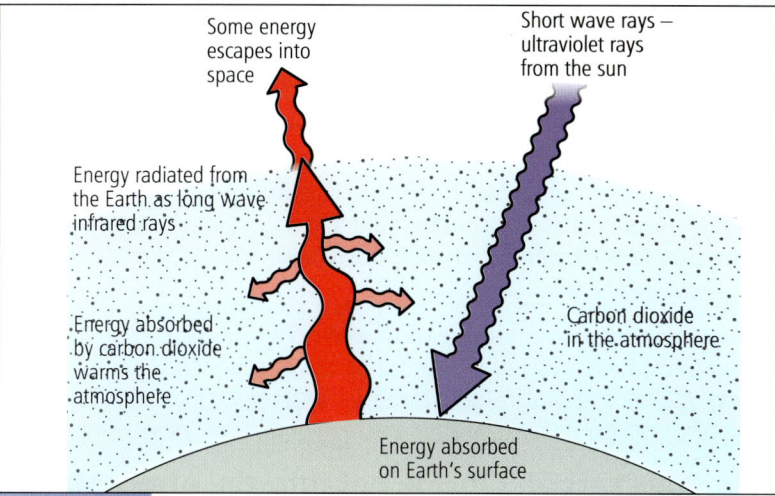

Figure 15.5.1 How the greenhouse effect works

Climate change and its results

A warmer atmosphere can affect our climate:

- The temperature of the air increases. This will lead to melting of the polar ice-caps. This in turn will increase sea levels and cause flooding of low-lying areas.
- There may be less rainfall in some areas. This will lead to the formation of more deserts and so less food production.
- There may be more violent weather in many areas – more storms and stronger winds. This will cause more destruction of property and crops.

Many scientists think that we are beginning to see these effects happening now.

STUDY TIP

It is important that you do not confuse the effects of different pollutants: carbon dioxide and methane are linked to global warming and sulfur dioxide is linked to acid rain.

DID YOU KNOW?

Some scientists predict that the average temperature near the Earth's surface may rise by as much as 6 °C by the year 2100.

SUMMARY QUESTIONS

1 Copy and complete using the words below:

   **absorb   atmosphere   global   greenhouse
   increase   methane   radiated**

   Carbon dioxide and _____ are _____ gases that increase _____ warming. The greenhouse gases _____ heat energy _____ from the Earth's surface. The more carbon dioxide there is in the _____, the greater the _____ in global warming.

2 What do you understand by the term *greenhouse gas*?

3 Methane is a greenhouse gas. State two sources of methane other than natural gas.

KEY POINTS

1 Methane and carbon dioxide are greenhouse gases.

2 Methane is produced as a result of digestion in cows and sheep, and from rice paddy fields.

3 Greenhouse gases cause an increase in global warming by absorbing heat energy radiated from the Earth's surface.

15.6 The carbon cycle

LEARNING OUTCOMES
- Describe the carbon cycle to include the processes of combustion, respiration and photosynthesis

The increased use of motor vehicles increases the concentration of carbon dioxide in the air

STUDY TIP
The two important regulating features of the carbon cycle are the uptake of carbon dioxide by photosynthesis and the production of carbon dioxide during respiration.

DID YOU KNOW?
There is fifty times more carbon in the oceans than there is in the air.

The sea, the atmosphere, the rocks and living things all contain carbon. We call these sources carbon reservoirs. The amount of carbon in each of these reservoirs has not changed much over millions of years. This is because there is a balance between the release and uptake of carbon from these reservoirs.

At the present time scientists are most worried about the carbon reservoir in the atmosphere. Most of the carbon in the atmosphere is present as carbon dioxide. We saw in Topic 15.5 that an increase in carbon dioxide in the atmosphere is linked to an increase in global warming.

How does carbon dioxide get into the atmosphere?

Most fuels produce carbon dioxide when they burn. This carbon dioxide escapes into the atmosphere.

$$CH_4(g) + 2O_2(g) \longrightarrow CO_2(g) + 2H_2O(l)$$
$$\text{methane} + \text{oxygen} \longrightarrow \text{carbon dioxide} + \text{water}$$

You can see from the equation that oxygen is also removed from the atmosphere.

An even larger amount of carbon dioxide is produced by **respiration**. This is the process by which living things get energy from food. It takes place in the cells of all animals, plants and microbes. The main source of energy in most organisms is glucose. We can summarise respiration as:

$$C_6H_{12}O_6 + 6O_2 \longrightarrow 6CO_2 + 6H_2O$$
$$\text{glucose} + \text{oxygen} \longrightarrow \text{carbon dioxide} + \text{water}$$

Smaller amounts of carbon get into the atmosphere from the breakdown of vegetation in swamps. This carbon may enter the atmosphere as methane or carbon dioxide. Carbon dioxide is also produced from the thermal decomposition of limestone and bacterial decomposition of dead plants and animals.

There is a huge amount of carbon dioxide dissolved in the world's oceans. Some of this is present as dissolved carbon dioxide, and some is present as hydrogencarbonate ions, HCO_3^-. When the water in the oceans gets warmer, carbon dioxide comes out of solution and enters the atmosphere.

How is carbon dioxide removed from the atmosphere?

Plants remove large amounts of carbon dioxide from the atmosphere by **photosynthesis**. We can summarise this as:

$$6CO_2 + 6H_2O \longrightarrow C_6H_{12}O_6 + 6O_2$$
$$\text{carbon dioxide} + \text{water} \longrightarrow \text{glucose} + \text{oxygen}$$

The glucose formed is used for plant growth and respiration. The energy for this reaction comes from sunlight. Chlorophyll, the green pigment in plants, traps the light and acts as a catalyst.

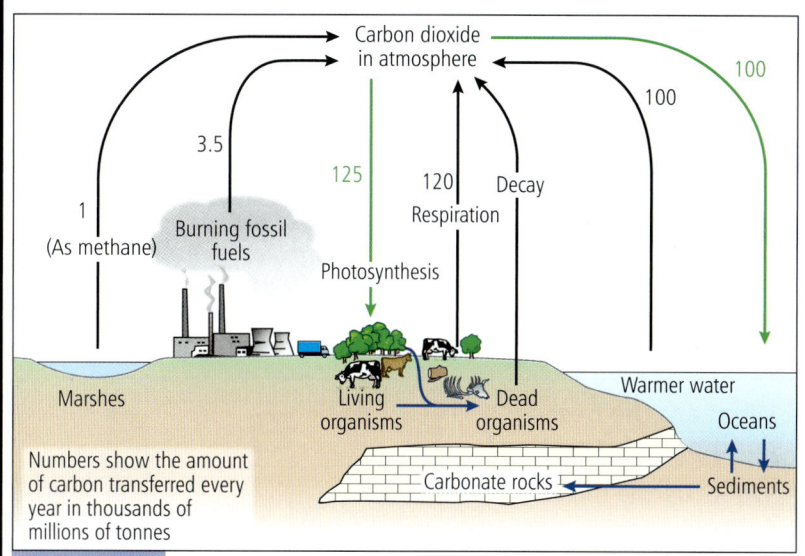

Figure 15.6.1 The main processes in the carbon cycle

Carbon dioxide is quite soluble in water and large amounts are removed from the atmosphere by the oceans.

Keeping the balance

Look at the equations for photosynthesis and respiration. You will notice that they are the opposite of one another. Respiration releases carbon dioxide to the air and photosynthesis removes carbon dioxide from the air. These two processes are roughly balanced so that the amount of carbon dioxide in the air remains fairly constant.

Some people are worried that the clearing of large areas of forests will reduce the amount of carbon dioxide removed from the air and make the carbon cycle unbalanced. The uptake of carbon dioxide from the atmosphere into the oceans is also roughly balanced by the carbon dioxide released back to the atmosphere from the oceans.

However, we have not taken into account the burning of fossil fuels. If fossil fuels were being formed from carbon dioxide as fast as they are being used, there would not be a problem. But fossil fuels are not being formed. So scientists are worried that the increased carbon dioxide concentration in the air will put the carbon cycle out of balance. This can lead to increased global warming.

More carbon dioxide could also be released into the atmosphere by warmer seas. Some scientists are worried that if the cycle becomes too unbalanced global warming will get out of control.

KEY POINTS

1 The carbon cycle keeps the concentration of carbon dioxide in the atmosphere approximately constant.
2 The amount of carbon dioxide produced in respiration is balanced by that which is absorbed in photosynthesis.
3 Burning fossil fuels and deforestation may unbalance the carbon cycle.

SUMMARY QUESTIONS

1 Copy and complete using the words below:

**atmosphere carbon
glucose oxygen
photosynthesis
uses water**

The main processes that keep the amount of _____ dioxide in the _____ constant are respiration and _____. Respiration uses _____ from the air to oxidise _____ to carbon dioxide and _____. Photosynthesis _____ carbon dioxide from the air to make glucose and oxygen.

2 Describe two processes that remove carbon dioxide from the atmosphere.

3 Describe two processes that add carbon dioxide to the atmosphere.

15.7 Preventing rust

LEARNING OUTCOMES

- State the conditions required for rusting
- Describe and explain methods of rust prevention involving coatings
- Describe and explain sacrificial protection in terms of the reactivity of metals

What causes rusting?

Chemicals in the air may attack metals causing the surface to get eaten away. We call this **corrosion**. The corrosion of iron and steel is called **rusting**. Only iron and steel rust.

PRACTICAL

What are the best conditions for rusting?

We set up the three tubes each with a clean iron nail at the bottom. We observe the nails over a period of time. We record how long it takes for each to rust.

The nail in the tube with just air and water rusts quickly. The nails in the other tubes do not rust.

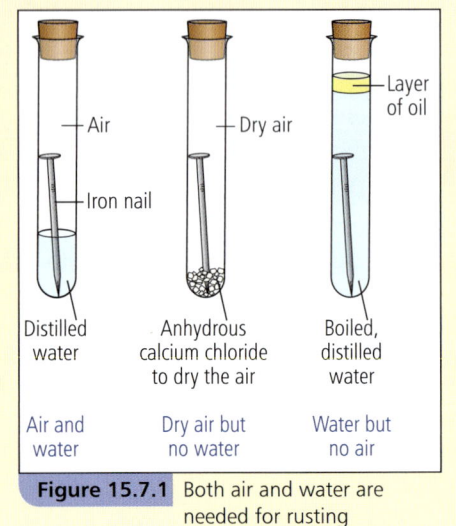

Tube 1: Air, Iron nail, Distilled water – Air and water
Tube 2: Dry air, Anhydrous calcium chloride to dry the air – Dry air but no water
Tube 3: Layer of oil, Boiled, distilled water – Water but no air

Figure 15.7.1 Both air and water are needed for rusting

For rusting to occur, both oxygen and water are needed. Rusting is an oxidation reaction:

$$4Fe(s) + 3O_2(g) + 2H_2O(l) \rightarrow 2Fe_2O_3.H_2O(s)$$
iron + oxygen + water → hydrated iron(III) oxide

There is no one simple formula for rust. The simplest formula that we can suggest is the formation of hydrated iron(III) oxide. The amount of water in rust varies with the conditions. You can see that rusting is a redox reaction: the iron increases its oxidation state from 0 to +3, and the oxidation state of oxygen changes from 0 to −2.

A layer of rust is very weak and it soon flakes away from the surface of the iron. The newly exposed surface then starts to rust. Rusting is speeded up by electrolytes such as salt. That is why ships rust very quickly unless they are treated to prevent rust.

Stopping the rust

Rusting destroys about 20% of the world's iron and steel every year. So it is very important to stop rusting. We can stop rusting by coating the iron or steel with a layer which keeps out air and water. We can use:

- Paint – for bridges and cars
- A plastic coating – garden furniture and wire netting for fencing are often covered with plastic

STUDY TIP

When considering methods that prevent rusting, 'removing water and air' are *explanations*, not methods.

- Metal plating – the iron is coated with another metal. This is often carried out by electroplating. Chromium plating is used for bathroom taps. Steel food cans are plated with tin. Iron for roofing is coated with zinc by dipping it into the molten metal – we call this **galvanising**
- Greasing and oiling – used for tools and the moving parts of machinery.

Galvanised steel is steel coated with zinc. The zinc forms a coating that stops water and oxygen from reaching the steel. But even when the zinc is scratched, the steel underneath will not rust. So there must be another way in which the zinc protects the steel. How does it do this?

Zinc is more reactive than iron. A more reactive metal loses electrons and forms positive ions more easily than a less reactive metal (See Topic 13.3). When the layer of zinc is scratched, a sort of electrochemical cell is set up in the presence of moisture. The electrons flow from the zinc to the iron.

$$Zn(s) \longrightarrow Zn^{2+}(aq) + 2e^-$$

The zinc is oxidised and forms zinc ions. So the zinc corrodes rather than the iron. We say the zinc is a *sacrificial* metal. So this method of using a more reactive metal to protect a less reactive metal from corrosion is called **sacrificial protection**. The more reactive metal corrodes rather than the iron or steel.

The iron remains protected because the electrons are accepted on its surface. This keeps the iron in a reduced state – it is not being oxidised to rust.

The electrons from the zinc react with hydrogen ions in the water to form hydrogen gas.

$$2H^+(aq) + 2e^- \longrightarrow H_2(g)$$

The sacrificial metal does not have to completely cover the surface of the iron for the method to work. Pipelines, oil rigs and ships can be protected from corrosion by attaching bars of zinc or magnesium in direct contact with the iron.

Ships' hulls have to be painted regularly, otherwise they will succumb to rust from contact with salt water and air.

Zinc strips protect the iron from rusting

DID YOU KNOW?

In 2003 the Kinzua bridge in Pennsylvania, USA, was destroyed in a storm because the bolts holding the structure to the ground had rusted away.

SUMMARY QUESTIONS

1 Copy and complete using the words below:

flakes   hydrated   oxygen   rust   steel   surface

Rusting occurs when _____ and water react with iron or _____. Rust is _____ iron(III) oxide. The layer of rust easily _____ off the _____ of the iron. The fresh iron surface then starts to _____.

2 Explain why iron plated with tin does not rust.

3 Explain why a piece of magnesium in contact with a steel pipeline prevents the pipeline rusting.

KEY POINTS

1 Both water and oxygen are needed for iron to rust.
2 You can prevent rusting by painting, plating, coating with plastic or greasing.
3 In sacrificial protection a more reactive metal is in contact with steel. The more reactive metal corrodes in preference to the steel.

SUMMARY QUESTIONS

1 Link each pollutant on the left with its effect on the right.

sulfur dioxide	smog as well as acid rain
nitrogen oxides	damage to the nervous system
methane	acid rain
lead compounds	increased global warming

2 Describe a chemical test for water.

3 Copy and complete using the words from the list below.

acid corroded fifth fossil oxidised rust sulfur sulfuric trioxide water

Oxygen forms one _____ of the atmosphere. In the presence of _____, oxygen reacts with iron to form _____. Iron is also _____ by _____ rain. Acid rain is formed when _____ fuels containing _____ are burnt. The sulfur dioxide formed is _____ in the atmosphere to form sulfur _____. This combines with rainwater to form a dilute solution of _____ acid.

4 Link each word on the left with the corresponding phrase on the right.

rusting	a gas in the atmosphere that absorbs heat energy
corrosion	the corrosion of iron
fossil fuels	a gas forming about 1% of the air
greenhouse gas	burning these causes acid rain
argon	a gas forming about 80% of the air
nitrogen	the 'eating away' of metal surfaces

5 State three ways of preventing rusting. For each method explain how rusting is prevented.

6 Catalytic converters remove nitrogen oxides from the exhaust gases in car engines.
   (a) Name two oxides of nitrogen.
   (b) Explain how nitrogen oxides are formed in car engines.
   (c) Explain how catalytic converters remove nitrogen oxides.

PRACTICE QUESTIONS

1 Which one of these statements about rusting is true?
   A Rust is hydrated iron(III) sulfate.
   B Oxygen is needed for rusting but water is not.
   C Both air and water are needed for rusting.
   D Stainless steel rusts faster than iron.

(Paper 1)

2 Which one of these statements about the carbon cycle is correct?
   A Burning fossil fuels is responsible for most of the carbon dioxide in the atmosphere.
   B Methane gets into the atmosphere by burning fossil fuels.
   C Removing many of the forests on Earth will decrease the amount of carbon dioxide in the atmosphere.
   D The amount of carbon dioxide put into the atmosphere by respiration is roughly the same as carbon dioxide removed by photosynthesis.

(Paper 2)

3 Clean air is about four-fifths nitrogen.
   (a) (i) State the names of three other gases present in unpolluted air. [3]
       (ii) For one of these gases, state the

7 Put these phrases about the separation of nitrogen and oxygen from liquid air into the correct order.
   A oxygen remains as a liquid at the bottom of the column
   B carbon dioxide is removed by bubbling through sodium hydroxide
   C the mixture of nitrogen and oxygen undergoes fractional distillation
   D gases with a very low boiling point such as argon are removed
   E nitrogen comes off at the top of the column as it has a lower boiling point
   F air is compressed

name of the gas and the approximate percentage of the gas in air. [1]
(b) Carbon monoxide is an atmospheric pollutant.
  (i) Suggest a source of carbon monoxide and explain how it is formed. [2]
  (ii) State an adverse effect of carbon monoxide on health. [1]
(c) Methane is also an atmospheric pollutant.
  (i) State two sources of methane. [2]
  (ii) Explain how methane may contribute to climate change. [2]
  (iii) State one effect of climate change. [1]

(Paper 3)

4 Water is an important raw material in the home and in industry.
(a) State one use of water (i) in industry (ii) in the home. [2]
(b) Describe a chemical test for water and state the result if the test is positive. [2]
(c) Water must be purified if it is to be used in the home. In the water purification process describe:
  (i) how insoluble substances are removed from the water [2]
  (ii) why chlorine is used. [1]
(d) Water plays a part in the formation of acid rain.
  (i) What do you understand by the term *acid rain*? [1]
  (ii) Explain how acid rain is formed. [3]

(Paper 3)

5 The concentration of carbon dioxide in the atmosphere is kept constant by respiration and photosynthesis.
(a) Explain what is meant by the term *photosynthesis*. [3]
(b) Write a symbol equation for the overall reaction in photosynthesis. [2]
(c) (i) Describe one other process other than photosynthesis that removes carbon dioxide from the atmosphere. [1]
  (ii) Describe one process, other than respiration, that adds carbon dioxide to the atmosphere. [1]
(d) Write a word equation for respiration. [2]

(e) Carbon dioxide is a greenhouse gas.
  (i) What do you understand by the term *greenhouse gas*? [2]
  (ii) Explain why the concentration of carbon dioxide in the atmosphere has increased over the past 100 years. [1]

(Paper 4)

6 Oxides of nitrogen are atmospheric pollutants.
(a) State one natural source of nitrogen oxides. [1]
(b) Nitrogen oxides and carbon monoxide are formed in car engines. Explain how these oxides are formed in car engines. [3]
(c) Catalytic converters are added to cars to remove oxides of nitrogen and carbon monoxide. Explain in simple terms how a catalytic converter works. [3]
(d) Unburned hydrocarbons from car engines can also react with nitrogen oxides:

$$4C_6H_{14} + 38NO_2 \longrightarrow 24CO_2 + 28H_2O + 19N_2$$

Identify the reducing agent and the oxidising agent in this reaction. Give reasons for your answers. [4]

(Paper 4)

7 Air is a mixture of gases.
(a) Describe the composition of clean dry air. [4]
(b) (i) Describe how nitrogen and oxygen can be separated from air. [2]
  (ii) The separation of nitrogen from oxygen depends on a particular physical property of these gases. Name this property. [1]
(c) Oxygen and water are needed for rusting of iron.
  (i) Describe one simple way of preventing oxygen and water from reaching the surface of the iron. [1]
  (ii) Rusting can also be prevented by sacrificial protection. Explain how sacrificial protection prevents rusting. [3]

(Paper 4)

16 The chemical industry

16.1 Fertilisers

LEARNING OUTCOMES

- Describe the need for nitrogen-, phosphorous- and potassium-containing fertilizers
- Describe the displacement of ammonia from its salts

We rely on many plants for our food. Plants need not only carbon dioxide, light and water for healthy growth, they also need nitrogen to make amino acids and proteins. There is plenty of nitrogen in the air but few plants can use this because nitrogen is an unreactive gas. Most plants take up nitrogen from the soil in the form of nitrates and ammonium salts. When farmers harvest their crops, the nitrogen is removed and not returned to the soil. After a few years the amount of nitrogen in the soil is reduced and new crops won't grow very well. Phosphorus and potassium which are also needed for plant growth are removed from the soil as well. What can we do to replace the loss of these essential elements from the soil?

We can add a **fertiliser**. A fertiliser is a substance added to the soil to replace the elements taken up by plants. Farmers also add fertilisers to the soil to increase the yields of their crops. They can use animal manure as a fertiliser. But only a limited amount of this is available. So artificial fertilisers containing nitrogen, phosphorus and potassium are added to the soil to replace the essential plant nutrients that have been lost. These fertilisers are called **NPK fertilisers** after the symbols for these three elements. A typical *compound fertiliser* may contain ammonium nitrate, ammonium phosphate and potassium chloride. A *single fertiliser* does not contain potassium and phosphorus.

Making fertilisers

Nearly 10% of the entire chemical industry is involved in making fertilisers. Many fertilisers contain ammonium salts. These are made by neutralising ammonia with acids:

$$NH_3(aq) + HNO_3(aq) \rightarrow NH_4NO_3(aq)$$
ammonia + nitric acid → ammonium nitrate

$$3NH_3(aq) + H_3PO_4(aq) \rightarrow (NH_4)_3PO_4(aq)$$
ammonia + phosphoric acid → ammonium phosphate

STUDY TIP

When you write equations for the formation of ammonium salts remember that no water is formed as a product. For example:
ammonia + sulfuric acid → ammonium sulfate.

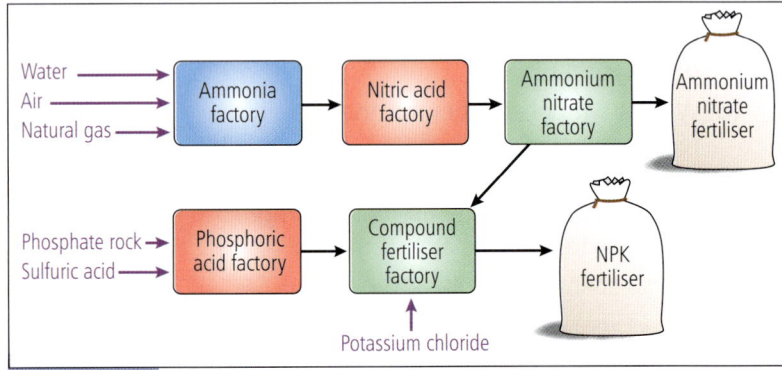

Figure 16.1.1 A flow chart for making an NPK fertiliser

The ammonia for these reactions is made by the **Haber process** (see Topic 16.2). The sulfuric acid for making phosphoric acid is made by the **Contact process** (see Topic 16.4). The potassium in fertilisers usually comes from salts such as potassium chloride which are mined.

PRACTICAL

Making ammonium sulfate fertiliser

1. You titrate ammonia with sulfuric acid using methyl orange as an indicator.
2. When the methyl orange turns pink, you record the burette reading.
3. Then you repeat the titration without the indicator, adding the same amount of acid as you did before.
4. You then pour the solution from the flask into an evaporating basin. You evaporate some of the water and allow the fertiliser to crystallise.

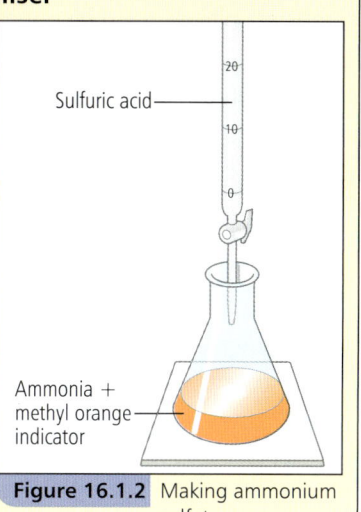

Figure 16.1.2 Making ammonium sulfate

Fertiliser being spread on fields by hand

Displacement of ammonia from ammonium salts

Many fertilisers are slightly acidic. But many crop plants do not grow well in acidic conditions. So farmers sometimes add lime (calcium oxide, CaO) to the soil to neutralise the acidity. Lime reacts with water in the soil to form slaked lime, calcium hydroxide. This is alkaline. If too much lime is put on the soil it reacts with ammonium salts in fertilisers to release ammonia:

$$2NH_4Cl(aq) + Ca(OH)_2(aq) \rightarrow 2NH_3(g) + CaCl_2(aq) + 2H_2O(l)$$

ammonium chloride | calcium hydroxide | ammonia | calcium chloride | water

The stronger alkali displaces ammonia from the ammonium salts. The ammonia escapes as a gas into the air and is lost from the soil.

DID YOU KNOW?

China now uses more synthetic fertilisers than any other country – about 50 000 000 tonnes per year.

SUMMARY QUESTIONS

1. Copy and complete using the words below:

   **elements   fertilisers   harvested
   phosphorus   potassium   soil**

   Plants need nitrogen, _____ and _____ for healthy growth. Artificial _____ are added to the _____ to replace the essential _____ lost when the crops are _____.

2. Describe how artificial fertilisers are made.
3. Ammonium nitrate is present in many fertilisers. Write a word equation for the reaction used to make ammonium nitrate.

KEY POINTS

1. Fertilisers are added to soil to replace the essential elements lost when plants are harvested.
2. NPK fertilisers contain nitrogen, phosphorus and potassium.
3. Fertilisers are often mixtures of salts containing ammonium, potassium, nitrate and phosphate ions.

16.2 Making ammonia

LEARNING OUTCOMES

- Describe the sources of the hydrogen and nitrogen used to make ammonia
- Describe and explain the essential conditions used in the Haber process

Millions of tonnes of ammonia are produced every year to make fertilisers, nitric acid, nylon and explosives. Over 80% of the ammonia produced is used to make fertilisers. But until 1908 nobody could make ammonia by combining nitrogen and hydrogen because nitrogen is very unreactive.

The problem was solved by a German chemist called Fritz Haber. He found that nitrogen can combine with hydrogen at high pressure and temperature. We call this method the **Haber process**.

The raw materials

The hydrogen for the Haber process is made either from natural gas or by 'cracking' ethane. The ethane is obtained from the fractional distillation of petroleum.

From cracking ethane using a high temperature and catalyst:

$$C_2H_6(g) \longrightarrow C_2H_4(g) + H_2(g)$$
$$\text{ethane} \longrightarrow \text{ethene} + \text{hydrogen}$$

From natural gas by reaction with steam in the presence of a nickel catalyst:

$$CH_4(g) + H_2O(g) \xrightarrow{\text{heat + Ni catalyst}} CO(g) + 3H_2(g)$$
$$\text{methane} + \text{steam} \longrightarrow \text{carbon monoxide} + \text{hydrogen}$$

The carbon monoxide, which can poison the catalyst used in the Haber process, is removed by reaction with more steam:

$$CO(g) + H_2O(g) \longrightarrow CO_2(g) + H_2(g)$$

Nitrogen for the Haber process is extracted from air, the oxygen having been removed by reaction with hydrogen.

The Haber process

The stages in the Haber process are:

- A mixture of nitrogen (1 volume) and hydrogen (3 volumes) is compressed.
- The compressed gases pass into a large tank called a converter. This contains trays of iron catalyst. The temperature in the converter is 450 °C and the pressure is 200 atmospheres. Nitrogen and hydrogen combine:

$$N_2(g) + 3H_2(g) \rightleftharpoons 2NH_3(g)$$

Under these conditions about 15% of the nitrogen and hydrogen are converted to ammonia.

- The mixture is passed to a cooling chamber. The ammonia condenses here and is removed.
- The unreacted nitrogen and hydrogen are returned to the converter. In this way they are not wasted.

A chemical plant for making ammonia

DID YOU KNOW?

The Haber process is more properly called the Haber–Bosch process. Carl Bosch did thousands of experiments to find the best catalyst and conditions to make ammonia.

STUDY TIP

You should know the conditions used in the Haber process and why these particular conditions are used by referring to the equilibrium reaction.

Figure 16.2.1 The Haber process

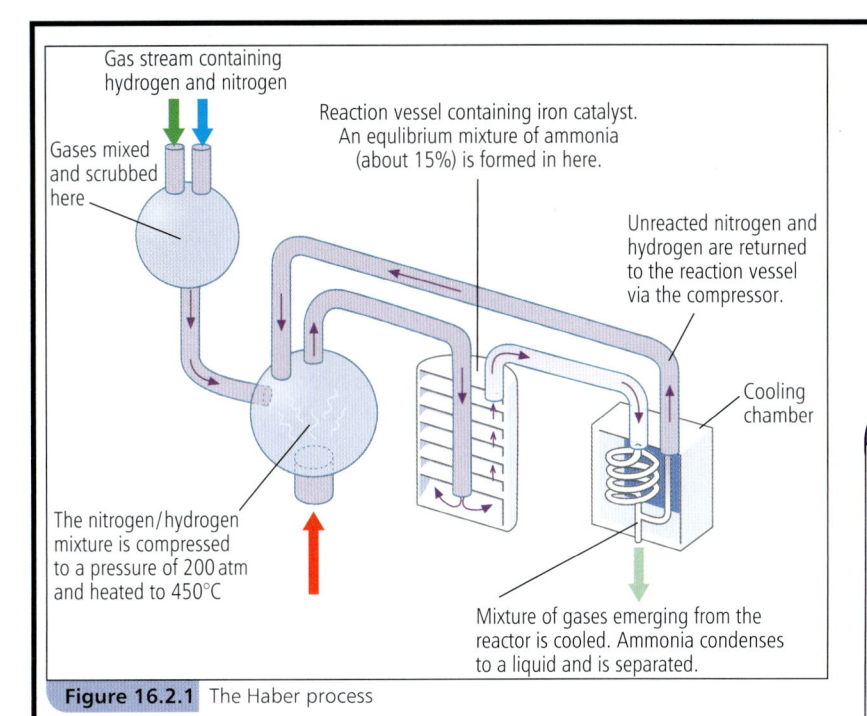

What are the best reaction conditions?

The reaction to make ammonia is an **equilibrium reaction**. The reaction is **exothermic** in the forward direction. Figure 16.2.2 shows how the equilibrium yield of ammonia varies with temperature and pressure. The **yield** is the percentage of ammonia in the equilibrium mixture.

You can see that the yield of ammonia increases with an increase in pressure. This is because, for a gas reaction, increasing the pressure shifts the equilibrium in the direction of lower volume. In this case it is to the right – in favour of an increased yield of ammonia. But high pressure is expensive. It costs a lot of money to make strong, safe vessels and a lot of expensive fuel is used to keep the pressure high. So we do not use a very high pressure.

The yield of ammonia decreases with increasing temperature. This is because an increase in temperature favours the endothermic reaction – in this case the endothermic reaction is the reverse reaction. But if we use too low a temperature the rate of reaction will be too slow. There is a conflict between the best equilibrium yield and the best rate of reaction. So we use a compromise temperature of 450 °C. This gives quite a good yield with a fast enough rate of reaction.

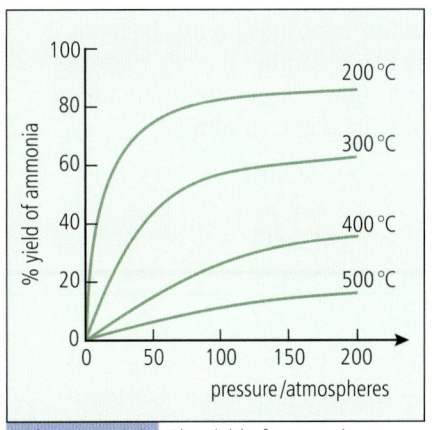

Figure 16.2.2 The yield of ammonia depends on the temperature and pressure

The catalyst has no effect on the equilibrium – it speeds up the rate of both the forward and the reverse reaction equally.

SUMMARY QUESTIONS

1 Copy and complete using the words below:

atmospheres   compressed converter   Haber   iron nitrogen   speed

Ammonia is made by the _____ process. Hydrogen and _____ are _____ and reacted together in a _____ at 450 °C and 200 _____ pressure. A catalyst of _____ is used to _____ up the reaction.

2 Explain why a compromise temperature of 450 °C is used in the Haber process.

3 How are the nitrogen and hydrogen for the Haber process obtained?

KEY POINTS

1 Nitrogen will not react with hydrogen at room temperature and pressure.

2 The conditions used in the Haber process are 200 atm pressure, a temperature of 450 °C and an iron catalyst.

3 A compromise temperature is used in the Haber process which gives quite a good yield at a fast enough rate of reaction.

205

16.3 Sulfur and sulfuric acid

LEARNING OUTCOMES

- Name some sources of sulfur
- Name some uses of sulfur and sulfur dioxide
- Describe the properties and uses of dilute and concentrated sulfuric acid

Sulfur: sources and uses

Sulfur is a yellow non-metallic element that is found uncombined in layers under the ground. Sulfur is mined in the United States, Mexico, Poland and Sicily in the south of Italy. It is also found combined with metals in ores such as zinc blende, ZnS, and galena, PbS. Natural gas and petroleum also contain sulfur, mainly as hydrogen sulfide, H_2S. Some natural gas from the South of France has a sulfur content of 15%.

About 90% of the sulfur extracted is used to manufacture sulfuric acid. The rest is used to make rubber tyres more flexible (vulcanising) and for making dyes.

Sulfur dioxide

Sulfur dioxide is a colourless poisonous gas. Large amounts of sulfur dioxide are added to the atmosphere every year by volcanoes and the burning of fossil fuels. Sulfur dioxide is highly acidic when it touches moist surfaces. It is the gas given off when sulfur burns:

$$S(s) + O_2(g) \longrightarrow SO_2(g)$$

It can also be made by roasting ores:

$$2ZnS(s) + 3O_2(g) \longrightarrow 2ZnO(s) + 2SO_2(g)$$
zinc sulfide + oxygen (in air) → zinc oxide + sulfur dioxide

These two reactions provide sulfur dioxide for the manufacture of sulfuric acid. Sulfur dioxide is used:

- as a bleach during the manufacture of paper from wood pulp. It is especially useful for bleaching materials such as silk, wool and straw which are destroyed by stronger bleaches such as chlorine.
- to preserve food and drinks. It does this by killing any bacteria present. Although sulfur dioxide can be added directly to drinks to preserve them, compounds called sulfites are more often added. These decompose to sulfur dioxide in acidic conditions:

$$SO_3^{2-}(aq) + 2H^+(aq) \longrightarrow SO_2(g) + H_2O(l)$$
sulfite ion + acid → sulfur dioxide + water

Sulfur crystals

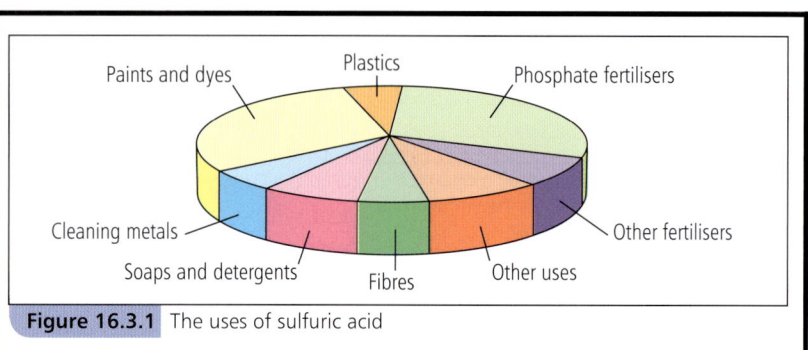

DID YOU KNOW?

The amount of sulfur dioxide produced by volcanoes is just over half the amount produced by burning fossil fuels.

Figure 16.3.1 The uses of sulfuric acid

Sulfuric acid

Sulfuric acid is one of the most important industrial chemicals. It has many uses (see Figure 16.3.1).

Sulfuric acid is a dibasic acid. This means that on reaction, two hydrogen ions can be formed per molecule.

$$H_2SO_4(aq) + 2NaOH(aq) \rightarrow Na_2SO_4(aq) + 2H_2O(l)$$
sulfuric acid + sodium hydroxide → sodium sulfate + water

Concentrated sulfuric acid is used as an acid drain cleaner and for cleaning metals. Concentrated sulfuric acid is an oxidising agent. It oxidises copper to copper (II) sulfate, and carbon to carbon dioxide. It is also a dehydrating agent, for example it removes water from ethanol to form ethane.

Dilute sulfuric acid is a strong acid and has typical acidic properties:

- It reacts with metals above hydrogen in the reactivity series to form a salt and hydrogen:

$$H_2SO_4(aq) + Zn(s) \rightarrow ZnSO_4(aq) + H_2(g)$$
sulfuric acid + zinc → zinc sulfate + hydrogen

- It reacts with metal oxides to form a salt and water:

$$H_2SO_4(aq) + MgO(s) \rightarrow MgSO_4(aq) + H_2O(l)$$
sulfuric acid + magnesium oxide → magnesium sulfate + water

- It reacts with metal hydroxides to form a salt and water:

$$H_2SO_4(aq) + 2KOH(aq) \rightarrow K_2SO_4(aq) + 2H_2O(l)$$
sulfuric acid + potassium hydroxide → potassium sulfate + water

- It reacts with carbonates to form a salt, water and carbon dioxide:

$$H_2SO_4(aq) + Na_2CO_3(aq) \rightarrow Na_2SO_4(aq) + H_2O(l) + CO_2(g)$$
sulfuric acid + sodium carbonate → sodium sulfate + water + carbon dioxide

> **STUDY TIP**
>
> Sulfuric acid has two hydrogen ions that can be replaced. Make sure that you remember this when writing symbol equations for the reaction of sulfuric acid with metals, metal oxides and metal carbonates.

SUMMARY QUESTIONS

1. Name two sources of sulfur.
2. Name two uses of sulfur dioxide.
3. Write symbol equations for the reaction of sulfuric acid with:
   a magnesium
   b potassium hydroxide.

KEY POINTS

1. Sulfur is used for vulcanising rubber and for making sulfur dioxide.
2. Sulfur dioxide is used to bleach wood pulp and as a food preservative.
3. Concentrated sulfuric acid is an oxidising agent and a dehydrating agent.
4. Dilute sulfuric acid reacts with metals, metal oxides, metal hydroxides and carbonates.

16.4 Manufacturing sulfuric acid

LEARNING OUTCOMES

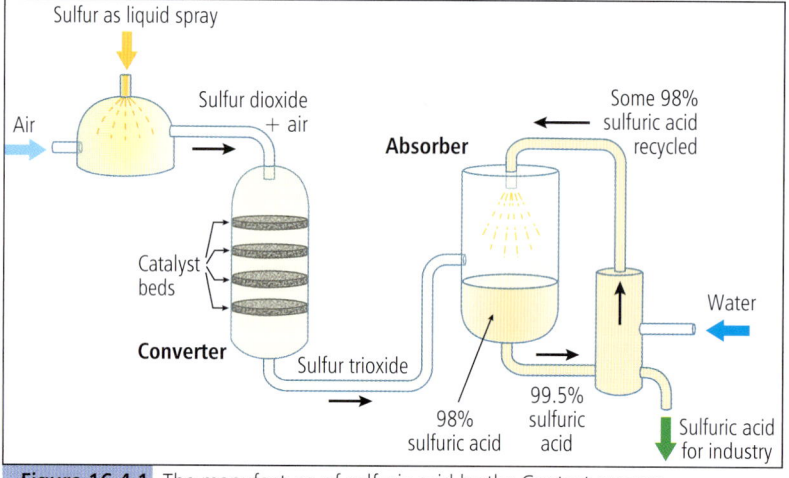

- Describe the manufacture of sulfuric acid by the Contact process
- Describe the essential conditions and reactions in the Contact process

Sulfuric acid has been an important chemical for many years. In 1843, the German chemist Liebig wrote 'We may judge the prosperity of a country by the amount of sulfuric acid it uses.' More sulfuric acid is made in the world than any other chemical. About 150 000 000 tonnes are made every year. It is used to make fertilisers, detergents, paints, fibres and to clean metals. It is also used in car batteries.

Most sulfuric acid is made in industry by the **Contact process**. Although the idea behind the Contact process was discovered in 1831, this method was not used until there was a greater demand for very concentrated sulfuric acid.

The Contact process

The raw materials for the Contact process are sulfur, air and water. The sources of sulfur are:

- sulfur from beneath the ground
- sulfide ores
- hydrogen sulfide from petroleum or natural gas.

A chemical plant for making sulfuric acid

Figure 16.4.1 The manufacture of sulfuric acid by the Contact process

The stages in the Contact process are:

- A spray of molten sulfur is burned in a furnace in a current of dry air. Sulfur dioxide is formed:

$$S(l) + O_2(g) \longrightarrow SO_2(g)$$

- The sulfur dioxide is cooled and reacted with excess air. This happens in a tower called a converter. The converter contains four layers of a catalyst, vanadium(V) oxide. The temperature in the converter is about 450 °C and the pressure is atmospheric pressure. In the converter sulfur dioxide is converted to sulfur trioxide:

$$2SO_2(g) + O_2(g) \rightleftharpoons 2SO_3(g)$$
$$\text{sulfur dioxide} + \text{oxygen} \rightleftharpoons \text{sulfur trioxide}$$

STUDY TIP

You will need to know the main reactions in the Contact process and be able to write the relevant equations. You also need to know why the particular conditions of temperature and pressure are used.

This reaction is exothermic. The heat is removed between each catalyst layer to keep the temperature quite low. The mixture of gases leaving the tower contains between 96% and 99.5% sulfur trioxide.

- The sulfur trioxide is absorbed into a 98% solution of sulfuric acid. This happens in a tower called an absorber. We do not dissolve the sulfuric acid directly into water. This is because when sulfur trioxide reacts with water a fine mist of sulfuric acid forms. This does not condense very easily. The sulfur trioxide dissolves in the 98% sulfuric acid to form a thick liquid called oleum:

$$SO_3(g) + H_2SO_4(l) \longrightarrow H_2S_2O_7(l)$$
sulfur trioxide   sulfuric acid   oleum

- The oleum is mixed with a little water to make concentrated 98% sulfuric acid:

$$H_2S_2O_7(l) + H_2O(l) \longrightarrow 2H_2SO_4(l)$$

Some of this 98% acid is returned to the absorber and the rest is run off to be used as concentrated sulfuric acid.

What reaction conditions do we use?

The reaction to make sulfur trioxide from sulfur dioxide is an **equilibrium reaction**. The yield of sulfur trioxide increases with an increase in pressure – in the direction of decreasing volume. But atmospheric pressure is used. There is no need to increase the pressure because the yield of sulfur trioxide is already very high at atmospheric pressure.

The reaction is exothermic. So the yield of sulfur trioxide decreases with increasing temperature. This is because an increase in temperature favours the endothermic reaction – the reverse reaction. But if we use too low a temperature the rate of reaction will be too slow. So we use a compromise temperature of 450 °C. This gives quite a good yield of sulfur trioxide with a fast enough rate of reaction.

The catalyst has no effect on the equilibrium – it speeds up the rate of the forward and the reverse reactions equally.

> **DID YOU KNOW?**
>
> The 'Father of Chemistry' Jabir ibn Hayyan discovered sulfuric acid about 1200 years ago. It is not certain whether he was Arabic or Iranian.

SUMMARY QUESTIONS

1 Copy and complete using the words below:

air Contact dioxide oleum sulfuric trioxide vanadium

In the _____ process sulfur is first burned to form sulfur _____. The sulfur dioxide reacts with more _____ in a converter packed with layers of _____ (V) oxide catalyst. The sulfur _____ formed is then absorbed in 98% _____ acid to form _____.

2 Describe two oxidation reactions that take place during the manufacturing of sulfuric acid.

3 Explain why atmospheric pressure is used in the Contact process to convert sulfur dioxide to sulfur trioxide.

KEY POINTS

1 In the Contact process, sulfur is first burned to form sulfur dioxide.

2 The sulfur dioxide is converted to sulfur trioxide at 450 °C in the presence of a catalyst of vanadium(V) oxide.

3 The sulfur trioxide produced in the Contact process is absorbed into 98% sulfuric acid.

16.5 The limestone industry

LEARNING OUTCOMES

- Describe the manufacture of lime in terms of thermal decomposition
- Name some uses of lime and slaked lime
- Name the uses of calcium carbonate

The main source of calcium carbonate is limestone. This rock was formed millions of years ago from the remains of the shells and skeletons of tiny sea creatures which died and fell to the bottom of the sea.

Lime

When we heat limestone strongly it breaks down into calcium oxide and carbon dioxide. This is an example of **thermal decomposition**. The common names for calcium oxide are **lime** or quicklime.

$$CaCO_3(s) \longrightarrow CaO(s) + CO_2(g)$$
calcium carbonate → calcium oxide + carbon dioxide

PRACTICAL

The thermal decomposition of limestone

You heat a tiny 'chip' of calcium carbonate very strongly. Calcium oxide is formed on its surface. You put the chip on a watch glass and add a few drops of universal indicator. There is a reaction and the indicator turns purple.

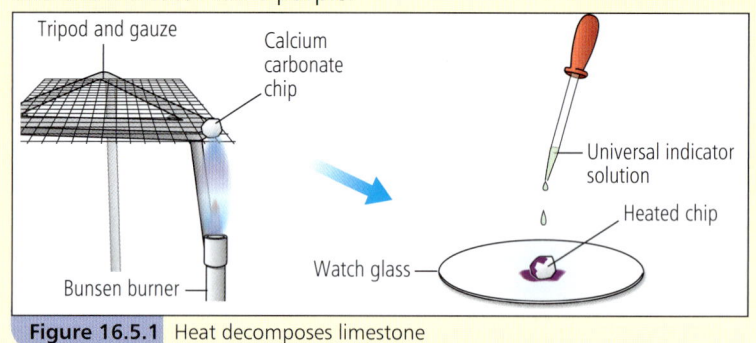

Figure 16.5.1 Heat decomposes limestone

STUDY TIP

It is a common error to suggest that oxygen is a reactant or product in the production of lime. The reaction is a thermal decomposition. Oxygen does not react with calcium carbonate and oxygen gas is NOT given off in the thermal decomposition.

Lime is usually made in a furnace called a lime kiln. In a rotary lime kiln, lumps of limestone are fed in at the top. The limestone is strongly heated by a current of hot air. The rotation of the kiln helps the limestone mix with the hot air. The limestone decomposes to calcium oxide (lime). This is removed at the bottom of the kiln.

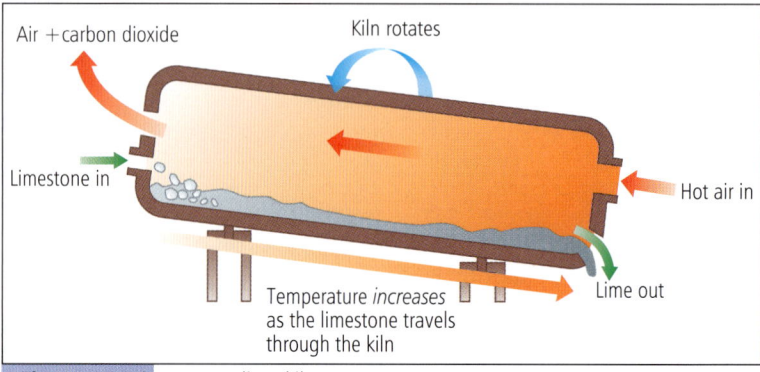

DID YOU KNOW?

The thermal decomposition of limestone to make lime is one of the oldest chemical reactions known. Some of the later pyramids in Egypt have lime mortar between the stone blocks.

Figure 16.5.2 A rotary lime kiln

Although the decomposition of calcium carbonate is a reversible reaction, the lime kiln is not a closed system (see Topic 9.1). The hot air carries the carbon dioxide out of the kiln so that the reverse reaction to re-form calcium carbonate cannot occur.

Slaked lime

Slaked lime is calcium hydroxide. This is made from lime by adding water. The water is added slowly because the reaction is very exothermic:

$$CaO(s) + H_2O(l) \longrightarrow Ca(OH)_2(s)$$
calcium oxide + water $\longrightarrow$ calcium hydroxide
(lime) (slaked lime)

A solution of calcium hydroxide in water is **limewater**. This is used as a test for carbon dioxide (see Topic 11.4).

Uses of limestone products

Limestone is used for building, in the extraction of iron, and in the construction of roads. In powdered form we use it to neutralise excess acidity in soils and lakes and to make glass.

Limestone is also used in the manufacturing of cement: the limestone is mixed with clay and heated in a furnace at about 1500 °C. It is then mixed with gypsum – calcium sulfate. This mixture is crushed to make cement. A mixture of cement with water and sand is called mortar.

Lime mortar is made from calcium oxide. It hardens when it reacts with carbon dioxide in the air.

Lime and slaked lime are used for treating excess acidity in soils and lakes.

When coal or petroleum fractions containing sulfur impurities are burned in power stations or to heat furnaces, sulfur dioxide is formed. This causes acid rain. The sulfur dioxide can be removed by passing it through a spray or slurry of calcium carbonate, calcium oxide or calcium hydroxide. This is called flue gas desulfurisation:

$$CaO(s) + SO_2(g) \longrightarrow CaSO_3(s)$$
calcium + sulfur $\longrightarrow$ calcium
oxide dioxide sulfite

The calcium sulfite formed is reacted with air and water to form calcium sulfate:

$$2CaSO_3(s) + O_2(g) + 4H_2O(l) \longrightarrow 2CaSO_4.2H_2O(s)$$
calcium sulfite + oxygen + water $\longrightarrow$ hydrated calcium sulfate

The calcium sulfate can be used to make plasterboard or to provide sulfur for making sulfuric acid. But it is more often just dumped in the ground because transporting it is too expensive.

Lime mortar was used to cement together the stone blocks of the pyramids

KEY POINTS

1. Calcium oxide (lime) is made by heating limestone in a rotary kiln.
2. Limestone is used in the extraction of iron and the manufacturing of cement.
3. Lime and slaked lime are used to treat acidic soil and to neutralise acidic waste products in industry.

SUMMARY QUESTIONS

1. Copy and complete using the words below:

   **carbon   decomposes
   lime   water**

   When limestone is heated it _____ to calcium oxide and _____ dioxide. Calcium oxide is also called _____. Slaked lime is made by adding _____ to lime.

2. State two uses of slaked lime and two uses of lime.

3. Describe how lime is made.

SUMMARY QUESTIONS

1. Link each compound on the left with its description on the right.

ammonium nitrate	this is quicklime, used for neutralising acidic soil
calcium oxide	an alkaline gas used to make fertilisers
calcium hydroxide	this is used in the manufacture of iron and cement
calcium carbonate	a solution of this chemical is called limewater
ammonia	a salt used as a fertiliser

2. Copy and complete using the words from the list below.

   acidity   ammonium   crops   elements   harvested   limestone   phosphate   reduce

   Farmers use several chemicals to make sure that their _____ grow well. Fertilisers such as _____ sulfate and ammonium _____ are added to the soil to replace the _____ removed when crops are _____. Quicklime and powdered _____ are added to the soil to _____ its _____.

3. State the names of a suitable acid and alkali you can use to make each of the following fertilisers:
   (a) ammonium sulfate
   (b) potassium phosphate
   (c) ammonium nitrate.

4. Write symbol equations for:
   (a) the reaction of nitrogen and hydrogen to make ammonia
   (b) the reaction of sulfur dioxide with oxygen to make sulfur trioxide
   (c) the thermal decomposition of calcium carbonate
   (d) the reaction of sulfuric acid with calcium carbonate.

5. Write symbol equations for the reaction of sulfuric acid with:
   (a) magnesium oxide
   (b) zinc
   (c) sodium hydroxide
   (d) sodium carbonate.

PRACTICE QUESTIONS

1. Which one of these statements about fertilisers is true?
   A   Ammonium nitrate can be used as a fertiliser.
   B   Fertilisers contain nitrogen, sulfur and iron.
   C   Fertilisers are added to the soil to make it more alkaline.
   D   Fertilisers are made by combining calcium with oxygen.

   (Paper 1)

2. Which one of these statements about the Haber process is correct?
   A   The raw materials are nitrogen dioxide, from the air and methane from the petroleum industry.
   B   The temperature used is about 450 °C because a higher temperature would reduce the yield of ammonia.
   C   The temperature is above 800 °C because a higher temperature gives a greater yield of ammonia.
   D   The pressure in the reaction vessel is 20 atmospheres because the reaction is exothermic.

   (Paper 2)

3. Fertilisers are spread on fields by farmers.
   (a) Why do farmers use fertilisers? [1]
   (b) State the names of the three elements most commonly found in fertilisers. [3]
   (c) Ammonium sulfate is a fertiliser. Describe how you can make ammonium sulfate in the laboratory from aqueous ammonia and sulfuric acid. [4]
   (d) Ammonium nitrate is also a fertiliser. Write a word equation to show how ammonium nitrate can be produced. [2]

   (Paper 3)

4. Calcium carbonate is used in the manufacture of lime.
   (a) Name a rock which contains largely calcium carbonate. [1]
   (b) (i) Describe how lime is manufactured from calcium carbonate rocks. [3]
       (ii) Copy and complete the equation for this reaction:
            $CaCO_3 \longrightarrow$ _____ + _____ [2]

(c) Explain why farmers sometimes put lime on their fields. *[2]*

(d) When lime reacts with water to form slaked lime, the reaction mixture gets very hot.
  (i) What is the name given to a chemical reaction which gives out heat? *[1]*
  (ii) Copy and complete the symbol equation for this reaction:
  CaO + _____ → _____ *[2]*
  (iii) Give one use of slaked lime. *[1]*

(Paper 3)

5 Ammonium phosphate is a fertiliser.
  (a) Suggest two substances that you could react together to make ammonium phosphate. *[2]*
  (b) What method would you use to make ammonium phosphate in the laboratory? *[1]*
  (c) Copy and complete the following sentences about the crystallisation of ammonium phosphate. Select words from the list below.

  **concentrated   dilute   dried
  phosphate   solution   water**

  A _____ of ammonium _____ is put into an evaporating basin and the _____ is evaporated until the _____ is very _____. The solution is then left to form crystals. The crystals are then removed and _____. *[4]*

  (d) Name another two fertilisers that can provide a plant with nitrogen in soluble form. *[2]*

(Paper 3)

6 The Haber process can be represented by the equation:

$N_2(g) + 3H_2(g) \rightleftharpoons 2NH_3(g)$ $\Delta H = -92 \, kJ/mol$

  (a) State the source of nitrogen for this process. *[1]*
  (b) State two sources of hydrogen for this process. *[2]*
  (c) Explain why the yield of ammonia rises as the pressure increases. *[2]*
  (d) How does the yield of ammonia vary with temperature? Explain your answer. *[3]*
  (e) Why is the reaction carried out at about 450°C rather than at a higher or lower temperature? *[3]*

(Paper 4)

7 A flow chart for the production of sulfuric acid is shown below.

A → Furnace → Converter → Absorber → Diluter
B ↗   ↑
                            C

  (a) State the name for the process used to make sulfuric acid. *[1]*
  (b) State the names of substances (i) A (ii) B and (iii) C in the flow diagram. *[3]*
  (c) Write a balanced equation for the formation of sulfur trioxide in the converter. *[2]*
  (d) State the name of the catalyst used in the converter. *[1]*
  (e) What conditions of temperature and pressure are used in the converter? Explain why these conditions are used. *[4]*
  (f) Explain why sulfur trioxide is not dissolved in water in the absorber. *[1]*

(Paper 4)

8 Sulfuric acid, H_2SO_4, is a dibasic acid.
  (a) What do you understand by the term *dibasic*? *[1]*
  (b) Sulfuric acid is a strong acid. Explain the meaning of *strong acid*. *[1]*
  (c) Write a balanced equation for the reaction of dilute sulfuric acid with calcium. *[2]*
  (d) Sulfuric acid reacts with several compounds of calcium. Write balanced equations for the reaction of sulfuric acid with two different calcium compounds. Include state symbols. *[6]*
  (e) Sulfuric acid is made from sulfur dioxide. Name two other uses of sulfur dioxide. *[2]*

(Paper 4)

17 Organic chemistry and petrochemicals

17.1 Organic chemistry

LEARNING OUTCOMES

- Describe a homologous series as a family of similar compounds with similar chemical properties due to the presence of the same functional group
- Name and draw the structures of methane, ethane, ethanol and ethanoic acid
- Describe the general characteristics of a homologous series
- Recall that compounds in a homologous series have the same general formula

Supplement

Organic chemistry

About two hundred years ago the Swedish chemist Jöns Jakob Berzelius divided chemicals into two main groups: organic and inorganic chemicals. Most organic chemicals burn or char (go black) when heated. Most inorganic chemicals just melt on heating.

All **organic compounds** contain carbon. They usually contain hydrogen and may contain other elements as well. Millions of organic compounds are known. So we have to make rules for naming them. Fortunately for us, many organic compounds can be put into groups. A group of organic molecules with similar chemical properties is called a **homologous series**. Two homologous series are **alcohols** and **carboxylic acids**. Here are the names and formulae of some compounds in these two homologous series:

alcohol homologous series		carboxylic acid homologous series	
methanol	CH_3OH	methanoic acid	$HCOOH$
ethanol	C_2H_5OH	ethanoic acid	CH_3COOH
propanol	C_3H_7OH	propanoic acid	C_2H_5COOH

You can see that all the alcohols have an -OH group and that all the carboxylic acids have a -COOH group. We call this group the **functional group**. A functional group is an atom or group of atoms that gives a compound particular properties. Carboxylic acids behave in a different way from alcohols but each carboxylic acid has very similar chemical properties.

More about homologous series

We can tell which homologous series a compound belongs to by the ending of its name. For example, the members of the alcohol homologous series all end in –ol. The table gives a list of some of these endings.

homologous series	name ending	functional group	example
alkane	-ane	—C—H	ethane, C_2H_6
alkene	-ene	C=C	ethene, C_2H_4
alcohol	-ol	—O—H	ethanol, C_2H_5OH
carboxylic acid	-oic acid	—C(=O)—O—H	ethanoic acid, CH_3COOH

Most organic compounds char or burn when heated in air. Most inorganic compounds just melt or vaporise.

We can show that a homologous series has the same general characteristics in several ways:

- We can give each homologous series a general formula which applies to all members of the homologous series. For example, all members of the alkane homologous series have the general formula C_nH_{2n+2}, where n is the number of carbon atoms. The alkane with five carbon atoms is called pentane. Its formula is $C_5H_{(2\times5)+2}$, which is C_5H_{12}. All members of the alkene homologous series have the general formula C_nH_{2n}.
- As the number of carbon atoms in a homologous series increases by one, the number of hydrogen atoms increases by two. For example: CH_3OH, C_2H_5OH, C_3H_7OH – each differs from the next by a CH_2 group.
- The members of a homologous series have very similar chemical properties because they all have the same functional group.
- The physical properties in a homologous series change in a regular way as the number of carbon atoms increases. For example, the boiling points of the alkanes (see Topic 18.1).

DID YOU KNOW?
Until Friedrich Wöhler made urea in 1828, scientists thought that compounds in the body differed from inorganic compounds because they had a special 'life force' in them.

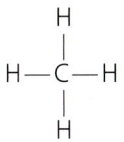

The full structural formula of methane, CH_4

Formulae of organic compounds

The full structural formula for methane shows how the atoms are bonded by covalent bonds. This type of formula is sometimes called the displayed formula.

The molecular formula shows the actual number of each type of atom in a compound without showing the bonds. The molecular formula for methane is CH_4.

The full structural and molecular formulae for ethane are:

full structural formula: H—C—C—H (with H atoms above and below each C)

molecular formula: C_2H_6

We sometimes abbreviate the structural formula to show each carbon atom with its attached hydrogen atoms one by one but without showing the single bonds. For ethane this type of structural formula is written CH_3CH_3.

STUDY TIP
When drawing the full structural formula of an organic compound you should show all atoms and all bonds. Don't forget that there is a bond in the alcohol functional group –O–H.

SUMMARY QUESTIONS

1 Copy and complete using the words below:

**atom   chemical   compound   ethane
functional   homologous**

Methane and ____ belong to the same ____ series. They have the same ____ group. A functional group is an ____ or group of atoms that gives a ____ its particular ____ properties.

2 Draw:
   a   the full structural formula of methane
   b   the molecular formula of ethanol.

3 State three characteristics of a homologous series.

KEY POINTS

1 A functional group is an atom or group of atoms that gives an organic compound its particular chemical properties.

2 A homologous series is a group of organic compounds with the same functional group and similar properties.

3 A homologous series has particular characteristics.

17.2 Hydrocarbons

LEARNING OUTCOMES

- Define the term *hydrocarbon*
- Name and draw the structures of alkanes having up to four carbon atoms
- Describe and identify structural isomerism

A **hydrocarbon** is a compound which contains only carbon and hydrogen atoms. The **alkanes** and the **alkenes** are two important homologous series of hydrocarbons. Alkanes have only single covalent bonds but alkenes can have one or more double bonds between their carbon atoms.

Naming alkanes

What do all compounds with names starting with meth- have in common? The answer is that they have only one carbon atom. A compound beginning with eth- has two carbon atoms in its chain. A compound with three carbon atoms has a name beginning with prop-. The prefix tells us how many carbon atoms there are in the longest chain. The compound on the right has a name starting with but- because the longest chain has four carbon atoms. The names of the first six alkanes are shown below. You will need to remember only the first four of these (or the first two if you are doing the core paper). But it is good to be familiar with the other names because you are certain to come across them.

Natural rubber is a very useful hydrocarbon which is obtained from the sap of certain trees

DID YOU KNOW?

The compound BHC (hexachlorocyclohexane) has eight different isomers but only one of the eight acts as an insecticide.

STUDY TIP

When drawing alkenes make sure that there are not too many hydrogen atoms around the carbon atoms that form the double bond. Check to see that each carbon atom has four bonds.

Alkyl groups

prefix	number of carbon atoms	name and molecular formula	full structural formula
meth-	1	methane, CH_4	H–C(H_2)–H (1 C, 4 H)
eth-	2	ethane, C_2H_6	H–C–C–H (each C with H's)
prop-	3	propane, C_3H_8	H–C–C–C–H
but-	4	butane, C_4H_{10}	H–C–C–C–C–H
pent-	5	pentane, C_5H_{12}	H–C–C–C–C–C–H
hex-	6	hexane, C_6H_{14}	H–C–C–C–C–C–C–H

When we remove a hydrogen atom from an alkane chain we have a group called an alkyl group. So the alkyl group from ethane, C_2H_6, is C_2H_5-. The alkyl group from butane, C_4H_{10}, is C_4H_9-. The general formula for an alkyl group is C_nH_{2n+1}. Alkyl groups are named after the hydrocarbons by changing the -ane ending of the hydrocarbon to -yl. So we call C_2H_5- an ethyl group and C_4H_9- a butyl group.

Structural isomers

The carbon chain in alkanes and other organic compounds can be branched.

2-methylpropane (branched chain)

butane (straight chain)

2-Methylpropane has four carbon atoms and has the same molecular formula as butane, C_4H_{10}. But it is not butane because the carbon atoms are arranged differently. Compounds with the same molecular formula but with a different structural formula are called **isomers**.

We say that the isomer of butane with the CH_3- group sticking out has a branched chain. Isomers may have the same chemical properties but they have different physical properties. The boiling point of straight-chained butane is 0 °C but the branched chain isomer has a boiling point of −12 °C.

The rules for naming branched-chain alkanes can be quite complicated. You do not have to learn these but it is useful to be able to recognise why we use numbers in the names of some organic compounds. Using the compound below as an example:

- You find the longest carbon chain and name the compound after the number of carbon atoms in the longest chain. There are four carbon atoms in the longest chain. So it is named after butane.
- You then look for the alkyl side chain. In this case it is a methyl group. So the compound is methylbutane.
- You then have to number the alkyl group side chain by counting the numbers of the carbon atoms from one end of the carbon chain. You count from the end of the carbon chain that gives you the lowest number. In this case counting from the left, the alkyl group is on the second carbon atom. So the compound is 2-methylbutane.

All members of the alkene homologous series with one C=C bond have the general formula C_nH_{2n}. These can also form isomers where the position of the double bond changes. These structural isomers are called position isomers:

KEY POINTS

1. A hydrocarbon is a compound containing carbon and hydrogen only.
2. The prefixes meth-, eth-, prop-, amongst others, tell us the number of carbon atoms in the main chain of an organic compound.
3. Structural isomers are compounds with the same molecular formula but different structural formulae.
4. Compounds with alkyl groups sticking out from the main carbon chain are called branched-chain compounds.

SUMMARY QUESTIONS

1. Copy and complete using the words below:

   **butane   chain   members
   number   pent-   prefixes
   prop-   three**

   The different _____ of a homologous series can be identified by the _____ meth-, eth-, _____, but-, _____ and so on. These prefixes show the _____ of carbon atoms in the main _____ of the compound. For example, _____ has four carbon atoms in its carbon chain and propane has _____.

2. Name:
   a. the straight-chained alkane with four carbon atoms
   b. the alkene with three carbon atoms.

3. Draw the full structural formula for the two isomers of butane.

17.3 Fuels

LEARNING OUTCOMES

- Name the fuels: coal, natural gas and petroleum
- Describe the fuels obtained from petroleum
- Describe the properties of molecules within a fraction

The **fossil fuels** coal, petroleum (crude oil) and natural gas all contain **hydrocarbons**. We cannot use petroleum (crude oil) as a fuel because it is a sticky black liquid that is difficult to set alight. When it does burn, it produces clouds of poisonous black smoke. Petroleum is a mixture of many types of hydrocarbons having different lengths of carbon chain. Some of the chains are branched and there may even be compounds with rings of carbon atoms.

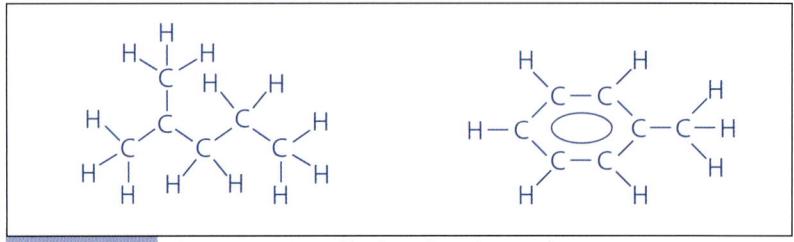

Figure 17.3.1 There is a variety of hydrocarbons in petroleum

Natural gas is methane

Fractional distillation is used to separate the hydrocarbon molecules in petroleum into groups that have similar boiling points. These groups of molecules are called **fractions**. Each contains hydrocarbons having a certain range of carbon atoms. Apart from the refinery gases, all these fractions are liquids. Many of these fractions are used as fuels:

fraction	number of carbon atoms	type of fuel
refinery gas	1–4	methane, propane, butane for gas cylinders
gasoline (petrol)	5–10	petrol for cars
kerosene (paraffin)	10–16	for aircraft
diesel oil (gas oil)	16–20	diesel for cars and larger vehicles
fuel oil	20–30	for ships and home heating

STUDY TIP

Don't get confused between petroleum and petrol. Petroleum is crude oil. Petrol, also known as gasoline, is a fraction obtained when we distil petroleum.

Each fraction has particular properties. The table below shows some of these properties. You can see that the properties show a trend as the number of carbon atoms increase.

fraction	number of C atoms	state	melting and boiling point	density	flammability
refinery gas	1–4	gas	increases ↓	increases ↓	increases ↓
gasoline	5–10	liquid			
kerosene	10–16	liquid			
diesel oil	16–20	liquid			
fuel oil	20–30	liquid			

The viscosity (glueyness) of the liquids also increases with the number of carbon atoms. For example, gasoline and kerosene flow easily, but fuel oil flows more slowly. The hydrocarbons within each fraction also show trends in boiling point, density and flammability.

- Methane (natural gas) – As well as being found under the ground, methane is also produced by the decomposition of materials in rubbish sites. (In some places, the gas can be piped out from the rubbish site and used for heating.)
- Hydrogen – A good fuel because it releases a lot of energy per gram and is non-polluting. However, it is usually made by using energy from other fuels (see Topic 7.5).

On burning in a good supply of oxygen, hydrocarbons fuels form carbon dioxide and water.

PRACTICAL

What is formed when fuels burn?

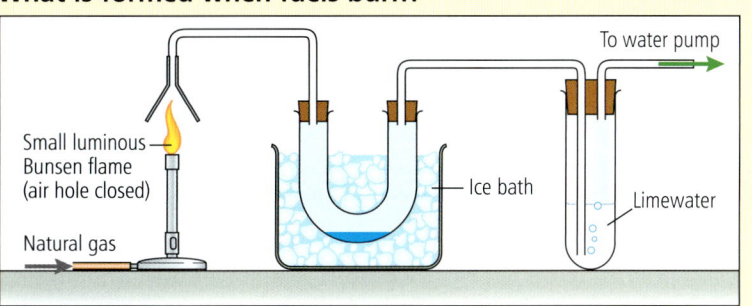

Figure 17.3.2 CO_2 and H_2O form when fuels burn

We test the products formed when a fuel burns using this apparatus. We burn the fuel under the funnel. The gases produced are sucked through the apparatus by a pump. Water collects in the U-shaped tube. You can test that this is water using white anhydrous copper(II) sulfate which turns blue. The limewater turns milky showing that carbon dioxide is produced.

DID YOU KNOW?

WHAT MAKES A GOOD FUEL?

There are several things we take into account when we choose a fuel for a particular job:

- How much heat does it gives out? Most hydrocarbon fuels give out a similar amount of energy per gram but hydrogen produces a lot more energy per gram.
- Is it polluting? Coal is very polluting, oil is less polluting and natural gas does not produce much pollution. But all these fuels produce the greenhouse gas carbon dioxide when burned.
- Is it easy to use? Solid fuels such as coal and wood are not as easy to use as liquid fuels.
- Is it readily available? Many people are worried that the supply of petroleum and natural gas will run out over the next 100 years. This means that we will have to use more biofuels.
- Is it cheap? The price depends on many things: how easy it is to extract and transport, how available the fuel is and politics.
- Is it easy and safe to transport? Many fuels are flammable so care has to be taken when transporting them and using them.

DID YOU KNOW?

Travelling by air produces nearly twenty times more greenhouse gases than travelling the same distance by train.

KEY POINTS

1 A good fuel releases a lot of heat energy per gram, is non-polluting and is easy to transport.
2 The fractional distillation of petroleum provides us with a variety of liquid and gaseous fuels.
3 The molecules within each fraction show trends in physical properties.
4 The products of the complete combustion of a hydrocarbon fuel are carbon dioxide and water.

SUMMARY QUESTIONS

1 Copy and complete using the words below:

   **excess   fractional   lighter
   petroleum   water**

   Many of the _____ fractions of hydrocarbons produced by the _____ distillation of _____ are useful fuels. When you burn a hydrocarbon in _____ air, carbon dioxide and _____ are formed.

2 Write word equations for the complete combustion of:
   a methane
   b carbon
   c hydrogen.

3 State three characteristics of a good fuel.

17.4 Petroleum

LEARNING OUTCOMES

- Describe the separation of petroleum into different fractions by fractional distillation
- Name some uses of these fractions

In an oil refinery the mixture of hydrocarbons in petroleum is separated into smaller groups. Each of these groups with a limited range of carbon atoms is called a **fraction**. For example, the gasoline fraction contains hydrocarbons with about five to ten carbon atoms.

The hydrocarbon fractions are separated by **fractional distillation**. We sometimes call this **fractionation**. Fractional distillation separates the hydrocarbons using the difference in their boiling points. Larger hydrocarbons have higher boiling points than smaller hydrocarbons.

The petroleum is first heated so that all the hydrocarbons are present as gases. The petroleum is then fed into a tall tower called a **fractionating column**. The column is kept hot at the bottom (about 350 °C) but it is cooler at the top. So there is a range of temperatures in the column.

Near the bottom of the column those hydrocarbons with higher boiling points condense. Hydrocarbons with lower boiling points are still gases. These move further up the column. As they move up the column, each hydrocarbon condenses at the point where the temperature in the column falls just below the boiling point of the hydrocarbon.

Petroleum undergoes fractional distillation in an oil refinery

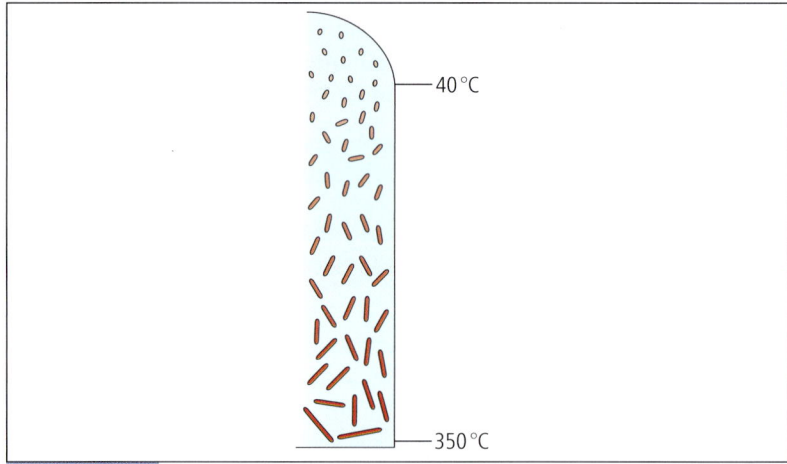

Figure 17.4.1 Lighter hydrocarbons with lower boiling points move further up the fractionating column

Hydrocarbons with similar boiling points are collected as fractions. Some of the hydrocarbons do not condense. They come off as gases at the top of the column. These are the *refinery gases* such as methane, ethane, propane and butane. In many oil refineries these are removed from the petroleum before fractionation.

The useful fractions

Fractional distillation separates petroleum into different fractions with a range of boiling points. Each fraction has a particular use.

DID YOU KNOW?

One quarter of the world's petroleum comes from Saudi Arabia.

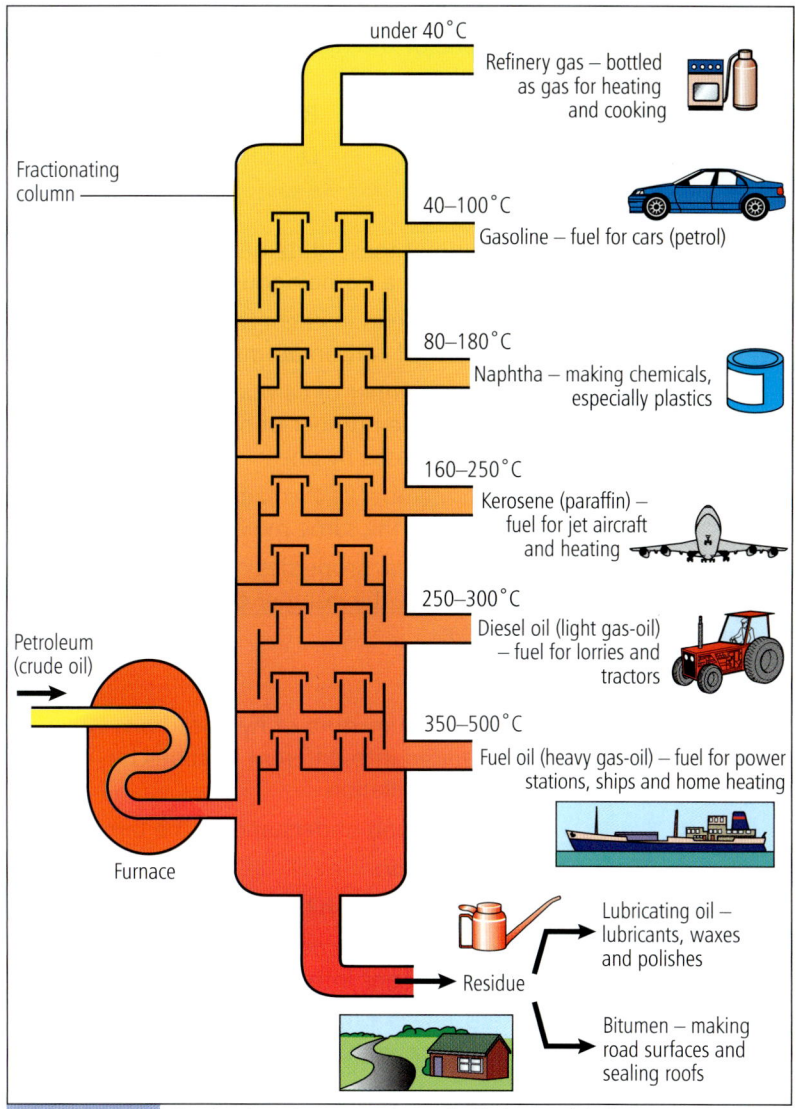

Figure 17.4.2 The fractions from petroleum distillation and their uses

STUDY TIP

You do not have to remember the boiling range or typical number of carbon atoms in each fraction. But you do have to know the uses of each fraction and where they condense in the fractionating column.

SUMMARY QUESTIONS

1. Copy and complete using the words below:

   **boiling   condense   fractionating
   fractions   higher   hydrocarbons**

   Petroleum is separated into different _____ in an oil refinery. Each fraction is a group of _____ with similar _____ points. The hydrocarbon molecules move up the _____ tower. Hydrocarbons with _____ boiling points _____ lower in the tower.

2. Draw a flow diagram to show the main stages in the fractional distillation of petroleum.

3. Give one use for each of these fractions:
   a fuel oil
   b kerosene
   c naphtha.

KEY POINTS

1. Petroleum is separated into different fractions by fractional distillation.

2. Each fraction has hydrocarbons with similar boiling points.

3. The hydrocarbons in petroleum are separated by fractional distillation because of the difference in their boiling points.

4. Each fraction obtained from petroleum has a particular use.

SUMMARY QUESTIONS

1. Match each petroleum fraction on the left with their use on the right.

naphtha	fuel for diesel engines
bitumen	jet fuel
light gas-oil	surfacing roads
kerosene	making chemicals

2. Copy and complete using the words from the list below.

   **alcohols   alkanes   alkenes   ethane
   functional   homologous**

   Methane and _____ belong to the _____ series called the _____. A homologous series is a family of similar compounds with the same _____ group. For example, _____ always have the –OH functional group and _____ have the C=C functional group.

3. Put the following fractions in order of decreasing boiling point:

   bitumen; fuel oil; kerosene; naphtha; refinery gas.

4. Match each word on the left with its description on the right.

methane	a group of molecules with a similar range of boiling points
coal	a thick liquid mixture of hydrocarbons
petroleum	a solid fuel that often contains sulfur
fraction	the main constituent of natural gas
hydrogen	a gaseous fuel that forms only water when it burns

5. Write the formula of the functional group present in:
   - (a) alcohols
   - (b) alkenes
   - (c) alkanes
   - (d) carboxylic acids.

6. (a) What do you understand by the term *isomer*?
   (b) Draw two isomers of:
      (i) an alkane having 5 carbon atoms
      (ii) an alkene having 4 carbon atoms
      (iii) an alcohol having 4 carbon atoms.

7. State three general characteristics of a homologous series.

PRACTICE QUESTIONS

1. Which one of the following molecules is an alkene:

   A CH_3COOH B $CH_3CH=CH_2$
   C $CH_3CH_2CH_3$ D $CH_3CH_2CH_2OH$

   (Paper 1)

2. Which one of these statements is a correct description of a homologous series?
   - A There is trend in chemical properties as the number of carbon atoms increases.
   - B As the number of carbon atoms increases by one, the number of hydrogen atoms also increases by one.
   - C Each member of the same homologous series has the same molecular formula.
   - D The physical properties change in a regular manner as the number of carbon atoms increases.

   (Paper 2)

3. Petroleum is a mixture of hydrocarbons which are separated into different fractions by fractional distillation.

   (a) What do you understand by the terms
      (i) fraction
      (ii) hydrocarbon? [2]

   (b) Explain how fractional distillation separates hydrocarbons into different fractions. [3]

   (c) Kerosene is a fraction obtained from the distillation of petroleum.
      (i) State one use of kerosene. [1]
      (ii) Name two other petroleum fractions. For each of these fractions give one use. [4]

   (d) Copy and complete the following sentences about petroleum fractionation using words from the list. (Not all words are used.)

   **condense   evaporate   fractions
   mass   higher   longer   lower   shorter**

   Hydrocarbon _____ higher in the distillation column have _____ hydrocarbon chains and _____ relative molecular _____ than hydrocarbons lower in the column. The fractions with higher boiling points _____ lower in the column. [5]

   (Paper 3)

4 When fuels burn they release energy.
   (a) What is the name given to a chemical reaction that releases heat energy? *[1]*
   (b) (i) State the name of:
      (1) a gaseous fuel
      (2) a liquid fuel
      (3) a solid fuel. *[3]*
   (ii) For each fuel that you have chosen in part **(i)**, state one disadvantage of that fuel. *[3]*
   (c) Many fuels are alkanes.
      (i) What do you understand by the term *alkane*? *[1]*
      (ii) Write the molecular formula for ethane. *[1]*
      (iii) Draw the full structural formula for methane, showing all atoms and bonds. *[1]*
   (d) The alkanes are a homologous series of compounds. What do you understand by the term *homologous series*? *[1]*

(Paper 3)

5 The diagram shows four organic compounds.

A: $H-C(H)(H)-C(H)(H)-H$ (ethane)
B: $H-C(H)(H)-O-H$ (methanol)
C: $H_2C=CH_2$ (ethene)
D: $H-C(H)(H)-C(=O)-O-H$ (ethanoic acid)

   (a) Which two of these compounds are hydrocarbons? *[1]*
   (b) Which compound is an alkene? *[1]*
   (c) Which compound is an alcohol? *[1]*
   (d) Name compound D. *[1]*
   (e) Write the molecular formula for compound B. *[1]*
   (f) Write the full structural formula for another member of the same homologous series as compound A. *[1]*
   (g) Write the formula for the functional groups present in **(i)** compound B **(ii)** compound D. *[2]*

(Paper 3)

6 The diagram shows a fractionation column for the separation of petroleum fractions.

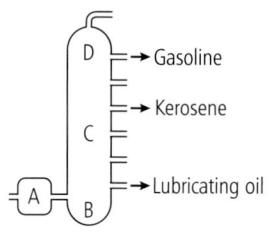

   (a) What do you understand by the term *petroleum*? *[2]*
   (b) Where in the diagram, A, B, C or D, is the temperature lowest? *[1]*
   (c) Where in the diagram is the petroleum turned to vapour. *[1]*
   (d) Which fraction labelled in the diagram has the lowest boiling point? *[1]*
   (e) State the name of two other fractions that are not shown on the diagram. For each of these fractions state **(i)** where they condense in the column and **(ii)** a use of the fraction. *[4]*

(Paper 3)

7 The alkanes and the alcohols are both homologous series.
   (a) What do you understand by the term *homologous series*? *[1]*
   (b) Write the general formula for the alcohol homologous series. *[1]*
   (c) Write the molecular formula for the fifth and sixth members of the alkane homologous series. *[2]*
   (d) (i) Draw the full structural formula for butanol, showing all atoms and bonds. *[1]*
      (ii) Draw the structure of another isomer of butanol. *[1]*
      (iii) What do you understand by the term *isomer*? *[1]*

(Paper 4)

18 The variety of organic chemicals

18.1 Alkanes

LEARNING OUTCOMES

- Describe the properties of alkanes
- Describe the bonding in alkanes
- Describe the reaction of alkanes with chlorine

Alkanes are hydrocarbons which have only single covalent bonds in their structure. We call them **saturated** hydrocarbons because no more atoms can be added to their molecules.

Physical properties of the alkanes

All alkanes are colourless gases, liquids or solids. The first four members of the alkane homologous series are gases at room temperature and pressure. Alkanes with 5 to 17 carbon atoms in their chains are liquids. Alkanes with more than 17 carbon atoms in their chains are solids. The boiling points of the alkanes vary in a regular way.

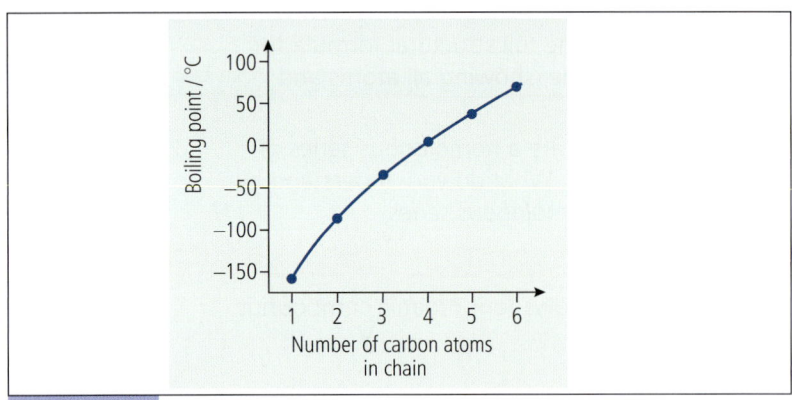

Figure 18.1.1 The boiling points of the alkanes vary in a regular way

You can see that as the carbon chain gets longer, the boiling points of the alkanes increase. The difference in the boiling point from one alkane to the next also gets smaller as the number of carbon atoms increases. You can predict the boiling point of other alkanes by following this trend. For example, using the graph, you might predict the boiling point of heptane (which has 7 carbon atoms in its chain) to be about 96–99 °C. The actual boiling point is 98 °C.

Chemical properties of the alkanes

Alkanes are generally unreactive compounds. They do not react with acids or alkalis. But they do burn and undergo a few reactions under special conditions. Alkanes burn with a clean blue flame if there is plenty of oxygen or air present. We describe this reaction as **complete** combustion. Carbon dioxide and water are formed:

$$C_3H_8(g) + 5O_2(g) \longrightarrow 3CO_2(g) + 4H_2O(l)$$
propane + oxygen ⟶ carbon dioxide + water

Candles are made from alkanes

DID YOU KNOW?

Female sand bees attract male sand bees by giving off a 'scent' made up of three alkanes.

If there is not enough oxygen present, combustion is **incomplete**. Carbon monoxide is formed and perhaps even soot (carbon particles).

$$2C_3H_8(g) + 7O_2(g) \rightarrow 6CO(g) + 8H_2O(l)$$
propane + oxygen → carbon monoxide + water

$$C_3H_8(g) + 2O_2(g) \rightarrow 3C(s) + 4H_2O(l)$$
propane + oxygen → carbon (soot) + water

The reaction of alkanes with chlorine

One chemical that alkanes will react with is chlorine – but only under certain conditions. Alkanes do not react with chlorine in the dark. However, if we mix chlorine with an alkane in a sealed tube and keep it in bright sunlight, the green colour of the chlorine disappears. A reaction has occurred. This is a **photochemical reaction** (see Topic 8.6). A chlorine atom replaces a hydrogen atom in the alkane. We call this type of reaction a **substitution reaction**.

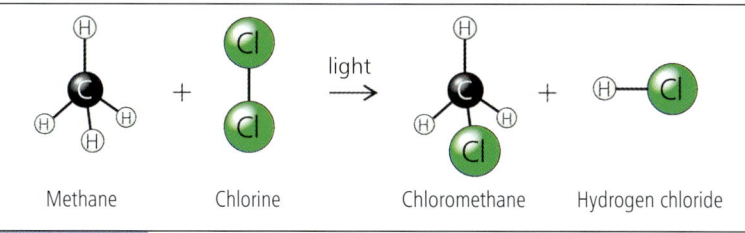

Figure 18.1.2 In sunlight, a chlorine atom replaces (substitutes) a hydrogen atom in methane

$$CH_4(g) + Cl_2(g) \xrightarrow{light} CH_3Cl(g) + HCl(g)$$
methane + chlorine → chloromethane + hydrogen chloride

You will notice that the acidic gas hydrogen chloride is produced. This turns damp blue litmus paper red.

If we use excess chlorine we can substitute more hydrogen atoms:

$$CH_3Cl + Cl_2 \rightarrow CH_2Cl_2 + HCl$$
$$CH_2Cl_2 + Cl_2 \rightarrow CHCl_3 + HCl$$

If enough chlorine is present all four hydrogen atoms can be replaced by chlorine atoms.

We can carry out similar reactions with other alkanes and other halogens:

$$C_3H_8 + Cl_2 \xrightarrow{light} C_3H_7Cl + HCl$$
$$CH_4 + Br_2 \xrightarrow{light} CH_3Br + HBr$$

STUDY TIP

You will be expected to be able to balance symbol equations for the combustion of alkanes. Remember to balance the oxygen.

KEY POINTS

1 The alkanes show trends in their physical properties.

2 Alkanes are generally unreactive except for burning.

3 Alkanes do not react with chlorine in the dark but they do in the presence of light.

SUMMARY QUESTIONS

1 Copy and complete using the words below:

**butane   ethane
gases   hydrocarbons
single   unreactive**

Alkanes are _____ having only _____ covalent bonds. They are _____ apart from burning. Methane, _____, propane and _____ are _____ at room temperature.

2 Describe how the boiling points of the straight-chained alkanes vary with temperature.

3 Write a balanced equation for the reaction of chlorine with propane in light to form the compound $C_3H_6Cl_2$.

225

18.2 Cracking alkanes

LEARNING OUTCOMES

- Describe the manufacture of alkenes and of hydrogen by cracking

All the fractions we get from the distillation of petroleum are useful. But some are more useful than others – there is a greater demand for them. We use more gasoline (petrol) and diesel than can be supplied by the fractional distillation of petroleum.

Oil companies solve this problem by breaking down larger hydrocarbons into smaller, more useful hydrocarbons. This is called **cracking**. Cracking is the thermal decomposition of alkanes. A catalyst is often used. Longer-chained alkanes are cracked to form a mixture of shorter-chained alkanes and alkenes. For example:

$$C_{10}H_{22} \xrightarrow[\text{catalyst}]{\text{heat}} C_5H_{12} + C_2H_4 + C_3H_6$$
$$\text{decane} \quad\quad\quad \text{pentane} \quad \text{ethene} \quad \text{propene}$$

From the cracking, we not only get shorter-chained alkanes which are useful for petrol, we also get alkenes. Alkenes are very useful for making a variety of chemicals including plastics.

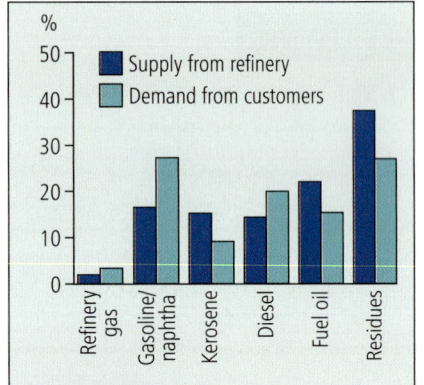

Figure 18.2.1 Supply of and demand for petroleum fractions

PRACTICAL

Cracking an alkane

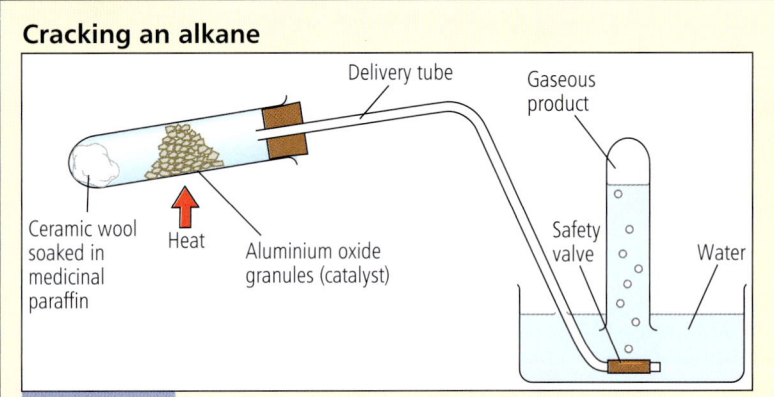

Figure 18.2.2 Aluminium oxide catalyses the cracking of alkanes

1. You can use medicinal paraffin as your alkane. You heat the aluminium oxide catalyst strongly and then heat the paraffin. The paraffin vapour passes over the aluminium oxide which is kept hot.

2. You collect the gases from the cracking in the test tube. You can tell if the gas contains alkenes by carrying out the bromine water test (see Topic 18.3).

Hydrogen can also be produced by cracking:

$$C_2H_6 \longrightarrow C_2H_4 + H_2$$
$$\text{ethane} \quad\quad \text{ethene} \quad \text{hydrogen}$$

Cracking petroleum fractions on a large scale

Cracking is often carried out on a large scale using a **catalyst**. The huge tank where this takes place is called a catalytic (cat) cracker. The vapour from the gas-oil or kerosene fractions is passed through a catalyst of silicon(IV) oxide and aluminium oxide at 400–500 °C.

STUDY TIP

When describing cracking you must state that (i) large hydrocarbon molecules are broken down to smaller ones (ii) and alkenes (iii) using a high temperature and (iv) a catalyst.

The catalyst is a fine powder which has to be continuously recycled to the cat cracker though a regenerator tank. This frees the catalyst from any carbon deposited on its surface.

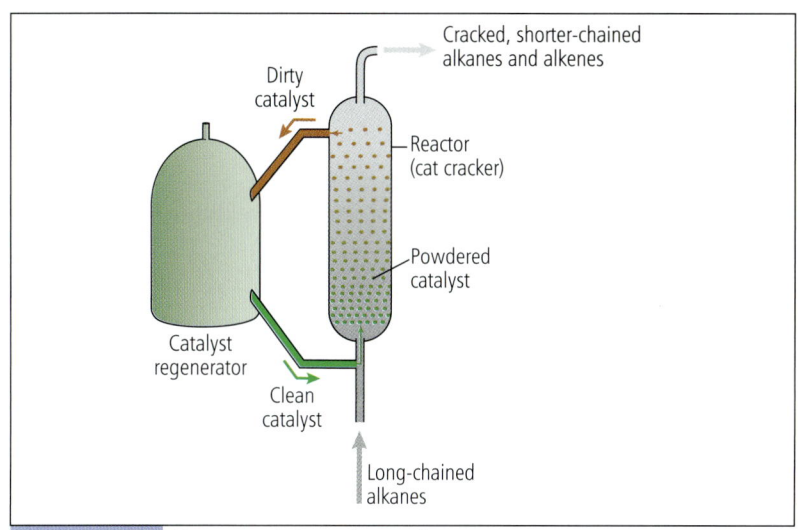

Figure 18.2.3 How a catalytic cracker works

A cat cracker like this one is used to split long-chain hydrocarbons into shorter-chain hydrocarbons and alkenes

The longer-chained alkanes in the gas-oil or kerosene fractions which are less useful are broken down to shorter-chained hydrocarbons:

- The shorter-chained alkanes are used for petrol and very small alkanes are used for fuel (liquid petroleum gases – LPG).
- The alkenes can be used to make a wide variety of chemicals, including plastics.
- Hydrogen may also be formed which can be used for making ammonia or as a fuel.

Catalytic cracking is not the only type of cracking. Long-chained alkanes can be cracked at a high temperature without a catalyst. A temperature between 450 °C and 800 °C is used. This type of cracking produces a greater percentage of alkenes.

DID YOU KNOW?

Thermal cracking was invented by the Russian engineer Vladimir Shukov in 1891.

SUMMARY QUESTIONS

1 Copy and complete using the words below:

alkanes   alkenes   aluminium   catalyst   high   long

We break down _____-chained alkanes into shorter-chained _____ and _____ by cracking them. We carry out cracking at a _____ temperature using an _____ oxide _____.

2 Complete the equations describing a cracking reaction:

decane ⟶ octane + _____

$C_{10}H_{22}$ ⟶ C_8H_{18} + _____

3 Why is cracking an important process in an oil refinery?

KEY POINTS

1 The demand for some petroleum fractions is greater than the supply.

2 Cracking is used to break long-chained alkanes to shorter-chained alkanes and alkenes.

3 Cracking is carried out using a high temperature and a catalyst.

18.3 Alkenes

LEARNING OUTCOMES

- Distinguish between saturated and unsaturated hydrocarbons from molecular structures and reaction with aqueous bromine
- Describe the addition reactions of alkenes with bromine, steam and hydrogen

Supplement

The **alkenes** are a homologous series of hydrocarbons whose names end in -ene. We call them **unsaturated hydrocarbons** because they have a C=C double bond. They do not have the maximum number of hydrogen atoms around each carbon atom – more atoms can be added to their molecules.

We can test to see if a hydrocarbon is unsaturated by using aqueous bromine (bromine water).

PRACTICAL

Is this compound saturated or unsaturated?

1 You take a test tube of the gas or liquid you want to test, for example the gas collected from an experiment on cracking.

2 You add a few drops of bromine water and shake the tube. Then you observe the colour of the liquid in the tube.

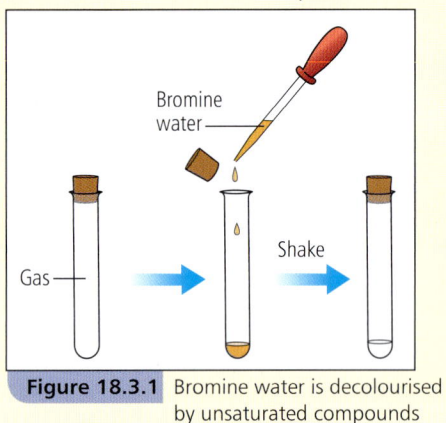

Figure 18.3.1 Bromine water is decolourised by unsaturated compounds

Butter contains many saturated fat molecules but vegetable oils contain a higher amount of unsaturated compounds

STUDY TIP

The test for an alkene is that it turns aqueous bromine colourless. Do not use the word 'clear' to mean colourless. Aqueous bromine stays the same yellow or orange colour when an alkane is added. A common error is to write 'no observations'.

Bromine water is yellow/orange in colour. If the bromine water is decolourised, the hydrocarbon is unsaturated. All alkenes decolourise bromine water. If the bromine water remains the same yellow or orange colour, the hydrocarbon is saturated.

Ethene, propene and butene are the first three compounds in the alkene homologous series. These three alkenes are all colourless gases. These are their structural formulae:

ethene, C_2H_4 propene, C_3H_6 butene, C_4H_8

Chemical properties of the alkenes

Combustion

The complete combustion of alkenes produces carbon dioxide and water:

$$C_2H_4(g) + 3O_2(g) \rightarrow 2CO_2(g) + 2H_2O(g)$$
ethene + oxygen → carbon dioxide + water

Addition reactions

Many of the reactions of alkenes are called **addition reactions**. In an addition reaction, two reactants add together to form only one product. Here are three examples.

1. The reaction of alkenes with bromine:

$$C_2H_4 + Br_2 \rightarrow C_2H_4Br_2$$
ethene   bromine   1,2-dibromoethane

You can see that the bromine has added across the double bond. Bromine does not react with saturated compounds because they do not have a double bond to 'open out'.

2. Hydrogen reacts with alkenes to form alkanes.

This addition reaction is also called a **hydrogenation** reaction. It can also be classed as a reduction reaction – one of the simple definitions of reduction is 'the addition of hydrogen to a compound' (see Topic 9.3). The reaction is carried out at 60 °C in the presence of a nickel catalyst.

$$C_2H_4 + H_2 \xrightarrow{\text{nickel/60 °C}} C_2H_6$$
ethene   hydrogen   ethane

This type of reaction is used to make margarine from unsaturated vegetable oils. The hydrogen reacts with the oil when it passes over the nickel catalyst at about 60 °C. Only some of the double bonds in the vegetable oil are changed to single bonds. This is enough to 'harden' the oil and make it into solid margarine.

3. Steam reacts with alkenes to form alcohols.

A high temperature (about 300 °C) and high pressure (70 atmospheres) are needed for this reaction. The steam is passed over a catalyst of concentrated phosphoric acid, H_3PO_4. This method gives a good yield of alcohol. Ethanol of high purity is made by this method.

$$C_2H_4 + H_2O \xrightarrow{H_3PO_4} C_2H_5OH$$
ethene   steam   ethanol

DID YOU KNOW?

A ripe banana put amongst some green tomatoes will help them ripen. The bananas give off ethene gas which ripens fruit.

KEY POINTS

1. Unsaturated hydrocarbons decolourise bromine water but saturated hydrocarbons do not.
2. Alkenes have a C=C double bond.
3. Alkenes undergo addition reactions with bromine, hydrogen and steam.

SUMMARY QUESTIONS

1. Copy and complete using the words below:

   bromine colourless hydrocarbon orange remains saturated shaken

   We can test whether a _____ is unsaturated or _____ by shaking it with _____ water. An unsaturated hydrocarbon turns _____ bromine water _____. The bromine water _____ yellow/orange when _____ with a saturated hydrocarbon.

2. How can you tell the difference between the two colourless hydrocarbons $CH_3CH=CH_2$ and $CH_3CH_2CH_3$?

3. Describe the addition reactions of alkenes with bromine, hydrogen and steam.

18.4 Alcohols

LEARNING OUTCOMES

- Describe the formation of ethanol by fermentation and by the addition of steam to ethene
- Describe the uses of ethanol
- Name and draw the structures of alcohols with up to four carbon atoms
- Describe how to make ethanoic acid

The **alcohols** are a homologous series having –OH as the functional group. Their names all end with -ol. Ethanol is:

molecular formula: C_2H_5OH

full structural formula:
```
    H   H
    |   |
H—C—C—O—H
    |   |
    H   H
```

Ethanol is a colourless liquid that boils at about 78 °C. It is miscible, meaning it mixes with water.

Ethanol can be manufactured from:

- ethene – by reacting steam and ethene at high pressure and temperature using a phosphoric acid catalyst (see Topic 18.3)
- glucose – by fermentation (see Topic 20.3).

Ethanol burns with a clean blue flame in excess air to form carbon dioxide and water:

$$C_2H_5OH + 3O_2 \rightarrow 2CO_2 + 3H_2O$$

The uses of ethanol

Ethanol has many uses:

- As a solvent: ethanol is used in perfumes and other cosmetics, in printing inks and in glues.
- As a fuel: ethanol can be mixed with petrol or used alone as a fuel for cars. It is less polluting than petrol and reduces the reliance on petrol and diesel.
- It is used to make other chemicals such as esters which are used in food flavourings and in many cosmetics.
- In some cultures the ethanol produced by fermentation is used for making alcoholic drinks. Excessive drinking of alcohol, however, leads to aggressive behaviour, depression and causes other medical problems.

Ethanol is a good solvent for perfumes

STUDY TIP

When writing a symbol equation for the combustion of an alcohol, when you balance the oxygen, remember that the alcohol contains oxygen too.

DID YOU KNOW?

The word alcohol first came into the English language in 1543 from the Arabic word 'al-ghul'.

More about the structure of alcohols

The general formula for alcohols is $C_nH_{2n+1}OH$. The formulae for the first four alcohols in this homologous series are:

CH_3OH methanol

C_2H_5OH ethanol

C_3H_7OH propan-1-ol

C_4H_9OH butan-1-ol

The number in the formula is used so that you can distinguish between different isomers of the alcohols. The numbers are not needed for methanol and ethanol though.

The isomers with the –OH group at the end are -1-ols. You will not be expected to name the different isomers but you may be asked to draw them. Some different (position) isomers of butanol are:

$$CH_3-CH_2-CH_2-CH_2-OH \qquad CH_3-CH_2-\underset{OH}{\underset{|}{CH}}-CH_3 \qquad CH_3-\underset{OH}{\overset{CH_3}{\underset{|}{\overset{|}{C}}}}-CH_3$$

butan-1-ol butan-2-ol 2-methylpropan-2-ol

Ethanol to ethanoic acid

Ethanol can be oxidised to ethanoic acid by:

1. oxidation in the air:

Acetobacter is a group of bacteria that causes wine to go sour. These bacteria are naturally present in the air and on surfaces around us. When we leave a solution of ethanol exposed to the air, enzymes from the bacterium speed up the conversion of ethanol to ethanoic acid. The reaction does not take place in the absence of oxygen.

$$C_2H_5OH + 2[O] \longrightarrow CH_3COOH + H_2O$$
ethanol   'oxygen'   ethanoic acid   water

We can write [O] in an equation when the oxidation reaction is complicated but we know that oxygen is involved. This is the reaction that makes vinegar. Vinegar is a solution of ethanoic acid.

2. acidified potassium manganate(VII):

Potassium manganate(VII) is a good oxidising agent, especially when sulfuric acid is added. We heat the ethanol with potassium manganate(VII) and sulfuric acid. We do this in a flask with a condenser in an upright position. We call this *refluxing*. This prevents the alcohol, which is very volatile, from escaping. The equation for the reaction can be represented the same as the one above. Other alcohols can be oxidised in the same way.

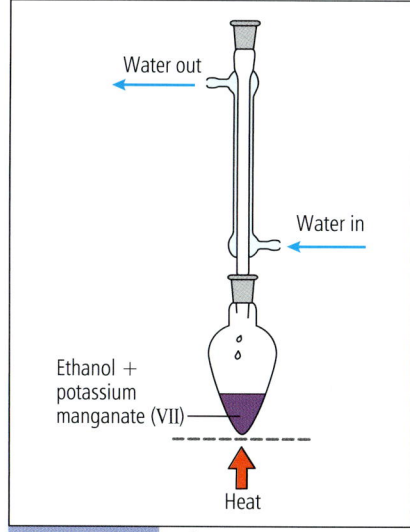

Figure 18.4.1 Refluxing a mixture

SUMMARY QUESTIONS

1 Copy and complete using the words below:

ethanol   ethene   fermentation   functional   steam

Alcohols have an –OH _____ group. The formula of _____ is C_2H_5OH. Ethanol can be made by reacting _____ with _____ or by _____.

2 Give three uses of ethanol.

3 Draw the complete structural formula for:
   a propan-1-ol
   b an isomer of propan-1-ol.

KEY POINTS

1 Alcohols are a homologous series with an –OH functional group.

2 Ethanol can be made by the addition of steam to ethene in the presence of a catalyst or by fermentation.

3 Propanol and butanol can exist as isomers with the —OH functional group in different positions.

4 Ethanoic acid can be made by the oxidation of ethanol.

18.5 Carboxylic acids

LEARNING OUTCOMES

- Describe the properties of aqueous ethanoic acid
- Describe the properties of ethanoic acid in terms of acid strength and reaction with ethanol
- Name and draw the structures of carboxylic acids with up to four carbon atoms
- Name and draw the structures of esters which can be made from unbranded alcohols and carboxylic acids containing up to four carbon atoms

The **carboxylic acids** are a homologous series with –COOH as the functional group. Their names all end in -oic acid. Their general formula is $C_nH_{2n+1}COOH$. The formulae for the first four carboxylic acids in this homologous series are:

HCOOH — methanoic acid
CH_3COOH — ethanoic acid
C_2H_5COOH — propanoic acid
C_3H_7COOH — butanoic acid

Carboxylic acids are colourless and have a sour taste like other acids. Like other acids they react with:

- metals to form metal salts and hydrogen
- hydroxides to form a salt and water
- carbonates to form a salt, carbon dioxide and water.

Chemical properties of carboxylic acids

Carboxylic acids are typical weak acids

Carboxylic acids are only partly ionised in water. The hydrogen of the –COOH group is the only one that is responsible for the acidity of carboxylic acids:

$$CH_3COOH \xrightleftharpoons{water} CH_3COO^- + H^+$$

ethanoic acid → ethanoate ion

Salts of ethanoic acid are named by changing the -oic to -oate.

Carboxylic acids show many of the reactions of a typical acid.

- They react with reactive metals to form a salt and hydrogen:

$$2CH_3COOH + Mg \rightarrow (CH_3COO)_2Mg + H_2$$
ethanoic acid → magnesium ethanoate

- They react with alkalis to form a salt and water:

$$CH_3COOH + NaOH \rightarrow CH_3COONa + H_2O$$
ethanoic acid → sodium ethanoate

- They react with metal carbonates to form a salt, water and carbon dioxide:

$$2CH_3COOH + Na_2CO_3 \rightarrow 2CH_3COONa + H_2O + CO_2$$
ethanoic acid → sodium ethanoate

The formation of esters

Carboxylic acids react with alcohols to form compounds called **esters**. The reaction is called esterification.

DID YOU KNOW?

The irritation caused by many insect bites is due to methanoic acid injected by the insect, whereas ethanoic acid is found in vinegar. Both acids have a highly pungent odour.

STUDY TIP

Remember that in naming the carboxylic acids, the carbon atom of the –COOH group is included. So CH_3COOH is ethanoic acid because compounds with two carbon atoms have names beginning with '-eth'.

In this reaction sulfuric acid acts as a catalyst. The $-\overset{\overset{O}{\|}}{C}-O-$ group formed is called an ester linkage. The water given off comes partly from the acid and partly from the alcohol:

$$CH_3C\overset{O}{\underset{O-H}{\diagdown}} + HOC_2H_5 \rightleftharpoons CH_3C\overset{O}{\underset{OC_2H_5}{\diagdown}} + H_2O$$

$$CH_3COOH + CH_3CH_2OH \xrightarrow{heat/H_2SO_4} CH_3COOC_2H_5 + H_2O$$
ethanoic acid   ethanol   ethyl ethanoate

Esters are named after the acid from which they are made. So the *ethanoate* comes last and the name of the alkyl group of the alcohol, in this case *ethyl*, comes first.

The naming of esters is based on the name of the carboxylic acid and alcohol used to make them.

- The name begins with the alkyl group from the alcohol.
- The name ends with the part coming from carboxylic acid, but –oic acid is changed to –oate.

$CH_3-C\overset{O}{\underset{O-CH_3}{\diagdown}}$ = methyl ethanoate

ethanoate   methyl
(from ethanoic (from
acid)   methanol)

So, $C_3H_7CO_2C_2H_5$ is ethyl butanoate and $HCO_2C_3H_7$ is propyl methanoate.

Each ester has its own typical fruity smell. For example, ethyl butanoate smells of pineapple. So we use esters in flavourings, shampoos and perfumes.

The flavour of most fruits is due to chemicals called esters

SUMMARY QUESTIONS

1 Copy and complete using the words below:

alcohols   functional   hydrogen   metals   weak

Carboxylic acids are _____ acids having the _____ group –COOH. They react with _____ to form salts and _____ and with _____ to form esters.

2 Write symbol equations for the reaction of ethanoic acid with:
   a sodium carbonate
   b ethanol.

3 Write the full structural formula for:
   a methanoic acid
   b butyl ethanoate.

KEY POINTS

1 Ethanoic acid can be produced by the oxidation of ethanol with acidified potassium manganate(VII).

2 Ethanoic acid is a weak acid. It reacts with metals, carbonates and hydroxides.

3 Carboxylic acids react with alcohols to form esters and water.

SUMMARY QUESTIONS

1 Match each word on the left with its description on the right.

ethanol	the breaking down of long-chained alkanes to alkenes and shorter-chained alkanes
cracking	organic compounds containing only single bonds
unsaturated	one of the products of the fermentation of glucose
addition	organic compounds containing C=C double bonds
saturated	a reaction in which two or more compounds combine to form only one compound

2 Copy and complete using words from the list below.

bromine decolourises double ethane ethene orange

Ethane can be distinguished from ethene by adding _____ water to each compound. Bromine water has an _____ colour. Ethene _____ bromine water but _____ does not. The bromine adds across the _____ bond in _____ and no other compound is formed.

3 Copy and complete these symbol equations:
  (a) $C_{10}H_{22} \rightarrow C_2H_4 + $ _____
  (b) $C_2H_4 + $ _____ $\rightarrow C_2H_4Br_2$
  (c) $C_2H_4 + H_2O \rightarrow$ _____
  (d) $C_5H_{12} + \_O_2 \rightarrow \_CO_2 + \_H_2O$
  (e) $C_4H_8 + $ _____ $\rightarrow \_CO + \_H_2O$

4 State the type of chemical reaction that occurs in each of these equations:
  (a) $C_3H_6 + H_2O \rightarrow C_3H_7OH$
  (b) $C_2H_5OH + 2[O] \rightarrow CH_3COOH + H_2O$
  (c) $C_7H_{16} + 11O_2 \rightarrow 7CO_2 + 8H_2O$
  (d) $CH_3COOH + C_2H_5OH \rightarrow CH_3COOC_2H_5 + H_2O$

5 What conditions and/or additional reagents are needed for the following reactions:
  (a) The formation of ethanol from ethene and steam
  (b) The conversion of ethene to ethane

PRACTICE QUESTIONS

1 Which one of these statements about cracking alkanes is correct?
  A Alkanes are cracked using a nickel catalyst and low temperatures.
  B Long-chained alkanes are broken down to shorter-chained alkanes and alkenes.
  C Short-chained alkanes are changed into long-chained alkanes and alkenes.
  D Hydrogen is removed from long-chained alkenes to form ethane.

(Paper 1)

2 What is the correct name of the ester with the formula $C_3H_7CO_2C_2H_5$?
  A ethyl butanoate
  B propyl propanoate
  C butyl ethanoate
  D ethyl propanoate

(Paper 2)

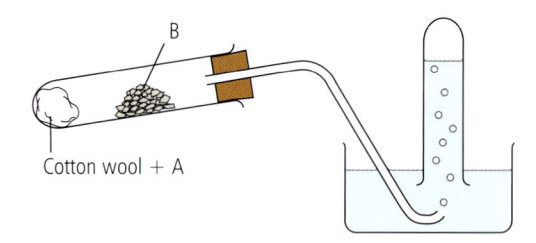

3 The laboratory apparatus used for cracking octane, C_8H_{18}, is shown above.
  (a) State the names of substances A and B. [2]
  (b) What do you understand by the term *cracking*? [1]
  (c) The formation of ethanoic acid from ethanol
  (d) The formation of ethyl ethanoate from ethanoic acid

6 Draw a flow diagram to show how you can make ethyl ethanoate using only ethanol as the organic reagent.

7 Write full structural formulae for:
  (a) propene
  (b) butanoic acid
  (c) two isomers of butane
  (d) the sixth member of the alcohol homologous series.

(c) (i) Copy and complete the equation for cracking octane:

$C_8H_{18} \rightarrow C_2H_4 +$ _____ [1]

(ii) State the name of the compound C_2H_4 and draw its full structural formula. [2]

(d) What conditions are used for cracking in industry? [2]

(e) Octane is a saturated hydrocarbon. What do you understand by the terms
(i) *saturated* and **(ii)** *hydrocarbon*? [2]

(Paper 3)

4 Ethene is an unsaturated hydrocarbon whereas ethane is a saturated hydrocarbon.

(a) Describe how you can distinguish between ethene and ethane using a chemical test. [3]

(b) Draw the full structural formula for ethane. [1]

(c) Ethene reacts with steam to form ethanol.
(i) What conditions are needed to carry out this reaction? [2]
(ii) Describe another method for making ethanol. [2]

(d) Copy and complete the equation for the complete combustion of ethanol:

$C_2H_5OH +$ ____$O_2 \rightarrow$ ____$CO_2 +$ ____H_2O [3]

(e) Describe how you can compare the energy given out by the combustion of two different alcohols. [3]

(Paper 3)

5 Ethanoic acid, CH_3COOH, is a typical weak acid.

(a) Draw the full structural formula for ethanoic acid. [2]

(b) Describe two ways in which ethanoic acid differs from a strong acid such as sulfuric acid. [2]

(c) Write balanced equations for the reaction of ethanoic acid with **(i)** magnesium and **(ii)** sodium carbonate. [4]

(d) Ethanoic acid can be made in the laboratory by the oxidation of ethanol with potassium dichromate.
(i) Name another oxidising agent that can be used to oxidise ethanol. [1]
(ii) What conditions are required for this oxidation? [2]

(iii) Write a symbol equation for this reaction. Use [O] to represent the oxidising agent. [2]

(Paper 4)

6 Alkanes and alkenes are both hydrocarbons.

(a) (i) Draw the full structural formula for an alkene with four carbon atoms. [1]
(ii) Draw the full structural formula for an isomer of the alkene you drew in part **(i)**. [1]

(b) Both alkenes and alkanes react with halogens.
(i) What essential condition is required for an alkane to react with chlorine? [1]
(ii) What type of chemical reaction is this? [1]

(c) (i) Write an equation to show the reaction of bromine with ethene. [2]
(ii) What observations are made during this reaction? [2]
(iii) What type of chemical reaction is this? [1]

(d) Alkenes react with steam to form alcohols.
(i) What conditions are needed for this reaction? [2]
(ii) Write an equation to show the reaction of propene with steam. [2]

(Paper 4)

7 Methane, methanol and methanoic acid each have one carbon atom.

(a) Write the full structural formula for
(i) methanol and **(ii)** methanoic acid. [2]

(b) Copy and complete the following sentences using suitable chemical terms:
Methane is generally _____ except for _____ and reaction with chlorine. In the presence of _____ and light methane undergoes a _____ reaction. [4]

(c) Methanoic acid reacts with ethanol to form an ester.
(i) What conditions are needed for this reaction? [1]
(ii) Draw the structure of the ester formed in this reaction. [1]

(Paper 4)

19 Polymers

19.1 What are polymers?

LEARNING OUTCOMES

- Understand the terms *macromolecule*, *monomer* and *addition polymerisation* with reference to the formation of poly(ethene)
- State some uses of plastics and man-made fibres
- Describe the pollution problems caused by non-biodegradable plastics

Plastics and proteins are **macromolecules**. Macromolecules are very large molecules made up of repeating units. One such unit that reacts to make a macromolecule is ethene. The ethene produced by cracking petroleum fractions is used for making a wide variety of chemicals: ethanol, antifreeze for cars, and **plastics**.

Polymers

We make plastics from lots of small molecules that join together to form long chains. We call the small molecules that join together **monomers**. The long-chain molecule formed by joining the monomers is called a **polymer**.

Figure 19.1.1 A model of polymerisation: the 'bead' monomers represent ethene. They join together to form poly(ethene).

A wide variety of articles are made from polymers

Figure 19.1.2 These 'bead' monomers represent propene. They join together to form poly(propene).

The process of joining monomers together to form polymers is called **polymerisation**. In Topic 18.3, we saw that ethene can take part in addition reactions. Polymers are often formed by addition reactions. We call this type of polymerisation, where no other substance is formed, **addition polymerisation**. Poly(ethene), commonly called polythene, is a plastic that is easy to shape and quite strong. It is made by joining ethene monomers together. How does this happen?

When alkene molecules react, one of the C=C bonds breaks and joins with its neighbouring molecule. When poly(ethene) is made, thousands of ethene molecules join together like this to form a long chain.

DID YOU KNOW?

Plastic trees have been 'planted' in some desert areas. When they are placed alongside real trees, they help trap moisture and improve the growth of the real trees.

Figure 19.1.3 Many ethene molecules join together by forming new bonds with the carbon atoms in the neighbouring molecule

Plastics: good or bad?

Plastics are particular types of polymers that can be moulded. The word plastic simply means something that can have its shape changed. Plastics have a wide range of uses:

- Poly(ethene) for bowls, buckets, dustbins and plastic bags
- Poly(propene) for milk crates and ropes
- Poly(chloroethene) (polyvinyl chloride, PVC) for insulation around electrical wires and for rainwater pipes and gutters
- Nylon for ropes, clothes and fishing nets
- Terylene for clothing.

But there is a down side to plastics. Many are non-**biodegradable**. This means that they are not broken down in the soil or water by micro-organisms when we throw them away. They just build up and cause a mess. If plastics get into drains they can block them and cause flooding. Plastics can kill wildlife by trapping small animals or blocking the digestive systems of animals and birds that eat the plastic along with their normal food. So how can we deal with unwanted plastics?

- Put them into landfill (waste) sites. These fill up very quickly and use up land that could be used for agriculture or housing.
- Burn them – we can use the heat produced to provide electricity or heating. But many plastics produce poisonous gases when they burn. PVC produces acidic hydrogen chloride. Plastics containing nitrogen may produce toxic hydrogen cyanide. Many plastics when burned at high temperatures also produce poisonous compounds called dioxins. It is very expensive to put filters on the furnaces used to burn plastics so this is rarely done.
- Recycling – some plastics can be melted and then moulded to make new articles. Not all plastics can be recycled. The ones that can be recycled have to be sorted out, which takes time and money.
- Cracking – some plastics can be melted, then cracked and then re-polymerised to make new articles.

Over the last few years more plastics have been made that break down in the environment, but the numbers of these are still small compared with non-biodegradable plastics.

STUDY TIP

When plastics are burned, poisonous or toxic gases are given off – not just harmful gases. A common error is to suggest that sulfur dioxide is given off when plastics burn. Few plastics contain sulfur.

SUMMARY QUESTIONS

1 Copy and complete using the words below:

   addition   ethene   molecules   monomer   polymer

   Small molecules such as _____ can join together by _____ polymerisation. We give the name _____ to the small _____ that join together to form a _____.

2 State two problems that arise from the disposal of plastics.

3 State one use for each of these polymers:
   a poly(ethene)
   b poly(vinyl chloride)
   c nylon.

KEY POINTS

1 Polymers are large molecules built up from small units (monomers).

2 In addition polymerisation the double bond between the carbon atoms changes to a single bond and the monomer molecules join together.

3 Nylon is used to make ropes and poly(ethene) is used to make bowls and plastic bags.

4 Each method of disposing of a plastic has its disadvantages.

19.2 More about polymer structure

LEARNING OUTCOMES

- Deduce the structure of a polymer from a given alkene
- Deduce the structure of a monomer from a given addition polymer

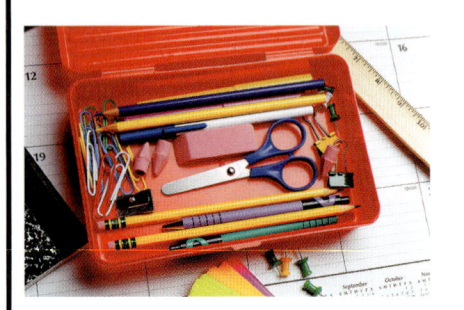

We can make many different types of **addition polymers**. Poly(propene), poly(phenylethene) and poly(chloroethene) are just three examples. Can you see how easy it is to name an addition polymer? You simply put the name of the monomer inside the brackets and put 'poly' in front.

However, some polymers are often called by their common names, so it is not always easy to deduce the structure from the name of the monomer. For example poly(tetrafluoroethene), which is used for non-stick pans, is commonly called 'teflon'.

From monomer to polymer

Propene is an alkene. The polymer formed from propene is called poly(propene). So how do we write the formula for a polymer when we are given the formula of the monomer?

One way of doing this is to show several repeating units:

- Write down the formulae for the number of monomer units you want, but change each double bond to a single bond.
- Draw single bonds between the monomer units.
- Put 'continuation bonds' at either end of the chain to show that the chain carries on in the same way.

We make lots of everyday articles from addition polymers

Figure 19.2.1 Three monomer units join to form part of a poly(propene) chain

We can also write the structure of the polymer much more simply:

- Draw the structure of the monomer but change the double bond to a single bond.
- Put 'continuation bonds' at either end of the 'molecule'.
- Put square brackets through the continuation bonds.
- Put an 'n' at the bottom right-hand corner. This means that the unit repeats itself n times.

DID YOU KNOW?

Poly(ethene) was discovered by accident. When the apparatus for an experiment on ethene leaked, a white solid was found, which had the empirical formula CH_2.

Figure 19.2.2 The repeating unit for poly(propene) from a large number (n) of propene monomers

Here is another example:

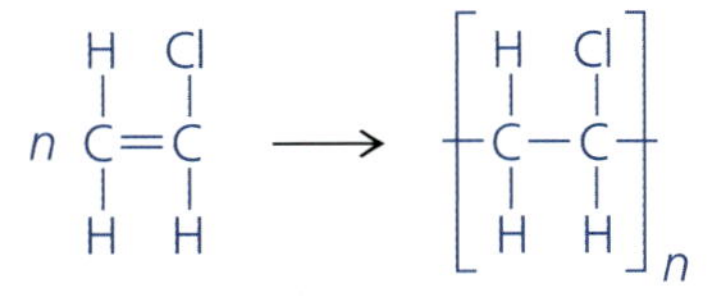

$$n \; \underset{\underset{H}{|}}{\overset{\overset{H}{|}}{C}} = \underset{\underset{H}{|}}{\overset{\overset{Cl}{|}}{C}} \longrightarrow \left[\underset{\underset{H}{|}}{\overset{\overset{H}{|}}{C}} - \underset{\underset{H}{|}}{\overset{\overset{Cl}{|}}{C}} \right]_n$$

Figure 19.2.3 Poly(chloroethene) from chloroethene

STUDY TIP

When writing the formula for an addition polymer don't forget: (i) include the continuation bonds and (ii) the double bond changes to a single bond.

From polymer to monomer

You can find the structure of the monomer from a diagram of the polymer:

- Identify the repeating unit in the polymer and draw this.
- Ignore the brackets, the continuation bonds and the 'n'.
- Make the single bond between the carbon atoms in the chain of the polymer into a double bond.

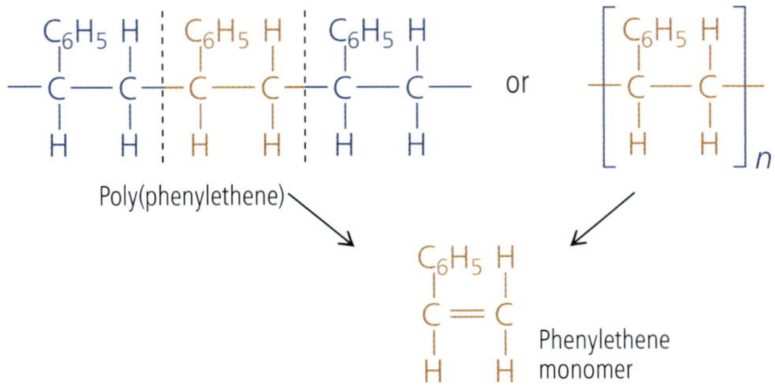

Figure 19.2.4 Deducing the monomer of poly(phenylethene)

SUMMARY QUESTIONS

1 Name the polymers formed from these monomers:
  a butene
  b fluoroethene
  c ethenyl ethanoate.

2 Draw part of the chain of poly(propene) to show three repeating units.

3 Deduce the structure of the monomer that forms this polymer:

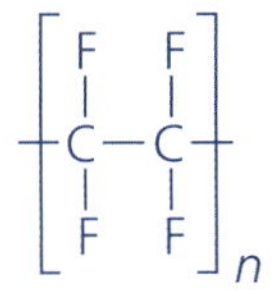

$$\left[\underset{\underset{F}{|}}{\overset{\overset{F}{|}}{C}} - \underset{\underset{F}{|}}{\overset{\overset{F}{|}}{C}} \right]_n$$

Figure 19.2.5

KEY POINTS

1 We can draw the structure of an addition polymer by joining several monomer units and changing the double bonds to single bonds.

2 We can draw a shorthand structure for a polymer by placing one unit in square brackets with continuation bonds.

3 We can deduce the structure of a monomer from a diagram of a polymer by identifying the repeating units.

19.3 Polyamides and polyesters

LEARNING OUTCOMES

- Understand the terms *polyamide* and *polyester*
- Understand condensation polymerisation
- Describe the formation of nylon and terylene
- Distinguish between addition polymers and condensation polymers

DID YOU KNOW?

Nobody seems to know the origin of the word 'nylon'. Some people think it is from 'no run' but others think it is from New York and London, where much of the research into it was done.

We make addition polymers by joining the monomers together to make a long chain. No other substance is formed. However, many polymers such as nylon and terylene are formed in a different way – by **condensation** polymerisation.

In condensation polymerisation, the monomers react together to form the polymer and another product. The other product is usually a small molecule such as water or hydrogen chloride. We say that the small molecule is *eliminated*. Condensation polymerisation usually involves two different monomers, each with a different functional group. These groups react together to form the polymer.

Polyamides

Nylon is a typical **polyamide**. The two types of monomer that form a polyamide are *carboxylic acids* and *amines*. An amine is a compound with an $-NH_2$ functional group. A carboxylic acid reacts with an amine to form an amide.

$$CH_3-\underset{\underset{O}{\|}}{C}-O-H \ + \ H-\underset{\underset{H}{|}}{N}-CH_3 \longrightarrow CH_3-\underset{\underset{O}{\|}}{C}-\underset{\underset{H}{|}}{N}-CH_3 \ + \ H_2O$$

A carboxylic acid | An amine | Amide linkage

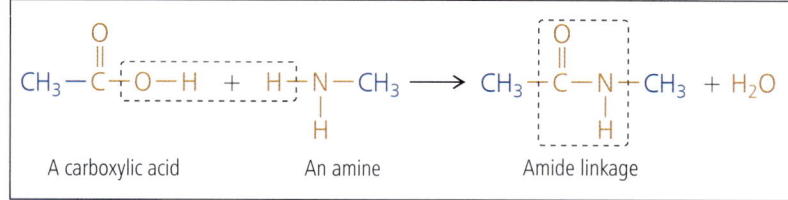

Figure 19.3.1 An amide linkage is formed when a carboxylic acid reacts with an amine

This type of reaction, where two molecules join together and a small molecule is eliminated, is called a **condensation reaction**.

We can represent the structures of the monomers that make a polyamide like this:

HOOC—▬—COOH H_2N—⬭—NH_2

A dicarboxylic acid   A diamine

Figure 19.3.3 Simplified structures of a dicarboxylic acid and a diamine

You can see that each monomer has two functional groups. We put the word di- in front of the names to show that there are two of the same functional groups per monomer. Because each end of the monomer has a reactive functional group, a long chain can be formed.

Nylon is a useful fibre for making strings and ropes

Figure 19.3.4 Nylon is formed from carboxylic acid groups reacting with amine groups

Polyesters

Terylene is a typical **polyester**. The two types of monomer that form a polyester are *carboxylic acids* and *alcohols*. An ester linkage is formed.

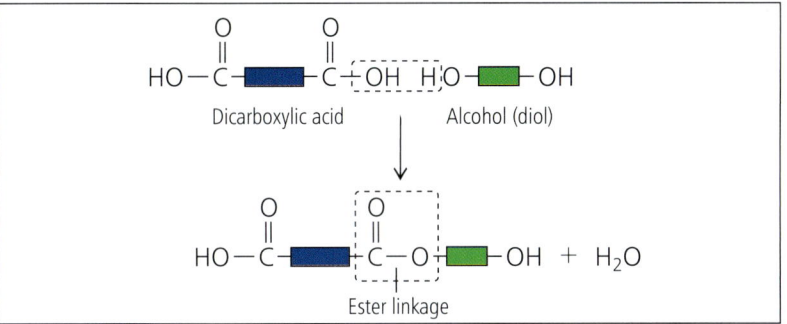

Figure 19.3.5 An ester link is formed when a carboxylic acid reacts with an alcohol

> **STUDY TIP**
> A common error in writing formulae for polyamides and polyesters when each monomer has only one type of functional group is to write all the
> $-\overset{O}{\underset{\parallel}{C}}-O-$ or $-\overset{O}{\underset{\parallel}{C}}-\underset{\underset{H}{|}}{N}-$
> groups in the same direction.

We can represent the equation for polyester formation like this:

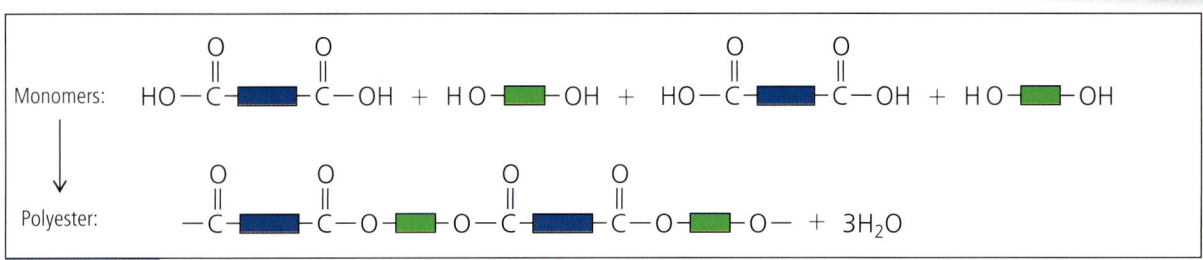

Figure 19.3.6 Terylene is formed from carboxylic acid groups reacting with alcohol groups

SUMMARY QUESTIONS

1 Copy and complete using the words below:

   **condensation   diamines   eliminated
   monomers   polyamide**

   Nylon is made by _____ polymerisation. The _____ are dicarboxylic acids and _____. These react to form a _____. A small molecule is _____.

2 Draw the structure of:
   a an amide linkage
   b an ester linkage.

KEY POINTS

1 In a condensation reaction the monomers join to form a polymer and a small molecule is eliminated.

2 Polyamides have an amide linkage
$-\overset{O}{\underset{\parallel}{C}}-\underset{\underset{H}{|}}{N}-$

3 Polyesters have an ester linkage
$-\overset{O}{\underset{\parallel}{C}}-O-$

241

SUMMARY QUESTIONS

1. Copy and complete using words from the list below:

   **chains ethene join monomers
   poly(ethene) polymer**

   At high temperature and pressure _____ molecules combine to form long _____ of _____. The long chain is called a _____. The small molecules that _____ to form the chain are called _____.

2. Match each word on the left with its description on the right.

monomer	a reaction where compounds combine and a small molecule is eliminated
addition	a molecule made by combining monomers
polymer	a simple molecule from which a polymer is made
condensation	a reaction where two or more molecules combine to form only one product

3. Match each polymer name on the left with its monomer unit on the right.

poly(chloroethene)	$C_6H_5CH=CH_2$
poly(phenylethene)	$CH_2=CH_2$
poly(ethenyl ethanoate)	$CH_3CH=CH_2$
poly(propene)	$CH_2=CHCl$
poly(ethene)	$CH_3COOCH=CH_2$

4. Write down one positive and one negative argument for each of the following ways of disposing of plastics:
   (a) burning the plastics
   (b) recycling the plastics
   (c) putting the plastics into a landfill site.

5. Which of these compounds can undergo addition polymerisation?
   (a) $CH_3CH=CH_2$ (b) $CH_3CH_2CH_3$
   (c) CH_3CH_2COOH (d) $C_6H_5NH_2$
   (e) $C_6H_5CH=CH_2$

6. State one use for each of these polymers:
   (a) poly(ethene)   (b) poly(propene)
   (c) nylon   (d) terylene

7. Draw simplified diagrams of the monomers used to make: (a) terylene (b) nylon.

PRACTICE QUESTIONS

1. Which one of these statements about poly(ethene) is true?
   A Poly(ethene) is formed by the addition polymerisation of ethane.
   B Ethene monomers are used to make poly(ethene).
   C Hydrogen is given off when poly(ethene) is made.
   D Poly(ethene) is made by condensation polymerisation.

   (Paper 1)

2. Which statement about the monomers W, X, Y and Z used to make polymers is correct?

   $CH_2=CH_2$ $HO-\square-OH$
   W   X

   $HO_2C-\square-CO_2H$ $H_2N-\square-NH_2$
   Y   Z

   A W and X polymerise to form an amide.
   B X and Y polymerise to form an ester.
   C X and Z polymerise to form an amide.
   D Y and Z polymerise to form an ester.

   (Paper 2)

3. At high temperature and pressure, ethene molecules combine to form a polymer.
   (a) Draw the full structural formula of an ethene molecule. [1]
   (b) State the general name given to a small molecule that combines to form a polymer. [1]
   (c) Give the name of the polymer formed from ethene molecules. [1]
   (d) What type of polymerisation occurs when ethene molecules are polymerised? [1]
   (e) What feature of ethene is responsible for its ability to polymerise? [1]
   (f) Draw a section of the polymer formed when ethene is polymerised. [2]

   (Paper 3)

4. Poly(propene) is a macromolecule that is made from propene monomers.
   (a) What do you understand by the term *macromolecule*? [1]
   (b) What type of reaction occurs when propene monomers combine to form poly(propene)? [1]

(c) Draw the structure of poly(propene) to show three repeating units. [2]

(d) Draw a simplified structure of poly(propene), showing one unit, using the letter 'n' to show the repetition of this unit. [2]

(e) Give one use of poly(propene). [1]

(Paper 4)

5 A simplified structure of nylon is shown below.

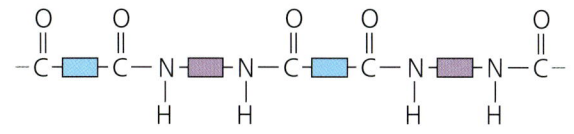

$$-\overset{O}{\underset{\|}{C}}-\boxed{}-\overset{O}{\underset{\|}{C}}-\underset{\underset{H}{|}}{N}-\boxed{}-\underset{\underset{H}{|}}{N}-\overset{O}{\underset{\|}{C}}-\boxed{}-\overset{O}{\underset{\|}{C}}-\underset{\underset{H}{|}}{N}-\boxed{}-\underset{\underset{H}{|}}{N}-\overset{O}{\underset{\|}{C}}-$$

(a) Nylon is made by condensation polymerisation. What do you understand by the term *condensation polymerisation*? [2]

(b) State the name given to the link between the monomer units in nylon. [1]

(c) Write the simplified formulae for the two monomers used to make nylon. [2]

(d) Give one use of nylon. [1]

(e) A polymer similar to nylon is made from two monomers having the structures shown below:

$H_2N-C_6H_4-NH_2$

$HOOC-C_6H_4-COOH$

(i) Draw the structure of this polymer to show one unit, using the letter 'n' to show the repetition of this unit. [3]

(ii) State the name of the small molecule eliminated when these two monomers combine. [1]

(Paper 4)

6 Some polymers contain halogen atoms. The structure of the polymer poly(dichloroethene) is shown below.

$$\left[\begin{array}{c}\underset{|}{\overset{H}{|}}\;\;\underset{|}{\overset{Cl}{|}}\\-C-C-\\\underset{|}{\overset{|}{H}}\;\;\underset{|}{\overset{|}{Cl}}\end{array}\right]_n$$

(a) What is the meaning of the letter 'n'? [1]

(b) (i) Draw the structure of the monomer used to make this polymer. [2]

(ii) State the name of this monomer. [1]

(c) Poly(dichloroethene) is formed by addition polymerisation. What do you understand by the term *addition polymerisation*? [1]

(d) The polymer 'Teflon' is obtained from the monomer, tetrafluoroethene:

$$\underset{F}{\overset{F}{\diagdown}}C=C\underset{F}{\overset{F}{\diagup}}$$

(i) State the chemical name of this polymer. [1]

(ii) Draw the structure of 'Teflon' to show three repeating units. [2]

(e) State one adverse effect of burning plastics containing halogen atoms. [1]

(Paper 4)

7 Polymer A has the structure:

$$-\overset{O}{\underset{\|}{C}}-C_{10}H_6-\overset{O}{\underset{\|}{C}}-O-CH_2-CH_2-O-\overset{O}{\underset{\|}{C}}-C_{10}H_6-\overset{O}{\underset{\|}{C}}-O-CH_2-CH_2-O-$$

(a) What type of linkage is formed when the monomers of this polymer combine? [1]

(b) (i) State the name of another polymer with this type of linkage. [1]

(ii) State one use of the polymer you suggested in part (i). [1]

(c) (i) Draw simplified structures of the two monomers used to make polymer A. [2]

(ii) What type of polymerisation occurs when polymer A is made? [1]

(Paper 4)

8 The simplified structure of the monomers used to make polymer B are shown below.

$$HO-\overset{O}{\underset{\|}{C}}-\boxed{}-\overset{O}{\underset{\|}{C}}-OH \qquad HO-\bigcirc-OH$$

(a) Draw the structure of the polymer formed when these two monomers combine. [3]

(b) (i) What type of linkage is present between the monomer units in polymer B? [1]

(ii) State the name of a polymer that contains this type of linkage. [1]

(iii) Give one use of the polymer you suggested in part (ii). [1]

(Paper 4)

243

20 Biological molecules

20.1 Natural macromolecules

LEARNING OUTCOMES

- Name proteins and carbohydrates as constituents of food
- Describe the structure of proteins
- Describe the hydrolysis of proteins to amino acids

Large molecules in food

We need food to give us energy and to keep making new cells in our body. The main classes of food molecules that give us energy are **carbohydrates** and **fats**. **Proteins** are needed for growth and other essential functions. Proteins and some carbohydrates, for example starch, are **macromolecules**. Fats are large molecules which usually contain three **ester linkages**.

The structure of proteins

Proteins have the same **amide linkage** as nylon. But instead of being made from two different monomers, proteins are made up from twenty. These monomers are **amino acids**. The structures of three amino acids are shown below.

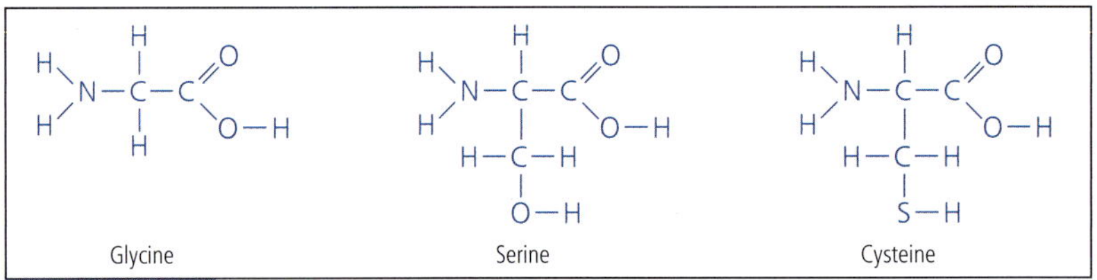

Figure 20.1.1 The structure of some amino acids

All amino acids have the **amine** ($-NH_2$) and **carboxylic acid** ($-COOH$) groups in common. But the side chains in each of the 20 amino acids are different. When proteins are made, an amide link is formed by the reaction of the amine group of one amino acid with the carboxylic acid group of the next amino acid. This is an example of a **condensation reaction**. Water is eliminated.

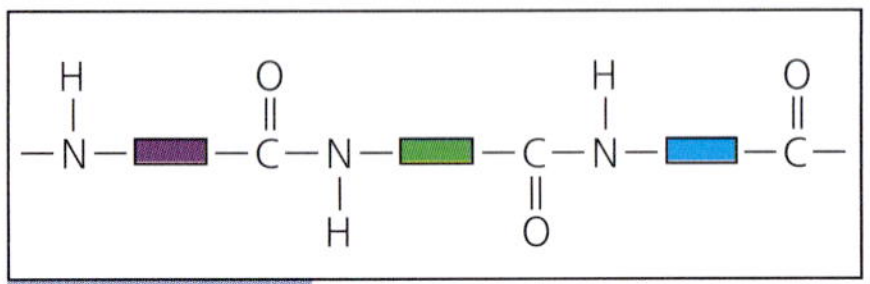

STUDY TIP

You should be able to recognise the repeating units in proteins as NHCOCH(R) and that this repeats along the chain.

Figure 20.1.2 Structure of a protein

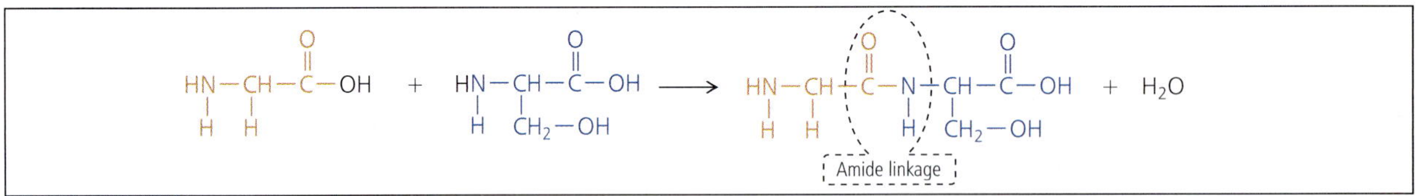

Figure 20.1.3 Forming an amide link between two amino acids

We can simplify the structure of an amino acid even further and show what happens when a protein is formed:

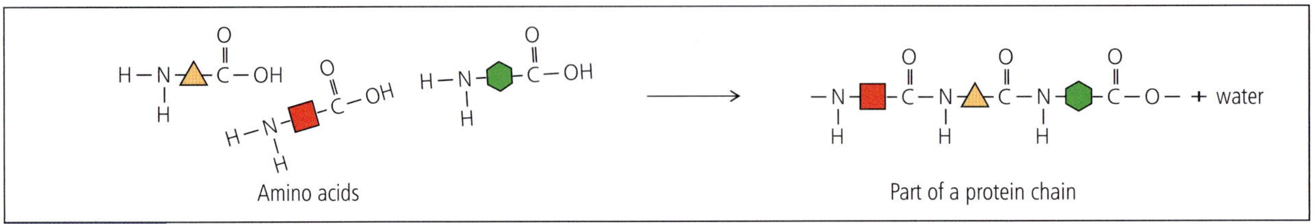

Figure 20.1.4 When proteins are made, amino acid monomers form amide linkages with one another by condensation polymerisation

The order of the amino acids in the protein does not follow any regular pattern.

From proteins to amino acids

In our bodies, proteins in our food are broken down to amino acids with the help of hydrochloric acid in the stomach. **Enzymes** catalyse this reaction. In the laboratory we can break down proteins to amino acids by heating them with hydrochloric acid.

$$\text{protein} + \text{water} \xrightarrow{\text{hydrochloric acid} + \text{heat}} \text{amino acids}$$

We call this a **hydrolysis** reaction. In a hydrolysis reaction, a compound reacts with water and breaks down to form two or more products. We often use acid or alkali to help the hydrolysis. If we use acid we call the reaction 'acid hydrolysis'.

Muscle tissue in animals is mainly protein

DEMONSTRATION

Hydrolysing a protein

The protein is heated with 6 mol/dm³ hydrochloric acid for several hours. The condenser is vertical to stop the loss of hydrochloric acid vapour. At the end of the experiment, you neutralise the excess hydrochloric acid with ammonia.

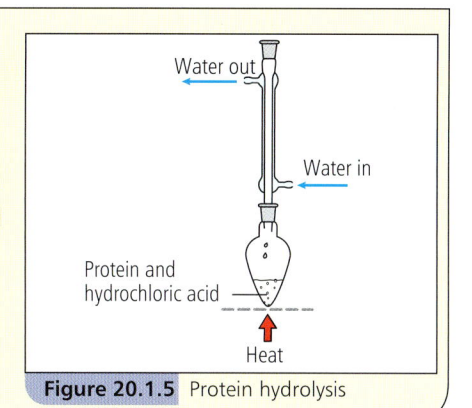

Figure 20.1.5 Protein hydrolysis

DID YOU KNOW?

The hydrochloric acid in your stomach has a pH between 1 and 2 to help unwind the food proteins so the peptide bonds can be hydrolysed.

When proteins are hydrolysed, the amide linkages are converted to carboxylic acid and amine functional groups – the reverse of the polymerisation reaction.

Figure 20.1.6 Hydrolysis of amide linkages in protein

You can use paper chromatography to identify the amino acids formed by the hydrolysis of a protein.

SUMMARY QUESTIONS

1 State the names of the two functional groups in amino acids that condense to form the amide linkages in proteins.

2 Explain why the formation of proteins from amino acids is a condensation reaction.

3 Draw a flow chart to describe how you can identify which amino acids are present in proteins.

KEY POINTS

1 Proteins and carbohydrates are the constituents of food.

2 Proteins have the same amide linkage as nylon but the monomer units are different.

3 Proteins can be hydrolysed to amino acids by heating with hydrochloric acid.

20.2 Mainly carbohydrates

LEARNING OUTCOMES

- Describe the formation of complex carbohydrates by the polymerisation of simple sugars
- Describe how chromatography can be used to identify the products of acid hydrolysis of carbohydrates

Figure 20.2.1 A glucose molecule

The word **carbohydrate** literally means carbon with water. So it is no surprise that the empirical formula for a carbohydrate is $C_x(H_2O)_y$. For example, the formula for glucose is $C_6H_{12}O_6$. But even simple carbohydrates are quite complex molecules, for example glucose. Fortunately we can simplify the structure of glucose to show only those groups which react to form glucose polymers.

We give the name *monosaccharides* to simple *sugars* such as glucose: *mono* means 'one' and *saccharide* means sugar.

We can join monosaccharide monomers together to make polymers called *polysaccharides*:

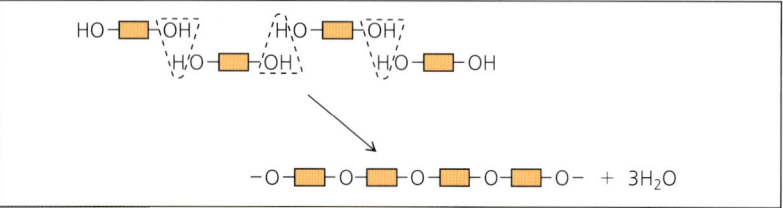

Figure 20.2.2 The monosaccharide monomers join to form a polysaccharide

You can see that this is a **condensation polymerisation**. The monosaccharide monomers have joined together and water has been eliminated. The linkage in these sugar polymers is called a *glycosidic* linkage. The general formula for a polysaccharide made from glucose is $(C_6H_{10}O_5)_n$ – glucose with water removed.

Two common sugar polymers found in plants are *starch* and *cellulose*:

- Starch is found in rice, pasta and potatoes. It provides us with most of the carbohydrate in our diet. Starch is made of glucose monomers arranged either in long chains or in branched chains.
- Cellulose is found in the cell walls of plants. It is also a polymer of glucose. But the glucose molecules are arranged in a different way. We can't digest cellulose but cows and sheep can!

The hydrolysis of carbohydrates

When we heat polysaccharides with hydrochloric acid they are hydrolysed to simple sugars. For example: starch is hydrolysed to glucose.

$$\text{starch} + \text{water} \xrightarrow{\text{hydrochloric acid, heat}} \text{glucose}$$

Figure 20.2.3 The hydrolysis of starch to glucose

DID YOU KNOW?

Scientists have found a way to produce hydrogen from cellulose. If this can be carried out on a large scale, the hydrogen could be used as a fuel for transport.

When starch is hydrolysed the glycosidic linkages break and simple sugars are formed. The hydrochloric acid acts as a catalyst in the reaction. In our bodies, starch is hydrolysed to glucose using several different **enzymes**.

Identifying hydrolysis products

The simple sugars formed by the hydrolysis of polysaccharides can be identified using paper chromatography. Because sugars are colourless, we have to use a **locating agent** so they show up on a **chromatogram**. We spray the chromatogram with a mixture of tin(II) chloride and a chemical called *N*-phenylenediamine. We then warm the chromatogram in an oven to develop the colour.

You cannot always identify a sugar or an amino acid from a chromatogram by simple chromatography. This is because some compounds have the same or similar R_f values. We overcome this by using two-dimensional chromatography: We carry out the chromatography as usual. We allow the paper to dry then turn it through 90° and carry out chromatography using a different solvent. Only then do we spray the chromatogram with a locating agent. The diagram shows how we separate the three amino acids aspartic acid (Asp), glutamic acid (Glu) and leucine (Leu) by this method.

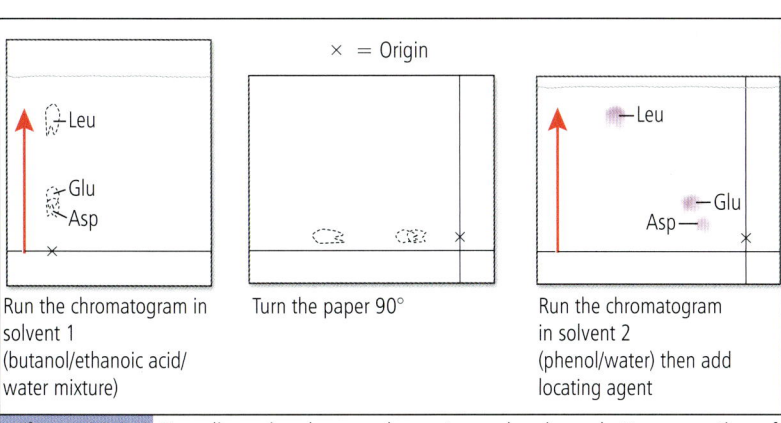

Many of the foods we eat contain carbohydrate

Figure 20.2.4 Two-dimensional paper chromatography gives a better separation of amino acids

STUDY TIP

You do not need to know the structure of carbohydrates but it is important to know how hydrolysis breaks down complex carbohydrates using simplified formulae.

SUMMARY QUESTIONS

1 Copy and complete using the words below:

   condensation   eliminated   glucose   water

   Starch is a _____ polymer formed when _____ molecules combine. In this reaction _____ molecules are _____.

2 Explain why the formation of starch from glucose is described as condensation polymerisation.

3 Draw a flow chart to describe how you can identify which sugars are present in a complex carbohydrate.

KEY POINTS

1 Carbohydrates have the general formula $C_x(H_2O)_y$.

2 The monomers that form complex carbohydrates can be represented by the formula HO—☐—OH.

3 Complex carbohydrates can be hydrolysed to simple sugars by concentrated hydrochloric acid.

4 We can use paper chromatography to identify sugars.

20.3 Fermentation

LEARNING OUTCOMES

- Describe enzymes as proteins that act as biological catalysts
- Describe the fermentation of simple sugars
- Compare two methods of producing ethanol: fermentation and the reaction of steam with ethene

Nearly all chemical reactions in living things are catalysed by **enzymes**. Enzymes are particular types of **protein** that act as biological **catalysts**. But unlike inorganic catalysts they are very sensitive to changes in temperature. They work best between 30 °C and 40 °C – at around body temperature. If the temperature is too low they work much more slowly. If the temperature is too high they lose their structure and don't work. We say they are *denatured*. Most enzymes will not work above a temperature of about 50 °C.

Bacteria and yeasts produce enzymes that catalyse **fermentation** reactions in organic materials. Fermentation is the breakdown of organic material with effervescence (bubbles) and release of heat. For thousands of years, ethanol has been made by fermentation. Ethanol is an excellent solvent and a good fuel, as well as being an important starting point for making other chemicals.

The fermentation of glucose

A wide range of vegetable material can be fermented. But the most commonly used substances for making ethanol are sugars and starch. The bacteria and yeast that cause fermentation are found on the surface of many plants, as well as in the air. When we carry out fermentation to produce ethanol, we use a particular form of yeast so that unwanted reactions do not occur.

The sugar from this sugar cane can be fermented to make ethanol

DID YOU KNOW?

Brazil is the world's largest producer of ethanol – much of it made from the fermentation of sugar cane waste. Since 1977 all fuel for cars in Brazil must contain at least 20% ethanol.

PRACTICAL

Fermenting sugars

1. You add moist yeast to the flask of warm glucose solution.
2. Then you shake the solution so that the yeast is suspended in the solution.
3. You keep the flask warm for about 30 minutes without shaking.
4. Then you 'pour' the gas from the flask into a boiling tube containing some limewater. The limewater turns milky. This shows that carbon dioxide is a product of fermentation.

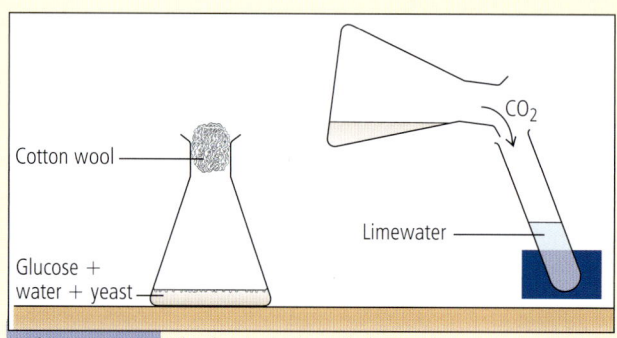

Figure 20.3.1 The fermentation of glucose by yeast

The fermentation of glucose produces ethanol and carbon dioxide:

$$C_6H_{12}O_6 \longrightarrow 2C_2H_5OH + 2CO_2$$
$$\text{glucose} \longrightarrow \text{ethanol} + \text{carbon dioxide}$$

The reaction takes place in the absence of air – it is **anaerobic**. The reaction will occur until about 14% ethanol is present in the mixture. If the ethanol concentration gets much higher than this the yeast will die. The ethanol is separated from the reaction mixture by **fractional distillation**.

Comparing methods

We can produce ethanol by fermentation or by reaction of ethene with steam. Each of these methods has advantages and disadvantages:

Ethanol from fermentation	Ethanol from ethene and steam
Simple method	More complex method
Needs a lot of very large tanks	Needs smaller-scale equipment to produce the same amount of ethanol
Uses a batch process: you have to start again from the beginning once you have removed the solution in the tank	A continuous process: the ethanol is removed continuously and the ethene and steam are fed into the apparatus continuously
Rate of reaction is slow	Rate of reaction is fast
Ethanol needs further purification by distillation	Produces ethanol of high purity
Uses renewable resources	The ethene is made from a non-renewable resource – petroleum

Ethanol as a fuel

Ethanol can be made by fermentation using a variety of plant sources. In Brazil, sugar cane waste is fermented to ethanol for use as a fuel for cars. It is either mixed with petrol to form a fuel called gasohol or used by itself. Ethanol is a cleaner fuel than petrol – it does not produce as much pollution.

The ethanol made by fermentation comes from a renewable resource. It is potentially 'carbon neutral': the carbon dioxide released into the atmosphere by burning the ethanol is balanced by the carbon dioxide absorbed from the atmosphere by the sugar cane during photosynthesis.

STUDY TIP

It is a common error to suggest that oxygen is required for the fermentation of glucose to ethanol.

KEY POINTS

1 An enzyme is a biological catalyst.

2 The fermentation of glucose to carbon dioxide and ethanol is catalysed by enzymes from yeast.

3 The production of ethanol by fermentation and hydration has both advantages and disadvantages.

SUMMARY QUESTIONS

1 Copy and complete using the words below:

   **carbon   catalyse
   enzymes   ethanol
   ferments   glucose   yeast**

   When you leave a solution of glucose with _____ for a few days it _____. The _____ is broken down to _____ and _____ dioxide. The yeast produces _____ that _____ this reaction.

2 How do enzymes differ from inorganic catalysts?

3 Give one advantage of making ethanol:

   a by fermentation

   b from ethene and steam.

SUMMARY QUESTIONS

1 Copy and complete using words from the list:

**biological   enzymes   inactivated
increase   inorganic   living   rate**

Catalysts are substances that _____ the _____ of a chemical reaction. The catalysts found in all _____ organisms are called _____. These _____ catalysts differ from _____ catalysts by being _____ at temperatures above about 45 °C.

2 Match each compound on the left with its descriptions on the right.

starch	a substance produced by the fermentation of sugars
glycerol	a polymer of glucose
amino acids	a compound that forms ester linkages with fatty acids
glucose	the monomers used to make proteins
ethanol	a compound with the formula $C_6H_{12}O_6$

3 Copy and complete using words from the list below:

**amide   amino   ester   fats   glycerol
glycosidic   monomers   polymers   starch**

Foods contain proteins, carbohydrates and _____. Complex carbohydrates such as _____ and cellulose are _____ of glucose. The linkages formed when glucose _____ polymerise are called _____ linkages. Proteins are polymers of _____ acids. The linkage between the amino acids in proteins is an _____ linkage. Fats are large molecules that have _____ linkages formed between _____ and fatty acids.

4 What do you understand by each of the following terms?
  (a) hydrolysis   (b) soaps
  (c) fermentation   (d) fats

5 (a) Draw a flow diagram to show how ethanol is made by fermentation and then purified to make the pure alcohol.
  (b) Make a list of the advantages and disadvantages of making ethanol by fermentation.

PRACTICE QUESTIONS

1 Which one of these statements about fermentation is correct?
  A The products of fermentation are ethene and carbon dioxide.
  B Fermentation is catalysed by the enzymes present in yeast.
  C The rate of fermentation increases rapidly at temperatures above 50 °C.
  D Fermentation cannot occur in the absence of oxygen.
  (Paper 1)

2 Which one of the following statements about the hydrolysis of proteins is correct?
  A Proteins can be hydrolysed by refluxing with concentrated hydrochloric acid.
  B The hydrolysis products of proteins are the amino acids glycine and alanine only.
  C When proteins are hydrolysed, the amide linkages are converted to carboxylic acid and alcohol functional groups.
  D You can use fractional distillation to identify the amino acids formed by hydrolysis of proteins.
  (Paper 2)

3 Ethanol can be made by leaving a mixture of glucose solution and yeast for several days.
  (a) What is the name given to this method of making ethanol? [1]
  (b) Describe one observation of the mixture that can be made during this process. [1]
  (c) The gas given off in this process turns limewater milky. Identify this gas. [1]
  (d) How can ethanol be removed from the reaction mixture? [1]
  (e) Enzymes in the yeast convert the glucose to ethanol. What do you understand by the term *enzyme*? [2]
  (f) Why does the reaction not work if it is carried out at 70 °C? [1]

(Paper 3)

4 The simplified structure of a protein is shown.

(a) (i) Write the formula for the amide link in the protein. *[1]*
(ii) What letter is usually used to represent the side groups in a protein? *[1]*
(iii) Nylon is a synthetic polymer with amide linkages. State two differences between the structures of nylon and proteins. *[2]*

(b) Proteases are enzymes that hydrolyse proteins to amino acids. What do you understand by the term *hydrolyse*? *[1]*

(c) The amino acids obtained by hydrolysis can be separated and identified using paper chromatography. Draw a labelled diagram of the apparatus used for paper chromatography. *[3]*

(d) The diagram below shows a chromatogram of the amino acids obtained by hydrolysis. Only some of the amino acids are shown.

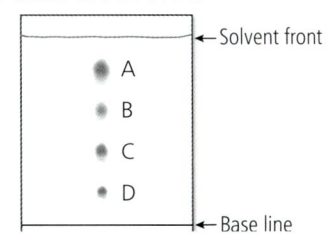

(i) Amino acids are colourless. Explain how the amino acids are made visible on the chromatogram. *[2]*
(ii) Which amino acid, A, B, C or D, is least soluble in the solvent? *[1]*
(iii) Calculate the R_f value of amino acid B. *[1]*

(Paper 4)

5 Simple sugars can be represented as HO—▭—OH.
(a) Simple sugars can be polymerised to form complex sugars called polysaccharides.
(i) Draw the structure of part of a polysaccharide chain showing three repeating units. *[2]*
(ii) State the name of a polysaccharide. *[1]*
(iii) What type of polymerisation occurs when simple sugars combine to form polysaccharides? *[1]*

(b) Polysaccharides can be hydrolysed to simple sugars.
(i) Describe how you can hydrolyse polysaccharides to simple sugars. *[2]*
(ii) Describe how paper chromatography can be used to identify simple sugars. *[4]*

(Paper 4)

6 The structures of glycine, serine and valine are:

$$H_2N-CH-COOH$$
$$|$$
$$H$$
Glycine

$$H_2N-CH-COOH$$
$$|$$
$$CH_2OH$$
Serine

$$H_2N-CH-COOH$$
$$|$$
$$CH(CH_3)_2$$
Valine

(a) To which group of compounds do these three compounds belong? *[1]*
(b) Glycine and valine can undergo a condensation reaction in which an amide linkage is formed.
(i) Draw the structure of an amide linkage. *[1]*
(ii) Name a synthetic polymer that contains an amide linkage. *[1]*
(c) Glycine can be polymerised to form poly(glycine).
(i) Draw a section of the chain of this polymer to show three repeating units. *[2]*
(ii) Poly(glycine) can be hydrolysed back to glycine. Describe how this hydrolysis is carried out. *[2]*

(Paper 4)

7 Ethanol can be produced by fermentation, or by the hydration of ethene.
(a) Describe three differences in the conditions used in these two reactions. *[6]*
(b) Describe two advantages and two disadvantages of making ethanol by fermentation. *[4]*

(Paper 4)

Alternative to Practical section

C1 Using and organising techniques, apparatus and material

LEARNING OUTCOMES

- Describe, explain or comment on experimental arrangements and techniques
- Identify sources of error and suggest possible improvements in procedures
- Suggest suitable techniques and apparatus for an investigation – see C4

APPARATUS

Your experience of practical work in chemistry could be gained from working as an individual, from group work or from teacher demonstration. The recognition of laboratory apparatus and standard equipment, such as Bunsen burners and tripods, is essential.

EXAMPLE 1

A solution of copper(II) sulfate was made by reacting excess copper(II) oxide with dilute sulfuric acid. The diagram shows the method used.

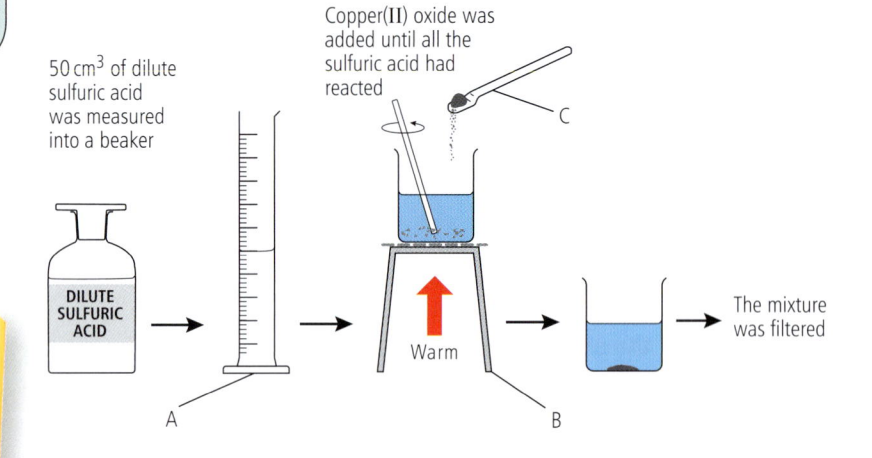

50 cm³ of dilute sulfuric acid was measured into a beaker

Copper(II) oxide was added until all the sulfuric acid had reacted

DILUTE SULFURIC ACID

Warm

The mixture was filtered

A   B   C

(a) Name the pieces of apparatus, A, B and C.
(b) Draw a labelled diagram to show how the mixture was filtered.

Cambridge IGCSE Chemistry 0620 Paper 6 Q1 June 2008

STUDY TIP

In (a) make sure that full and correct names are used. 'Cylinder', 'stand' and 'spoon' are not precise. 'Measuring cylinder', 'tripod' and 'spatula' are the correct names of the apparatus.

In (b) use a ruler and a sharp pencil and label the diagram clearly. A filter funnel and filter paper both need to be included in the answer.

SAFETY

Carry out experiments using safe and efficient procedures. These involve the use of protective clothing, such as goggles, and the awareness of the dangers of flammable and toxic chemicals.

Understanding experimental set-ups and identifying problems

Information will be provided to set the context as in the question in Example 2.

EXAMPLE 2

Sulfur dioxide gas is denser than air and soluble in water. A sample of sulfur dioxide can be prepared by adding dilute hydrochloric acid to sodium sulfite and warming the mixture.

Study the diagram of the apparatus used.

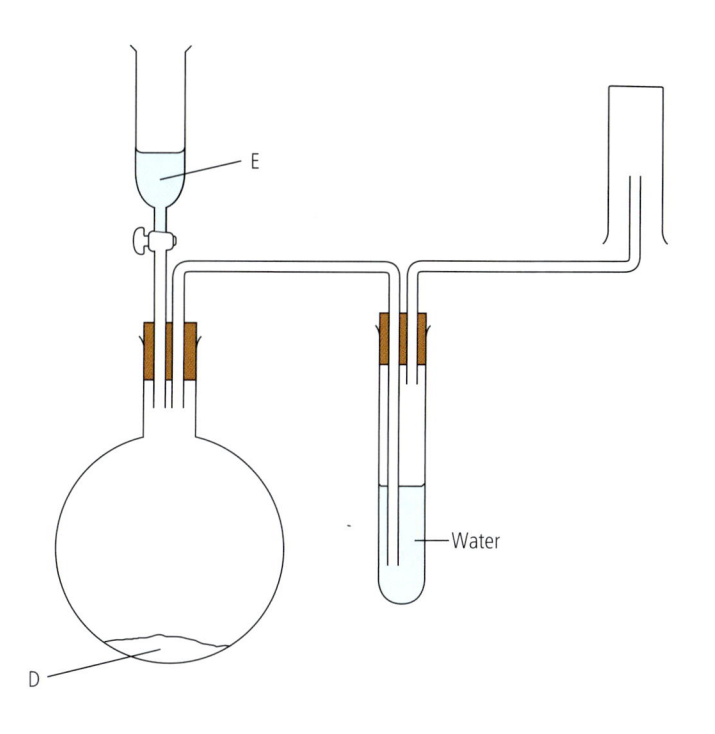

(a) Name the chemicals used, D and E.
(b) Where on the diagram is heat applied?
(c) Identify two mistakes in the diagram.

Cambridge IGCSE Chemistry 0620 Paper 6 Q3 June 2008

REDUCING ERRORS

Using apparatus that gives greater precision or repeating the experiments and taking the average of a set of results can help to reduce errors. Improved techniques can also reduce errors. Suggested improvements to reduce the sources or error in experiments should be meaningful and detailed.

- For example, vague statements such as 'Repeat the experiment to obtain more accurate results' and 'Use a more accurate measuring cylinder' are not sufficient.

Good answers are:

- Repeat the experiment and take the average of the results.
- Use a burette/graduated pipette instead of a measuring cylinder.

> **STUDY TIP**
>
> In (a), make sure that you label the chemicals correctly, e.g. sodium sulfite and not sodium sulfate. In the answer to (b) an arrow needs to be positioned with its point touching the flask underneath the solid. In (c) do not write vague answers. e.g. 'There is no lid on the collecting vessel.' Identifying clearly that the gas should be collected by downward delivery and should not be collected through water are the correct conclusions drawn from the supplied information.

C2 Observing, measuring and recording

LEARNING OUTCOMES

- Record readings from diagrams of apparatus
- Complete tables of data
- Plot graphs

Topic 1.5, **Apparatus for measuring**, includes information about accuracy when measuring volumes. Burette diagrams are made to be read to one decimal place.

In examination papers, diagrams of apparatus such as thermometers, burettes, measuring cylinders, gas syringes, etc. are used. These show readings at various stages of an experiment and you are expected to fill in a blank table of results.

EXAMPLE

A student carried out an experiment to measure the temperature changes during the reaction of two solutions, X and Y.

The results are shown in the table. Use the thermometer diagrams to record the maximum temperatures reached.

Volume of solution Y added to 25 cm³ of solution X/cm³	Thermometer diagram	Maximum temperature/°C
0	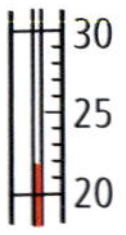 30 / 25 / 20	
10	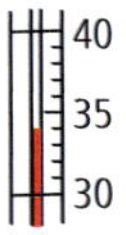 40 / 35 / 30	
20	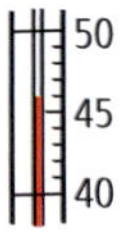 50 / 45 / 40	
30	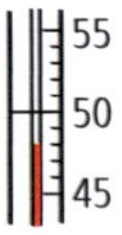 55 / 50 / 45	
40	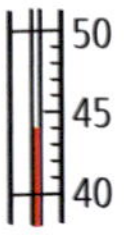 50 / 45 / 40	
50	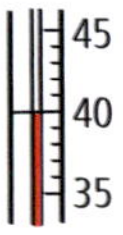 45 / 40 / 35	

STUDY TIP

Note the readings after checking the scale used in the diagrams.

STUDY TIP

Make sure that you don't misread the temperatures recorded when 10 cm³ and 40 cm³ of Y have been added. Incorrect readings would be '36 °C' and '46 °C'.

Cambridge IGCSE Chemistry 0620 Paper 6 Q3 June 2006

GRAPHS

Examination questions requiring you to draw a graph will include the grid. Axes may be labelled and scales included. When a blank grid is provided, an appropriate scale needs to be chosen and axes labelled clearly with units. Each point should be plotted as a small cross using a sharp pencil. Best-fit lines should be drawn leaving out any anomalous points, e.g. the point at 20 seconds in Experiment 4 in the upper graph. Dot-to-dot line graphs are examples of poor practice. The graphs will be smooth curves or straight lines.

Example of good practice

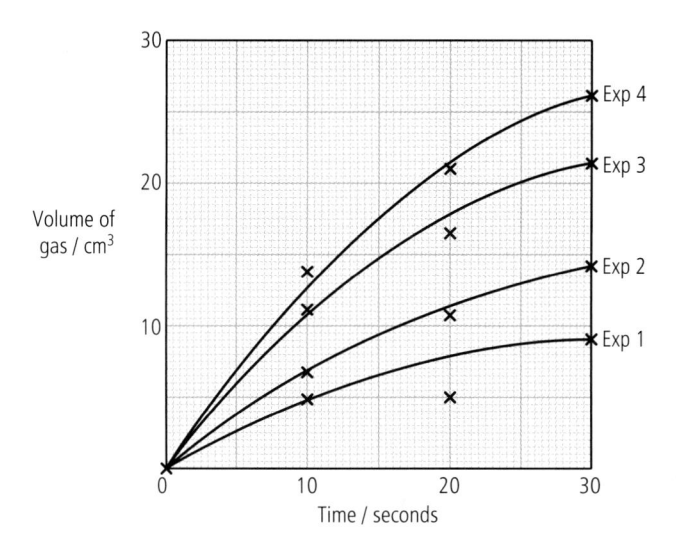

Example of poor practice

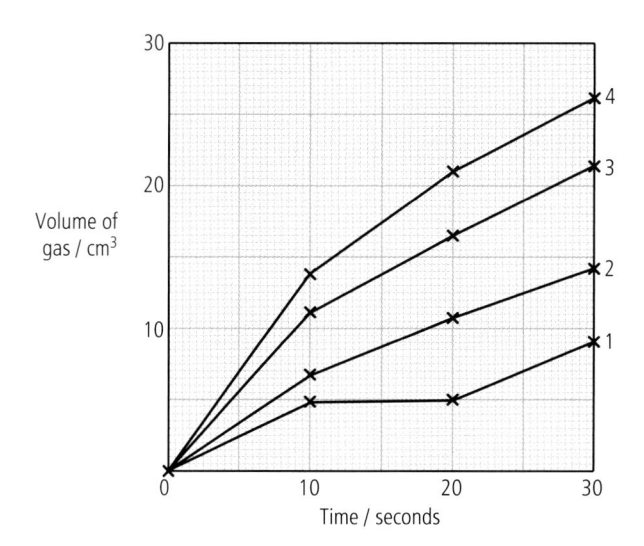

Cambridge IGCSE Chemistry 0620 Paper 6 Q4a June 2006

This is an example of poor practice because the points are joined by straight lines. There should be a continuous smooth curve going through or close to each point (and ignoring anomalous points).

C3 Handling experimental observations and data

LEARNING OUTCOMES

- Draw conclusions from information given
- Interpret and evaluate observations and data
- Draw conclusions from tests for gases and ions
- Interpret graphical information

STUDY TIP

Example 1 involves applying experience of common practical procedures to an unfamiliar situation. It also tests knowledge and understanding of chromatography. When drawing chromotography apparatus, make sure that the place where you put the original spot of colour is above the level of the solvent.

Making conclusions involves the interpretation of results using knowledge and understanding of chemical facts and concepts, and practical techniques.

EXAMPLE 1

A sample of orange fruit jam was investigated to check the three colourings present.

Step 1 The jam was boiled with water.

Step 2 The mixture was filtered.

Step 3 The filtrate was concentrated.

Step 4 The concentrate was analysed by chromatography.

(a) What was the purpose of Step 1?
(b) Why was the mixture filtered?
(c) How was Step 3 carried out?
(d) Draw a diagram to show the possible paper chromatogram obtained in Step 4.

Cambridge IGCSE Chemistry 0620 Paper 6 Q2 June 2006

Study tip – Looking for patterns in data

Two common examples:

- Two titrations were carried out using two different solutions of hydrochloric acid, A and B, and the same volume of aqueous sodium hydroxide. $25.0\,cm^3$ of the acid solution A neutralised the same volume of sodium hydroxide as $50.0\,cm^3$ of the acid solution B.

  Conclusion: The acid solution A is twice as concentrated as the acid solution B.

- Rates of reaction – Using experimental and graphical information to interpret data – see Topic 8.2.

EVALUATION OF AN EXPERIMENT

Suggest possible improvements to the method and/or apparatus used. Questions to ask are:

- Was it a fair test?
- Which variables are kept constant and which variables will be changed?
- What would be the point of repeating the experiment?

EXAMPLE 2

This example shows a sample answer to a question concerning qualitative analysis. It is very important to learn the tests for gases and ions.

A mixture of two compounds was tested. One compound was a water-soluble zinc salt and the other was insoluble. The tests and some of the observations are in the following table. Complete the observations in the table.

tests	observation
(a) One measure of the mixture was heated gently then strongly.	condensation at the top of the tube
The gas released was tested with cobalt chloride paper.	paper turned pink
The rest of the mixture was added to about 25cm³ of distilled water in a boiling tube. The contents of the tube were shaken and filtered. The following tests were carried out.	

Tests on the filtrate
The solution was divided into 2cm³ portions in four test-tubes.

(b) (i) Drops of aqueous sodium hydroxide were added to the first portion of the solution	White precipitate ✓ ✓
Excess aqueous sodium hydroxide was added.	white precipitate disappears leaving a clear solution ✓ [3]
(ii) Using the second portion test (b)(i) was repeated using aqueous ammonia instead of aqueous sodium hydroxide.	white precipitate produced, soluble in excess leaving a clear solution ✓ ✓ [3]
(iii) To the third portion of solution was added hydrochloric acid and barium nitrate solution.	white precipitate
(iv) To the fourth portion of solution was added nitric acid and silver nitrate solution.	no visible reaction

Tests on the residue

(c) Some of the residue was placed into a test tube. Dilute hydrochloric acid was added and the gas given off was tested with limewater.	rapid effervecence limewater turned milky

(d) What does test (a) indicate?

no chlorine. ✗ [1]

(e) What conclusions can you draw about the filtrate?

white precipitate produced with barium nitrate and HCl leads us to believe it is a sulfate.
The compound is zinc sulfate. No chlorine is present ✓ ✗ [2]

(f) What does test (c) indicate?

Indicates that the residue is a carbonate producing carbon dioxide ✓ ✓ [2]

Cambridge IGCSE Chemistry 0620 Paper 6 Q5 June 2006

STUDY TIP

The answers in (b) show that the knowledge of cation tests is good. The effect of aqueous ammonia and sodium hydroxide on a solution containing Zn^{2+}(aq) is described correctly.

STUDY TIP

In the answers to (b)(i) and (ii) it is best to use the word 'colourless' instead of 'clear' when referring to a solution that is not coloured.

STUDY TIP

The anion tests in (e) and (f) are also recognised. However, in (d), the use of cobalt chloride to test for the presence of water was confused with the test for chlorine gas.

C4 Planning investigations

LEARNING OUTCOMES

- Suggest suitable techniques and apparatus for an investigation
- Describe tests for gases and ions

Planning an investigation

In examination, you may be asked a question about planning an investigation.

Details of the experiments should include:

- apparatus/equipment to be used
- conditions to be employed, e.g. heat
- measurements and observations to be made and recorded
- comparison of experiments to ensure a fair test
- interpretation of results to make conclusions.

EXAMPLE 1

Diesel is a liquid fuel obtained from crude oil. Biodiesel is a fuel made from oil obtained from the seeds of plants such as sunflowers.

Using the apparatus below plan an experiment to investigate which of these two fuels produces more energy.

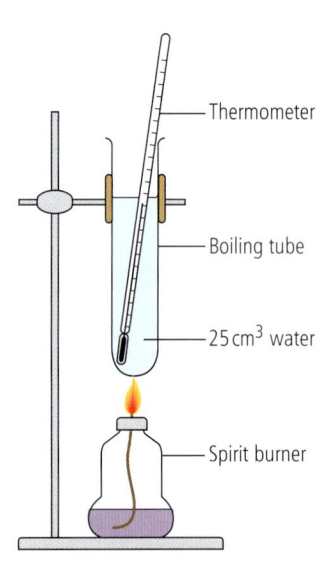

Diagram showing apparatus: Thermometer, Boiling tube, 25 cm³ water, Spirit burner

Good answers involve heating the water with each fuel in turn and measuring initial and final temperatures of the water over a specified time interval.

Cambridge IGCSE Chemistry 0620 Paper 6 Q7 November 2007

STUDY TIP

When answering questions about the energy released from fuels make sure that:

- the fuel is not mixed with the water
- the thermometer is in the water, not the fuel
- every step in the procedure is clearly described
- a clearly labelled diagram is drawn
- when comparing different fuels, the test is fair e.g. the same distance of the spirit burner from the test tube; the same mass of water used.

EXAMPLE 2

This example shows a sample response to a planning question. This is an excellent response.

The diagram shows two bottles of liquid oven cleaner.

> **STUDY TIP**
>
> All measurements and observations to be made are clearly recorded. The idea of a fair test is clearly realised and the comparison of the results to draw appropriate conclusions made. This student also notes safety precautions and suggests repeating the experiment to check reliability.

The oven cleaners contain sodium hydroxide solution. Plan an investigation to show which oven cleaner contains the highest concentration of sodium hydroxide.

> Titration. Place 25 cm³ of oven cleaner in two conical flasks. Use a volumetric ✓ pipette to measure the amount of alkali. Use a burette filled with 1 mol/dm³ of hydrochloric acid. ✓ Add 3 drops of methyl orange ✓ into the flask with alkali. Let the acid run into the flask until the colour of alkali changes from yellow to salmon-pink. ✓ This is the end point, record the amount of acid used. If less acid was used than alkali – acid is stronger and alkali is less concentrated. If more ✓ acid was used than alkali – alkali is more concentrated. The experiment could be repeated to check the reliability of results and care must be taken when using acid or alkali. Goggles should be worn. ✓ The results from both flasks with the different oven cleaners are compared.

[6]

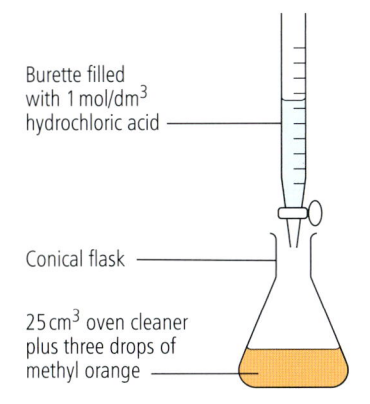

Cambridge IGCSE Chemistry 0620 Paper 6 Q6 June 2006

Revision checklist

This list is to help you check that you have completed and understood all the sections of the course. When you think you have understood each statement below put a tick in the appropriate box in pencil. When you start to revise for the examination you can rub out the tick and repeat the process.

(Supplement shaded in blue here)

Unit 1 Particles and purification

I can:

describe the distinguishing properties of solids, liquids and gases	✓
describe solids, liquids and gases in terms of particle separation, arrangement and types of motion	✓
describe change of state in terms of melting, boiling, evaporation, freezing, condensation and sublimation	✓
explain changes of state in terms of the kinetic theory	
describe qualitatively the pressure and temperature of a gas in terms of the motion of its particles	
describe and explain diffusion and show an understanding of Brownian motion	
describe, explain and state evidence for Brownian motion	
describe how rate of diffusion depends on molecular mass	
name apparatus for measuring mass, time, temperature and volume	
describe paper chromatography and interpret chromatograms	
calculate R_f values from chromatograms	
understand the use of locating agents in chromatography	
identify and assess the purity of substances from melting and boiling point data	
understand the importance of purity in medicines and foodstuffs	
describe methods of purification using solvents or by filtration or crystallisation	
understand the use of simple distillation and fractional distillation	
suggest suitable purification techniques for a given substance	

Unit 2 Atoms, elements and compounds

I can:

state the relative charges and masses of protons, neutrons and electrons	
define proton number and nucleon number	
describe isotopes as radioactive and non-radioactive	
describe some medical and industrial uses of radioisotopes	
understand the Periodic Table in terms of proton number and electronic structure	

Revision checklist

describe electronic structure in terms of shells of electrons	
understand the importance of the noble gas electronic structure	
understand the difference between elements, compounds and mixtures	
describe the differences between metals and non-metals	
describe the structure of alloys	

Unit 3 Structure and bonding

I can:

describe the formation of ions by electron loss and electron gain	
draw the structure of ions derived from elements in Groups I and VII	
draw the structure of ions derived from elements in other groups	
describe the lattice structure of ionic compounds	
describe the covalent bond as a pair of shared electrons	
describe the covalent structures of H_2, Cl_2, H_2O, HCl, CH_4 and NH_3	
understand the importance of the noble gas electronic structure in compounds	
describe the covalent structures of N_2, CO_2, CH_3OH, CO_2 and other compounds	
distinguish ionic and covalent compounds by differences in their physical properties	
explain the differences in physical properties of ionic and covalent compounds	
relate the structures of graphite and diamond to their properties	
relate the structures of graphite and diamond to their uses	
describe the structure of silicon dioxide and relate its structure to its properties	
describe a simple model for metallic bonding and relate this to metallic properties	

Unit 4 Formulae and equations

I can:

use the symbols for the elements to write formulae for simple compounds	
work out the formula of compounds from diagrams	
work out the formula of an ionic compound from the charges on the ions	
write word equations and simple balanced equations	
write balanced equations including the use of state symbols	
write ionic equations and ionic half equations when given sufficient information	
define *relative atomic mass*, *relative molecular mass* and *relative formula mass*	
do simple chemical calculations involving reacting masses	

Revision checklist

Unit 5 Chemical calculations

I can:

do simple chemical calculations based on ratios	
define the Avogadro constant and the mole	
understand how to use the mole concept to calculate reacting masses	
calculate stoichiometric reacting masses	
understand how to do calculations involving the concept of limiting reactants	
understand how to do calculations involving the molar gas volume	
calculate percentage yield and percentage purity	
calculate the empirical formula and molecular formula of a compound	
understand how to do calculations involving solution concentration in mol/dm^3	

Unit 6 Electricity and chemistry

I can:

define the terms *electrode, anode, cathode, electrolyte* and *electrolysis*	
describe the products of electrolysis of a molten salt between inert electrodes	
describe the electrolysis of concentrated HCl, concentrated NaCl and dilute sulfuric acid	
understand that metals or hydrogen are formed at the cathode during electrolysis	
understand that non-metals are formed at the anode during electrolysis	
describe the manufacture of chlorine and sodium hydroxide in a diaphragm cell	
predict the electrolysis products of dilute or concentrated halide solutions	
describe electrolysis in terms of the redox reactions at the electrodes	
describe the electrolysis of copper(II) sulfate using graphite and copper electrodes	
describe the refining of copper	
describe electroplating and name some of its uses	
describe the extraction of aluminium from bauxite	
describe the electrode reactions and the use of cryolite in extracting aluminium	
describe the use of steel-cored aluminium and copper electricity cables	

Revision checklist

Unit 7 Chemical changes

I can:

understand the differences between physical and chemical changes	☐
understand the meaning of the terms *endothermic* and *exothermic*	☐
interpret energy level diagrams	☐
draw and label energy level diagrams	☐
describe bond breaking as endothermic and bond making as exothermic	☐
describe the importance of fuels, especially coal, natural gas and petroleum	☐
describe the use of hydrogen as a fuel	☐
describe radioactive isotopes such as uranium-235 as a source of energy	☐
understand the importance of fuel cells and the main chemical reactions involved	☐
describe how simple cells can produce electricity	☐
relate the voltage produced in a simple cell to the reactivity of the electrodes	☐

Unit 8 Rate of reaction

I can:

describe a practical method for investigating the rate of reaction involving a gas	☐
work out a suitable method for investigating the rate of a reaction	☐
understand how to interpret data in experiments involving rate of reaction	☐
describe how particle size and catalysts affect the rate of reaction	☐
describe how fine powders can cause explosions when they combust	☐
describe the effect of concentration and temperature on the rate of reaction	☐
explain the effect of particle size, concentration and temperature on reaction rate	☐
describe the process of photosynthesis	☐
describe the use of silver salts in photography	☐
for the two reactions above, describe the effect of light on the rate of reaction	☐

Revision checklist

Unit 9 Chemical reactions

I can:

describe reversible reactions, especially the effect of heat upon hydrated salts	
understand the concept of equilibrium	
explain how changing the concentration of reactants or products affects equilibrium	
explain how change in pressure affects the equilibrium of a reaction involving gases	
explain how a change in temperature affects an equilibrium	
explain oxidation and reduction in terms of oxygen loss or gain	
understand the use of oxidation state in naming compounds	
explain redox reactions in terms of electron transfer	
define oxidising agent and reducing agent	
identify redox reactions by changes in oxidation number	
describe the colour changes involved using acidified $KMnO_4$ as an oxidant	
describe the colour changes involved using potassium iodide as a reductant	

Unit 10 Acids and bases

I can:

describe solutions as acidic, neutral or alkaline in terms of pH	
describe the use of universal indicator to measure pH	
describe the effect of acids and alkalis on litmus	
describe the reaction of acids with metals, bases and carbonates	
describe the reactions of bases with acids and ammonium salts	
explain why acids and bases are used to control soil pH	
describe acids and bases in terms of proton transfer	
explain the difference between strong and weak acids and bases	
describe the differences between acidic and basic oxides	
describe two other forms of oxide as neutral and amphoteric	

Revision checklist

Unit 11 Making and identifying salts

I can:

describe how to make a soluble salt from an acid and an alkali by a titration	
describe how to make a soluble salt from an acid and a metal or metal oxide	
describe how to make an insoluble salt by precipitation	
suggest a method of making a salt from given starting materials	
describe tests for ammonia, carbon dioxide, chlorine, hydrogen and oxygen	
describe tests for aluminium, calcium, chromium(III), copper(II), iron(II), iron(III) and zinc ions	
describe flame tests for lithium, sodium, potassium and copper(II)	
describe a test for ammonium ions	
describe tests for carbonate, chloride, bromide, iodide, nitrate, sulfate and sulfate ions	

Unit 12 The Periodic Table

I can:

describe the Periodic Table in terms of groups and periods	
use the Periodic Table to predict properties of the elements	
describe the change from metallic to non-metallic character across a period	
describe how the number of outer shell electrons in an atom is related to metallic or non-metallic character	
describe the physical properties of Li, Na and K and their reactions with water	
predict the properties of other alkali metals when given sufficient data	
describe the physical properties of Cl, Br and I and their reactions with halide ions	
predict the properties of other halogens when given sufficient data	
state the uses of the noble gases	
explain the lack of the reactivity of noble gases in terms of electronic structure	
identify trends in the properties of the alkali metals, halogens and noble gases	
identify trends in other groups of elements when given sufficient information	
describe the typical properties of the transition elements and their compounds	

Revision checklist

Unit 13 Metals and reactivity

I can:

describe the general physical and chemical properties of metals	
explain why metals are used in the form of alloys	
identify alloys from diagrams of their structure	
describe the order of reactivity K, Na, Ca, Mg, Zn, Fe and Cu	
describe the relative reactivity of K, Na, Ca, Mg, Zn, Fe and Cu with water or steam	
describe the relative reactivity of K, Na, Ca, Mg, Zn, Fe and Cu with dilute HCl(aq)	
deduce an order of reactivity from experimental results	
explain metal reactivity in terms of ease of formation of metal ions	
describe the reactivity series by reference to the reaction of metals with metal ions	
describe the reactivity series in terms of the reaction of metal oxides with carbon	
explain why aluminium is apparently unreactive	
describe the action of heat on metal hydroxides and nitrates	
link the thermal decomposition of hydroxides and nitrates to the reactivity series	

Unit 14 Metal extraction

I can:

use the reactivity series to explain why some metals are extracted using carbon	
use the reactivity series to explain why some metals are extracted by electrolysis	
describe the extraction of zinc from zinc blende	
name an ore of aluminium and an ore of iron	
name the raw materials required for extracting iron in a blast furnace	
describe the reactions occurring in the blast furnace, including formation of slag	
describe the conversion of iron into steel	
explain why basic oxides and oxygen are needed to convert iron into steel	
describe why the steel alloys are used for particular jobs rather than pure iron	
explain why alloys have particular properties by reference to their structure	
name some uses of aluminium, mild steel and stainless steel	
name some uses of zinc and copper	

Revision checklist

Unit 15 Air and water

I can:

describe a chemical test for water and its result	
discuss the problems of an inadequate supply of water	
describe the purification of the water supply, especially filtration and chlorination	
name some uses of water in the home and in industry	
describe the composition of unpolluted air	
describe the separation of oxygen and nitrogen from liquid air	
name carbon monoxide, sulfur dioxide and lead as atmospheric pollutants	
name the sources of the atmospheric pollutants mentioned above (CO, SO_2, Pb)	
describe the adverse effects of these pollutants on health	
describe the role of sulfur dioxide in producing acid rain	
name a source of nitrogen oxides and name them as atmospheric pollutants	
describe the adverse effects of nitrogen oxides on buildings and on health	
explain how catalytic converters remove nitrogen oxides from car exhausts	
describe carbon dioxide and methane as greenhouse gases and explain how they contribute to climate change	
describe the sources of carbon dioxide (including respiration) and methane	
describe global warming in simple terms and how it leads to climate change	
describe how the carbon cycle regulates the amount of carbon dioxide in the atmosphere	
describe the conditions required for rusting	
describe methods of rust prevention involving coating with a protective layer	
describe sacrificial protection in terms of the reactivity series	
describe galvanising as a method of rust prevention	

Revision checklist

Unit 16 The chemical industry

I can:

understand why farmers need to add fertilisers to the soil	
describe fertilisers in terms of the elements they contain	
describe how to make a simple fertiliser in the laboratory	
describe the displacement of ammonia from its salts under alkaline conditions	
name the sources of the raw materials used in the manufacture of ammonia	
describe the essential conditions to manufacture ammonia by the Haber process	
name some sources and uses of sulfur	
name some uses of sulfur dioxide	
describe the chemical properties and uses of concentrated and dilute sulfuric acid	
name the sources of the raw materials used in the manufacture of sulfuric acid	
describe the essential conditions for the manufacture of sulfuric acid	
describe the manufacture of quicklime from limestone	
name the uses of quicklime, slaked lime and limestone	

Unit 17 Organic chemistry and petrochemicals

I can:

describe the meaning of *homologous series*	
describe the general characteristics of a homologous series	
describe the meaning of *functional group*	
name and draw the structures of the alkanes	
name the type of compound present from the endings -ane, -ene, -ol and -oic acid	
describe and draw structural isomers	
name the compounds formed when fuels burn	
name methane as the main constituent of natural gas	
describe petroleum as a mixture of hydrocarbons	
describe how petroleum is separated into useful fractions by fractional distillation	
describe the properties of molecules within a fraction	
name the fractions: refinery gas, gasoline, naphtha, kerosene, diesel, fuel oil	
name the residues: lubricating oils and bitumen	
name the uses of the fractions from petroleum fractionation	

Revision checklist

Unit 18 The variety of organic chemicals

I can:

describe the bonding in alkanes	☐
describe the properties of the alkanes (combustion)	☐
describe the substitution reaction of alkanes with chlorine	☐
distinguish saturated from unsaturated hydrocarbons by using bromine water	☐
distinguish saturated from unsaturated hydrocarbons from their structure	☐
describe the addition reactions of alkenes with bromine, hydrogen and steam	☐
describe the manufacture of alkenes and hydrogen by cracking	☐
describe the properties of ethanol	☐
describe the structure of ethanol and name its uses	☐
name and draw the structures of alcohols with up to four carbon atoms	☐
describe the oxidation of ethanoic acid by fermentation	☐
describe the oxidation of ethanol by acidified potassium manganate(VII)	☐
describe ethanoic acid as a weak acid	☐
name carboxylic acids with up to four carbon atoms	☐
describe the reaction of ethanoic acid with ethanol to form an ester	☐
name and draw formulae of esters made from alcohols and carboxylic acids, each containing up to four carbon atoms	☐

Unit 19 Polymers

I can:

describe the polymerisation of ethene to form poly(ethene)	☐
understand the terms *monomer*, *polymer* and *polymerisation*	☐
understand the meaning of the term *macromolecule*	☐
name some uses of plastics and man-made fibres	☐
describe the pollution problems involved in the disposal of non-biodegradable plastics	☐
explain the difference between condensation and addition polymerisation	☐
deduce the structure of a polymer from a given alkene monomer	☐
deduce the structure of the monomer from the given structure of an addition polymer	☐
describe the formation of nylon and terylene	☐
understand the terms *condensation polymerisation*, *amide linkage* and *ester linkage*	☐

Revision checklist

Unit 20 Biological molecules

I can:

name the main constituents of food as carbohydrates and proteins	
describe in simple terms the structure of amino acids	
describe in simple terms the structure of proteins, especially the amide linkage	
describe the hydrolysis of proteins to amino acids	
describe in simple terms the structure of complex carbohydrates	
describe the hydrolysis of complex carbohydrates to simple sugars	
describe how chromatography is used to identify amino acids and simple sugars	
describe enzymes as biological catalysts	
describe how simple sugars are fermented to produce ethanol	
describe the advantages and disadvantages of manufacturing ethanol by fermentation and hydration of ethene	

Glossary

A

Acid A substance that forms hydrogen ions when dissolved in water. Acidic solutions have a pH less than 7.

Acidic oxide An oxide that reacts with a base to form a salt and water

Acid rain Rain with an acidity below about pH 5.0 due to the reaction of sulfur dioxide (formed by burning fossil fuels) with rainwater

Activation energy The minimum energy needed to break bonds to start a chemical reaction

Addition polymerisation The formation of polymers from monomers (having double bonds) where no other substance other than the polymer is formed

Addition reaction A reaction where two substances react together to form only one product

Air The mixture of gases always present in the atmosphere

Alcohol A compound with an –OH functional group and general formula $C_nH_{2n+1}OH$

Alkali A soluble base. An alkaline solution has a pH above 7.

Alkali metal The Group I elements in the Periodic Table. They have one electron in their outer shell.

Alkane A hydrocarbon with the general formula C_nH_{2n+2}

Alkene A hydrocarbon containing one or more C=C bonds and having the general formula C_nH_{2n}

Alloy A mixture of two or more metals or (less often) a metal with a non-metal

Amphoteric oxide An oxide which reacts with both acids and alkalis

Amide linkage The –CONH– group formed when proteins and polyamides such as nylon are formed from their monomers

Amino acid The monomers that combine to form proteins

Anhydrous A substance without water combined with it

Anion A negative ion (which moves to the anode during electrolysis)

Anode The positive electrode

Aqueous solution A solution made by dissolving a substance in water

Atmosphere (air) The layer of gases that surround the Earth

Atmosphere (unit) A unit of pressure used in chemistry

Atom The smallest part of an element that can take part in a chemical change

Atomic number The number of protons in the nucleus of an atom

Avogadro's constant The number of defined particles (ions, atoms, molecules) in one mole of those particles

B

Balanced equation An equation with the same number of each type of atom on both sides of the equation

Base A compound that reacts with acid to form salts. Bases are proton acceptors.

Basic oxide An oxide that reacts with an acid to form a salt

Battery – see Electrochemical cell

Bauxite The main ore of aluminium

Binary compound A compound containing just two elements, e.g. NaCl

Biodegradable Naturally decaying in the environment with the help of bacteria and fungi

Blast furnace A furnace into which air is blown and in which a metal oxide is reduced using carbon

Bond energy The energy needed to break a mole of given bonds

Bonding The way the atoms or ions are held together in a substance

Brass An alloy of copper and zinc

Brownian motion The random bombardment of molecules on small suspended particles leading to a random irregular motion

Burette A piece of glassware for delivering a variable volume of liquid accurately (usually up to 50 cm³)

Burning An exothermic reaction where substances combine and a flame can be seen

C

Carbohydrate The general name for simple and complex sugars. Carbohydrates have the general formula $C_x(H_2O)_y$.

Carbon cycle The series of reactions that keeps the amount of carbon dioxide in the atmosphere fairly constant and moves carbon from place to place in the Earth, water and atmosphere

Carboxylic acid A weak acid that has the general formula $C_nH_{2n+1}COOH$

Catalyst A substance that speeds up a chemical reaction without being chemically changed or used up

Catalytic converter A piece of equipment put on a car exhaust to remove nitrogen oxides and carbon monoxide

Cathode The negative electrode

Cation A positive ion (which moves to the cathode during electrolysis)

Cell A container in which electrolysis is carried out (Also see Electrochemical cell)

Celsius A scale used to measure temperature (degrees Celsius, °C)

Cement A building material made by heating limestone with clay

Centrifugation Separation of a solid from a liquid by spinning at high speed

Charge Electrical charge can be positive or negative

Chemical property A property involving the formation of a new substance

Chlorination The addition of chlorine, especially in water purification

Chlorophyll A green plant pigment that catalyses photosynthesis

Chromatogram A piece of paper showing the separation of substances after chromatography has been carried out

Glossary

Chromatography The separation of a mixture of substances using filter paper and a solvent

Climate change The change in weather patterns that is often linked to increase in global warming

Collision theory Using the idea of colliding particles to explain how reaction rates change with temperature and concentration

Combustion Burning (usually in a reaction with oxygen gas)

Compound A substance containing two or more types of atoms chemically combined

Concentration The amount of one type of substance dissolved in another. It is measured in mol/dm^3 or g/dm^3.

Condensation The change of a gas to a liquid

Condensation reaction A reaction where two or more substances combine and a small molecule is eliminated (given off)

Condenser A piece of apparatus for cooling a vapour and converting it to a liquid

Conductor A substance that allows heat or electricity to flow through it

Contact process The industrial process for making sulfuric acid

Controlled variable A variable that is kept constant during an investigation

Corrosion The 'eating away' of the surface of a metal by a chemical, e.g. by acids

Covalent bond A bond formed by the sharing of electrons between two atoms

Covalent compound A compound having covalent bonds. They can be simple molecules or giant structures.

Cracking The breaking of an organic compound into smaller molecules by heat (or heat and a catalyst)

Crude oil A mixture of hydrocarbons present under the Earth's crust as a black sticky liquid. Sometimes called petroleum

Cryolite A compound added to the melt when aluminium oxide is electrolysed. It helps dissolve the aluminium oxide and lowers the temperature of the electrolyte.

Crystallisation The formation of crystals when a saturated solution is left to cool

Crystallisation point The point at which crystals will form very quickly when a drop of saturated solution is placed on a cold surface

D

Decimetre cubed Unit of volume (dm^3) often used in chemistry

Decomposition The breakdown of a substance into two or more products

Delocalised electrons Electrons that are not associated with any particular atom

Density The mass of a substance divided by its volume

Diatomic Containing two atoms

Diffusion The random movement of particles (usually molecules or ions) in solution or in gases leading to complete mixing of the particles

Displacement A reaction in which one atom or group of atoms replaces another in a compound

Distillation A method of separating a liquid from a mixture by boiling the mixture then condensing the vapours

Ductile Can be pulled out into wires

E

Electrical conductivity The flow of electric current through a substance. Conduction is due to moving electrons in metals and moving ions in ionic solutions/melts.

Electrochemical cell A source of electrical power containing two different metals dipping into an electrolyte

Electrode A rod of metal or graphite that leads an electric current into or out of an electrolyte

Electrolysis The breakdown of an ionic compound using electricity

Electrolyte A substance that conducts electricity when molten or dissolved in water

Electron A negatively charged particle found in electron shells outside the nucleus

Electronic structure The number and arrangement of electrons in the electron shells of an atom

Electron shell The energy levels at different distances from the nucleus where the electrons are found

Electroplating A process that uses electricity to coat one metal with another

Element A substance containing only one type of atom

Empirical formula A chemical formula that shows the simplest ratio of atoms in a compound

Endothermic A reaction or process that absorbs (takes in) energy

End point The point in a titration when the indicator changes colour showing that the reaction is complete

Energy level – see Electron shell

Energy level diagram Shows the energy change from reactants to products for exothermic or endothermic reactions

Enzyme A biological catalyst

Equilibrium In a reversible reaction, the point where the forward and backward reactions are taking place at the same rate

Ester A compound formed when a carboxylic acid reacts with an alcohol

Ester linkage The –COO group found in esters, fats and oils

Evaporation The change from liquid to vapour state below the boiling point

Exothermic A reaction or process that releases (gives out) energy

F

Fermentation Often refers to the breakdown of glucose by yeast in the absence of oxygen to form ethanol and carbon dioxide. Other fermentations can also occur.

Glossary

Fertiliser A substance added to the soil to replace essential elements lost when crops are harvested

Fibre A polymer that can be drawn into threads

Filtrate The liquid that runs through the filter paper when carrying out filtration

Filtration Separating a solid from a liquid by using a filter paper

Flue gas desulfurisation The removal of sulfur dioxide from the waste gases produced in furnaces using bases

Fossil fuel Fuel (coal, oil, natural gas) formed from the remains of tiny, dead, sea creatures and plants over millions of years

Fraction A group of molecules with similar boiling points, distilling off at the same place in a fractionation column

Fractional distillation (fractionation) The separation of different substances in a liquid by their different boiling points

Fractionating column A tall column used for fractional distillation

Free electrons – see Delocalised electrons

Freezing The change of state from liquid to solid at the melting point

Fuel A substance that releases energy (usually when burned)

Fuel cell A cell where hydrogen and oxygen undergo reaction to produce an electric current

Functional group A group of atoms that is responsible for the characteristic reactions of a homologous series

G

Galvanising Coating a metal (usually iron) with a protective layer of zinc

Gas syringe A piece of glassware for measuring the volume of gases given off in a reaction

General formula A formula that applies to all the compounds in a particular homologous series

Giant covalent structure A structure with a continuous three-dimensional network of covalent bonds

Giant ionic structure A structure with a continuous three-dimensional network of ionic bonds

Global warming The warming of the atmosphere due to greenhouse gases trapping infrared radiation from the Earth's surface

Graphite A form of carbon that conducts electricity and is used for inert electrodes

Greenhouse effect The process by which heat energy is absorbed by the atmosphere after being radiated from the Earth's surface

Greenhouse gas A gas such as methane or carbon dioxide that absorbs infrared radiation from the Earth's surface

Group A vertical column of elements in the Periodic Table

H

Haber process The industrial process for making ammonia

Half equation Equation showing the oxidation or reduction parts of a redox reaction separately. Often used for reactions at the electrodes in electrolysis.

Halide A substance containing an ion formed from a halogen atom

Halogen An element in Group VII of the Periodic Table

Hematite A common ore of iron

Homologous series A group of organic compounds having the same general formula and similar chemical properties

Hydrated compound A compound containing water

Hydrocarbon A compound containing only carbon and hydrogen

Hydrogenation The addition of hydrogen to an unsaturated compound

Hydrolysis The breakdown of a compound by reaction with water. Acids or alkalis speed up hydrolysis.

I

Immiscible (liquids) Liquids which do not mix

Incomplete combustion Occurs when there is not enough oxygen to react completely with the substance burned

Independent variable The variable that you decide to change in an investigation

Indicator A substance that has two different colours depending on the solution in which it is placed. It often changes colour according to the pH of the solution.

Inert Does not react

Inert electrode An electrode that does not react during electrolysis

Inorganic compound Compounds that generally just melt or vaporise on heating without charring or burning

Insulator A substance that does not conduct electricity (or heat)

Ion An atom or group of atoms that has become positively or negatively charged

Ionic bond The electrostatic attraction between the positive and negative ions in the crystal lattice

Isomers Substances with the same molecular formula but a different arrangement of atoms

Isotopes Atoms of the same element which have the same proton number but a different nucleon number

K

Kinetic particle theory The idea that the arrangement and motion of the atoms can explain states of matter, diffusion and rates of reaction

L

Lattice A continuous regular arrangement of particles

Le Chatelier's principle When you alter the conditions in an equilibrium reaction, the reaction moves in the direction to oppose the change

Glossary

Limewater Calcium hydroxide solution used to test for carbon dioxide

Limiting reactant The reactant that is not in excess

Litmus An indicator used to test if a substance is acidic or alkaline

Locating agent A compound that reacts with a colourless substance on chromatography paper to form a coloured spot

Lone pair A pair of electrons not involved in bonding

M

Macromolecule A very large molecule, e.g. a polymer or a giant covalent structure

Malleable Can be hammered into shape

Mass number The number of protons plus neutrons in the nucleus of an atom

Melt (electrolytic) A molten electrolyte

Melting The change in state from solid to liquid

Metallic bonding Metal ions are held together by a 'sea' of delocalised electrons

Mixture Two or more substances mixed together but not chemically combined

Molar gas volume The volume occupied by one mole of any gas – 24 dm³ at room temperature

Molarity The concentration of a solution in mol/dm³

Molar mass – see Relative molecular mass

Mole The relative formula mass of a substance in grams

Molecular formula A formula showing the type and number of each element present in a molecule

Molecule A particle made up of two or more atoms held together by covalent bonds

Moles per decimetre cubed The unit of concentration – mol/dm³

Monomer A small molecule that can combine to form a polymer

Monatomic Consisting of one atom

N

Neutral Neither acidic nor alkaline. A neutral solution has pH 7

Neutralisation The reaction of an acid with a base to form a salt and water

Neutral oxide Oxides that are neither acidic nor basic

Neutron A particle in the nucleus of the atom which has no charge

Noble gas An element in Group 0 of the Periodic Table

Noble gas configuration This is obtained when atoms in molecules ions or have a complete outer shell of electrons

NPK fertilisers Fertilisers containing nitrogen, phosphorus and potassium

Nucleon A particle (proton or neutron) in the nucleus of an atom

Nucleon number The total number of protons plus neutrons in the nucleus of an atom

Nucleus The central part of an atom containing protons and neutrons

O

Ores Rocks containing a metal or metal compound from which a metal can be extracted

Organic compound Compounds containing carbon and usually hydrogen and perhaps other elements

Outer electrons – see Valency electrons

Oxidation The addition of oxygen to, removal of electrons from or increase in oxidation state of a substance

Oxidation state/number A number that describes how oxidised an atom is

Oxidising agent A substance that removes electrons from or adds oxygen to another substance; it oxidises another substance

P

Percentage composition The percentage by mass of each element in a compound

Percentage purity The mass of a given substance in a mixture divided by the overall mass of substance present, expressed as a percentage

Percentage yield The actual yield divided by the theoretical yield expressed as a percentage

Period A horizontal row of elements in the Periodic Table

Periodic Table An arrangement of elements in order of increasing atomic number such that some elements with similar properties are arranged in groups

Petroleum – see Crude oil

pH scale A scale of numbers that describes how acidic (or alkaline) a substance is

Photochemical reaction A reaction that depends on the presence of light

Photosynthesis The process by which plants make glucose (and oxygen) from carbon dioxide and water in the presence of sunlight

Physical property A property which does not involve a chemical reaction e.g. melting point

Pipette – see Volumetric pipette

Plastics Polymers that can be moulded

Pollutant A substance that contaminates (makes less pure) the air, water or soil

Polyamide A polymer with –CONH– linkages.

Polyester A polymer with –COO– linkages.

Polymer A substance made up from a huge number of small molecules that have combined

Polymerisation The chemical reaction combining monomers to form a polymer

Precipitate A solid formed when two solutions are mixed

Glossary

Protein A polymer made up from amino acids

Proton A positively charged particle in the nucleus of an atom. Also used as a term for a hydrogen ion

Proton acceptor The definition of a base

Proton donor The definition of an acid

Proton number The number of protons in the nucleus of an atom

Pure There is only one substance present; often used incorrectly to suggest that there are no pollutants present

Purification methods Methods to separate the substance you want from mixtures, e.g. distillation, filtration

Q

Qualitative analysis A way of identifying substances by carrying out chemical tests based on observations

Quicklime The common name for calcium oxide; also called lime

R

Radioactivity Radiation or sub-atomic particles given out from the nuclei of unstable atoms

Rate of reaction The amount of product converted to reactants in a given time

Reacting mass The amount of one reactant (in grams or moles) needed to react exactly with another reactant

Reactivity series A list of elements (usually metals) in their order of reactivity

Recycling The processing of used materials to form new products

Redox reactions Reactions in which reduction and oxidation occur together. The electrons lost by one substance are gained by another.

Reducing agent A substance that adds electrons to or removes oxygen from another substance

Reduction The removal of oxygen from, addition of electrons to or decrease in oxidation state of a substance

Refining Purifying a substance; used for purifying metals, e.g. copper or for separating petroleum into its fractions

Relative atomic mass The mass of an atom on a scale where an atom of carbon-12 weighs exactly 12 units

Relative formula mass The average mass of naturally occurring atoms of an element on a scale, where the 12C atom has a mass of exactly 12 units

Relative mass The mass of one particle compared with another; used when the actual mass is very small, e.g. masses of protons, neutrons and electrons

Relative molecular mass The sum of the relative atomic masses of all the atoms in a molecule

Renewable resource Something that can be replaced if we keep using it, e.g. using wood as a fuel

Residue The solid obtained on the filter paper when carrying out filtration

Respiration The reactions that release energy in all living things. The overall reaction is glucose combining with oxygen to form carbon dioxide and water.

Reversible reaction A reaction in which the products can react together to re-form the original reactants

R_f value In chromatography, the distance moved by a particular compound from the baseline compared with the distance moved by the solvent front

Rust Hydrated iron(III) oxide formed when iron reacts with air and water. The term only applies to corrosion of iron and steel.

S

Sacrificial protection A more reactive metal is placed in contact with a less reactive metal. The more reactive metal corrodes and saves the less reactive metal from corrosion.

Salts Compounds formed when hydrogen in an acid is replaced by a metal or an ammonium ion

Saturated (hydrocarbon) A hydrocarbon with only single bonds. It has the maximum amount of hydrogen possible.

Sea of electrons Term used for the delocalised electrons in metallic bonding

Separating funnel Piece of glassware used to separate two immiscible liquids

Slag The waste material formed when metals are extracted in a furnace

Slaked lime The common name for calcium hydroxide

Solubility The amount of solute that dissolves in a given quantity of solvent

Solute A substance that dissolves in a solvent

Solution A mixture of a solute in a solvent

Solvent A substance that dissolves another substance

Solvent extraction A method of extracting a substance that is more soluble in one solvent than another

Spectator ions Ions that do not take part in a reaction

State The three states of matter are solid, liquid and gas.

State symbols The letters (s), (l), (g) or (aq) placed after each formula in an equation to indicate the state of the reactants and products

Steel An alloy of iron containing controlled amounts of carbon and other metals

Stoichiometry The ratios of the reactants and products shown in a balanced chemical equation

Strong acid/alkali Acids or alkalis that are completely ionised when dissolved in water

Structural formula A formula that shows the structure of a molecule or ion. A full structural formula (sometimes called a displayed formula) shows all atoms and bonds.

Glossary

Sublimation The change of state from solid to gas or from gas to solid without a liquid being formed

Substitution reaction A reaction in which one atom or group of atoms replaces another

T

Thermal decomposition The breakdown of a compound into two or more substances by heat

Titration A method for finding the amount of a substance in a solution

Transition element A block of elements between Groups II and III in the Periodic Table

U

Universal indicator A mixture of indicators which is used to measure pH

Unsaturated (hydrocarbon) A hydrocarbon with one or more double (or triple) bonds

V

Valency The combining power of atoms, e.g. sodium has a valency of 1, oxygen has a valency of 2

Valency electrons The electrons in the outer shell of an atom

Volatile Easily vaporised

Volumetric flask A flask with a graduation mark for making up solutions accurately

Volumetric pipette A pipette used to measure out solutions accurately. A pipette for putting drops of liquid into a test tube is called a teat pipette.

W

Water of crystallisation The water combined with a compound in a crystal

Weak acid/alkali Acids and alkalis that are only slightly ionised when dissolved in water

Word equation An equation with the reactants and products written as their chemical names

Y

Yield The amount of product obtained in a reaction

Z

Zinc blende A common ore of zinc containing zinc sulfide

Index

A
acid rain 190–1
acids 126, 130, 167
   a soluble salt from an acid and alkali 140–1
   acids, bases and protons 132
   how changing the concentration of acid affects the rate of reaction 108
   litmus test 128
   pH scale 126
   reaction with carbonates 129
   reaction with metal hydroxides or aqueous ammonia 129
   reaction with metal oxides 129
   reaction with metals 128–9
   strong and weak acids 133, 232
   using universal indicator 127
addition polymerisation 236, 238
air 188
   carbon cycle 196–7
   gases 188–9
   global warming 194–5
   percentage of oxygen in 188
air pollution 190-3
alcohols 214, 230
   structure 230–1
   uses of ethanol 230
alkali metals 154-5
alkalis 126
   a soluble salt from an acid and alkali 140–1
   reacting ammonium sulfate with an alkali 131
   reaction with ammonium salts 130
alkanes 216
   chemical properties 224–5
   cracking 226–7
   naming 216
   physical properties 224
   reaction with chlorine 225
alkenes 216, 228
   chemical properties of the alkenes 229
alkyl groups 216–17
allotropes 42
alloys 164, 180
   steel alloys 182
   tin, lead and solder 164
   uses 165, 182–3
aluminium 171, 182
aluminium extraction 84-5
amide linkage 244
amino acids 244, 245
ammonia 129
   displacement of ammonia from ammonium salts 203
   Haber process 204–5
   using the litmus test 144
ammonium salts 203
   reaction of alkalis with ammonium salts 130

ammonium sulfate 203
   reacting with an alkali 131
anhydrous substances 116
anions 78, 148
   identifying carbonate ions 148
   identifying chlorides, bromides and iodides 148
   identifying nitrates 148–9
   identifying sulfates 149
   identifying sulfites 149
anodes 75
apparatus 252
   measuring 254
   reducing errors 253
aqueous solutions 54
   electrolysing concentrated aqueous sodium chloride 76–7
   electrolysis of dilute aqueous solutions 78–9
   reaction of acids with aqueous ammonia 129
atmosphere 188
atoms 22
   atomic number 23, 152
   molecules with three or more types of atoms 38
Avogadro constant 60

B
bases 130-3
bauxite 176
biodegradability 237
blast furnaces 177, 178
   chemical reactions in 178–9
boiling 4
   boiling point 14–15
bonds 92
   bond energies 92
brass 183
brine 77
bromides 148
Brownian Motion 6
burettes 11, 140

C
carbohydrates 244, 246
   hydrolysis of 246–7
carbon 170
carbon cycle 196-7
carbon dioxide 145, 196–7
carbon monoxide 190
carbonates 129
   carbonate ions 148
   reaction of acids with 129
   thermal decomposition 173
carboxylic acids 214, 232, 244
   chemical properties 232
   formation of esters 232–3
catalysts 107, 161, 226, 248
catalytic converters 193
cathodes 75, 146
cations 78

   flame tests 147
   testing for cations 146–7
cells 70
centrifugation 16
centrifuges 16
changes of state 4
   explaining 4–5
   heating curve for salicylic acid 5
chemical bonds 28
chemical calculations 58–9
   Avogadro constant and the mole 60
   finding the empirical formula 68–9
   finding the molecular formula 69
   gas volumes 64–5
   how much product? 62–3
   mole 60–1
   percentage by mass 64
   reacting masses 61
   relative atomic mass 58
   relative formula mass 59
   relative molecular mass 58–9
   simple chemical 59
   yield and purity 66–7
chemical changes 90
chemical equations 52–5
   half equations 78
   ionic equations 54–5
   using state symbols 54
chemical formulae 48–9
   empirical formula 50, 68–9
   molecular formula 50, 69
   simple rules for naming compounds 49
   structural formula 50
   working out the formula from diagrams 50
   working out the formula of an ionic compound 50–1
chemical properties 31
chemical symbols 48
chlorides 148
chlorination 187
chlorine 145
   reaction of alkanes with 225
chromatograms 12, 247
chromatography 12–13
   locating agent 13, 247
   making a chromatogram 12
climate change 195
coal 94
collision theory 108–9, 111
combustion 190
   alkenes 229
   incomplete 190
compounds 28
   double and triple bonds 38–9
   ionic or covalent? 40–1
   organic compounds 214–15
   saturated or unsaturated? 228
   simple rules for naming compounds 49

Index

working out the formula of an ionic compound 50–1
concentration 108, 118
condensation 4
condensation polymerisation 240, 246
condensation reactions 240, 244
conductors 86–7
constants 104
Contact Process 203, 208–9
  what reaction conditions do we use? 209
controlled variables 104
copper 166, 183
  electroplating 82
copper(II) oxide 120
  making copper(II) sulfate from copper(II) oxide 139
  reducing copper(II) oxide with carbon 170
  reduction of copper(II) oxide 120
copper refining 80
  changing the electrodes 81
  electrolysis with copper electrodes 81
  electrolysis with inert electrodes 81
copper sulfate 139
  how much copper sulfate can we get from malachite? 66
corrosion 83, 198
covalent bonding 36, 40, 41
  compounds with double and triple bonds 38–9
  diamond and graphite 42–3
  electronic structure of simple molecules 36–7
  giant covalent structures 42
  molecules with three or more types of atoms 38
cracking alkanes 226–7
crystallisation 16–17

D

decanting 16
decomposition 74
  thermal decomposition 172–3
denitrification 192
density 30, 154
diamond 42–3
diffusion 7
  comparing speeds 9
  explaining 8–9
  speed of diffusion and molecular mass 7
discharge series 76–7
displacement reactions 157, 168
distillation
  fractional distillation 18–19, 189, 218, 219, 220–1
  simple distillation 18
ductility 160

E

electrical conductivity 30
electrochemical cells 96
  getting the best voltage 97
  how does a simple electrochemical cell work? 96
electrodes 74
electrolysis 74–5, 176
  copper refining 80–1
  dilute aqueous solutions 78–9
  electrolysing brine 77
  electrolysing concentrated aqueous sodium chloride 76–7
  from ions to atoms 78–9
  hydrochloric acid 77
  molten lead(II) bromide 74–5
  predicting the products of electrolysis 75
electrolytes 74
  acidic electrolytes 99
  alkaline electrolytes 99
electrons 22
  arranging 26
  covalent bonds 36
  delocalised electrons 43
  electron configuration 26
  electron shells 22, 26
  electron transfer 121
  electronic structure 26
  lone pairs 36
  valency 26, 39, 152
electroplating metals 82
  copper 82
  how electroplating works 82–3
  uses 83
elements 27
  making sodium chloride 29
  simple rules for naming compounds 49
  transition elements 160–1
empirical formula 50, 68–9
endothermic reactions 90–1
energy levels 22, 26
  diagrams 91
enzymes 245, 248
equilibrium 116
  changing the concentration 118, 119
  changing the direction of a reaction 118
  changing the pressure 119
  changing the temperature 117
  equilibrium reaction 205, 209
esters 232–3
  ester linkages 244
ethanol 230
  comparing production methods 249
  ethanoic acid 231
  fuel 249
  production 248–9
evaporation 4
exothermic reactions 90, 205
experimental observations and data 256
  evaluation 256–7
experimental set-ups 252–3

explosive reactions 106–7

F

fats 244
fermentation 248
  glucose 248–9
fertilisers 202
  displacement of ammonia from ammonium salts 203
  making ammonium sulfate fertiliser 203
  making fertilisers 202–3
  NPK fertilisers 202
filtrates 16
filtration 16
forces of attraction 4
fractional distillation 18–19, 189, 218, 219
  fractionating columns 220
  fractionation 220
  petroleum 200–1
fractions 18, 218, 219, 220
freezing 4
fuel cells 98
  how does a fuel cell work? 98–9
  model fuel cell 98
  what are the advantages of fuel cells? 99
fuels 94–5
  comparing the energy released 94
  energy from radioactivity 95
  ethanol as a fuel 249
  fossil fuels 94, 218–19
  what is formed when fuels burn? 219

G

galvanising 199
gas syringes 11
gases 2, 144
  compressing gases 3
  diffusion 8–9
  following change in the volume of gas given off 103
  gases in the air 188–9
  greenhouse gases 194
  heating gases 3
  identifying carbon dioxide 145
  identifying chlorine 145
  identifying hydrogen 144
  identifying oxygen 144
  identifying sulfur dioxide 145
  molar gas volume 65
  natural gas 94
  noble gases 26–7, 158
  using the litmus test for ammonia 144
  volumes of gases 11, 64–5
global warming 193, 194
  climate change and its results 195
  greenhouse effect 194–5
  greenhouse gases 194

Index

glucose 248–9
gold 166
graphite 42–3
  why does graphite conduct electricity? 43
graphs 255
greenhouse effect 194–5
greenhouse gases 194

H
Haber process 203, 204–5
  raw materials 204
  what are the best reaction conditions? 205
hematite 176, 178
half equations 78
halogens 156
  reactions with halide ions 156–7
  reactions with Group I elements 156
  trends in physical properties 156
homologous series 214–15
hydration 116
hydrocarbons 216, 218
  alkyl groups 216–17
  saturated hydrocarbons 224
  structural isomers 217
  unsaturated hydrocarbons 228
hydrochloric acid 128
  electrolysis 77
hydrogen 144
hydrogenation 229
hydrolysis 245
  carbohydrates 246–7
  identifying hydrolysis products 247

I
indicators 128, 140
  universal indicator 126, 127
insulators 87
intermolecular forces 40
iodides 148
ionic bonding 34–5, 40, 41
ionic equations 54–5
ions 2
  carbonate ions 148
  compound ions 51
  dot-and-cross diagrams 32
  forming a stable structure 34–5
  how are ions formed? 32
  ions with multiple charges 35
  spectator ions 55, 143
iron 166, 178
  extracting iron 178–9
  steelmaking 180–1
  why do we add limestone? 179
isotopes 24–5
  uses of isotopes 25

K
kinetic particle theory 4, 6
  Brownian Motion 6
  diffusion 7

L
lattices 40
lead 164
  electrolysis of molten lead(II) bromide 74–5
  lead compounds 190
light-sensitive reactions 112
  how light affects the rate of photosynthesis 112
  photography 113
  photosynthesis 112
limestone 179, 210
  lime 210–11
  limewater 211
  slaked lime 211
  thermal decomposition 210
  uses of limestone products 211
limiting reactants (reagents) 62–3, 105
liquids 2
  volumes of liquids 10–11
litmus test 128, 144
locating agent 13, 247

M
macromolecules 42, 236
  large molecules in food 244–5
magnesium 166
  reaction with hydrochloric acid 128
malleability 160
mass 11
mass number 22, 24
measurements 10
  accuracy 11
  apparatus 11
  volumes of gases 11, 64–5
  volumes of liquids 10–11
measuring cylinders 11
melting 4
  melting point 14–15
metal extraction 176
  extracting iron 178–9
  extracting zinc 177
  reactivity series 176
metal hydroxides
  reaction of acids 129
  thermal decomposition 172
metal oxides 170–1
metals 30–1
  alkali metals 154–5
  alloys 164–5
  competing for oxygen 170
  electroplating 82–3
  explaining reactivity 169
  explaining the use of reducing agents 171
  metallic bonding 44–5
  modelling metallic structure 44–5
  properties 45
  reacting iron wool with steam 166
  reacting with oxygen 166
  reacting with water or steam 166–7
  reaction with dilute acid 167
  reactivity series 138, 166–9, 176
  recycling 183
  reducing copper(II) oxide with carbon 170
  reducing metal oxides with carbon 170
  thermal decomposition 172–3
  thermit reaction 168
  transition elements 160–1
  uses 182–3
  why does aluminium seem unreactive? 171
mixtures 29
mole 60
  calculations using the mole 60–1
  molar mass 60
molecules 2, 40
  diatomic molecules 52, 156
  electronic structure of simple molecules 36–7
  molecular formula 50, 69
  molecules with three or more types of atoms 38
monatomic substances 158
monomers 236, 238–9

N
neutral substances 126
neutralisation reactions 129, 130, 132
neutrons 22, 24
nitrates 148–9
nitrites 172–3
nitrogen oxide 192–3
noble gases 158
  chemical properties 158
  noble gas structure 26–7
  physical properties 158
  uses of the noble gases 158
non-metals 30–1
NPK fertilisers 202
nuclear fuels 95
nucleon number 22, 24
nucleons 22
nucleus 22

O
ores 176
organic chemistry 214–15
organic compounds 214
  formulae 215
  homologous series 214–15
oxidation 78, 113, 120
  oxidation numbers 122
  oxidation states 121–3
  oxidising agents 120
oxides 134
  acidic oxides 134
  amphoteric oxides 135
  basic oxides 134
  neutral oxides 135
oxygen 144, 170
  reacting metals 166

Index

P
paper chromatography 12–13
particles 2
particulates 190
percentage by mass 64
percentage purity 67
Periodic Table 23, 152
   Group I alkali metals 154–5
   Group VII halogens 156–7
   groups 26, 152
   outer electrons and the Periodic Table 153
   periods 26, 152
   transition elements 160–1
   trends across a period 153
   trends down the groups 152
   trends in other groups 159
petroleum 94, 220
   cracking petroleum fractions on a large scale 226–7
   fractional distillation 220–1
pH scale 126
   comparing the pH of household products 126
photochemical reactions 112, 148, 225
photography 113
photosynthesis 112, 196
   how light affects the rate of photosynthesis 112
physical changes 90
physical properties 30
planning investigations 258–9
plastics 236
   good or bad? 237
polyamides 240–1
polyesters 241
polymerisation 236
polymers 236–7
   from monomer to polymer 238–9
   from polymer to monomer 239
potassium manganate 123
precipitation reactions 103
   making an insoluble salt 142
   precipitates 142
   soluble or insoluble compounds? 142
   what happens in a precipitation reaction? 143
pressure 119
product 63
proteins 244, 248
   from proteins to amino acids 245
   hydrolysing 245
   structure 244–5
protons 22
   proton acceptors 132
   proton donors 132
   proton number 23, 152
purity 14
   distillation 18–19
   how do we know if a substance is pure? 14–15
   how do we purify mixtures? 15
   impurity 14
   percentage purity 67
   which method of purification? 19

Q
qualitative analysis 148

R
radioactivity 25
random movement 6
rates of reaction 102-10
reacting masses 61
reactivity series 138, 166–9, 176
redox reactions 113, 120–1, 157, 193
   analysing 122
   colour changes 123
   electron transfer 121
   oxidation states 121
   reacting iron(II) with potassium manganate(VII) 123
   reducing and oxidising agents 120
   reduction of copper(II) oxide 120
reduction 78, 113, 120
   reducing agents 120, 171
relative atomic mass 58
relative formula mass 59
relative molecular mass 58–9
residues 16
respiration 196
reversible reactions 116
   equilibrium 116–17
   heating hydrated salts 116
rusting 198
   sacrificial protection 199
   stopping the rust 198–9
   what are the best conditions for rusting? 198

S
safety 252
salicylic acid 5
salts 128
   how do we make salts? 138
   making copper(II) sulfate from copper(II) oxide 139
   precipitation method 142–3
   salts from insoluble bases 139
   salts from metals 138
   titration method 140–1
silicon (IV) oxide 43
slag 180
sodium chloride 76–7
   making sodium chloride from its elements 29
soil acidity 130–1
solder 164
solids 2
   separating a solid from a solution 16–17
solutes 16
solutions 16
aqueous solutions 54, 76–7, 78–9, 129
solution concentration 70
solvents 12, 16
   solvent extraction 17
state symbols 54
steam 166–7
steel 180
   steel alloys 182
   steelmaking 180–1
stoichiometry 61
structural formula 50
structural isomers 217
substitution reactions 225
sulfates 149
sulfites 149
sulfur 206
sulfur dioxide 145, 206
   acid rain 190–1
sulfuric acid 207
   Contact process 208–9
surface area 106
symbol equations 52–3

T
temperature 11, 110, 119
thermal decomposition 172
   carbonates 173
   limestone 210
   metal hydroxides 172
   nitrates 172–3
three states of matter 2
time 11
tin 164
titrations 70–1, 140
transition elements 160–1

V
valency 26, 39, 152
   combining power 39
variables 104
volatility 130
volumetric flasks 11
volumetric pipettes 11, 140

W
water 186
   reacting metals with 166–7
   testing for 186
   purification 186–7
   water of crystallisation 116
word equations 52

Y
yield 66, 205

Z
zinc 183
   extracting 177
   zinc blende 176